Blockchain, Internet of Things, and Artificial Intelligence

Blockchain, Internet of Things, and Artificial Intelligence

Edited by
Naveen Chilamkurti
T. Poongodi
Balamurugan Balusamy

CRC Press
Taylor & Francis Group
Boca Raton London New York

CRC Press is an imprint of the
Taylor & Francis Group, an **informa** business

A CHAPMAN & HALL BOOK

First edition published 2021
by CRC Press
6000 Broken Sound Parkway NW, Suite 300, Boca Raton, FL 33487-2742

and by CRC Press
2 Park Square, Milton Park, Abingdon, Oxon, OX14 4RN

CRC Press is an imprint of Taylor & Francis Group, LLC

Library of Congress Cataloging-in-Publication Data
Names: Chilamkurti, Naveen, 1974- editor. | Poongodi, T., editor. |
Balusamy, Balamurugan, editor.
Title: Blockchain, internet of things, and artificial intelligence / Naveen
Chilamkurti, T. Poongodi, Balamurugan Balusamy.
Description: First edition. | Boca Raton : C&H/CRC Press, 2021. | Includes
bibliographical references and index.
Identifiers: LCCN 2020045938 (print) | LCCN 2020045939 (ebook) | ISBN
9780367371531 (hardback) | ISBN 9780429352898 (ebook)
Subjects: LCSH: Blockchains (Databases) | Internet of things. | Artificial
intelligence--Social aspects.
Classification: LCC QA76.9.B56 B5624 2021 (print) | LCC QA76.9.B56
(ebook) | DDC 005.74--dc23
LC record available at https://lccn.loc.gov/2020045938
LC ebook record available at https://lccn.loc.gov/2020045939

ISBN: 978-0-367-37153-1 (hbk)
ISBN: 978-0-367-72448-1 (pbk)
ISBN: 978-0-429-35289-8 (ebk)

Typeset in Palatino
by SPi Global, India

Contents

Preface

The significance of the integration of blockchain, IoT, and AI in the fourth revolution driving the quality of life in the modern world has been recognized. The decentralized ledger and ultra-secure immutability, self-sovereign identity, and consensus mechanisms in blockchain along with AI algorithms, improve security. Moreover, interaction among users over the internet may create a huge impact using a smart contract in blockchain, and the integrated platform will likely to be revolutionary in many areas. Blockchain plays a significant role in providing a secured platform with IoT devices and AI algorithms. This book explores the concepts and techniques of blockchain, IoT, and AI. Also, the possibility of applying blockchain for providing security in various domains is discussed. This book highlights the application of integrated technologies in enhancing data models, better insights and discovery, intelligent predictions, smarter finance, smart retail, global verification, transparent governance, and innovative audit systems. This article will stimulate the minds of the IT professionals, researchers, and academicians towards fourth revolution technologies.

Editors

Naveen Chilamkurti is the reader and associate professor, and currently the Cybersecurity Discipline head, of Computer Science and Information Technology at La Trobe University, Australia. He completed PhD at La Trobe University in 2005. He is also the inaugural editor-in-chief for *International Journal of Wireless Networks and Broadband Technologies*, launched in July 2011. He is the author of various journal articles and multiple books. Dr. Chilamkurti was the head of the Department of Computer Science and Information Technology (Bundoora and Bendigo Campus) from April 2014 to 2016.

T. Poongodi is an associate professor in the School of Computing Science and Engineering, Galgotias University, Delhi, NCR, India. She completed PhD in information technology (information and communication engineering) at Anna University, Tamil Nadu, India. She is a pioneer researcher in the areas of big data, wireless ad-hoc network, Internet of Things, network security, and blockchain technology. She has published more than 50 papers in various international journals, national and international conferences, and book chapters in CRC Press, IGI Global, Springer, Elsevier, Wiley, and De-Gruyter and edited books in CRC, IET, Wiley, Springer, and Apple Academic Press.

Balamurugan Balusamy worked as associate professor for 14 years at Vellore Institute of Technology. He completed his bachelor's, master's, and PhD degrees at top premier institutions. His passion is teaching and he adapts different design thinking principles while delivering his lectures. He has worked on 30 books on various technologies and visited over 15 countries for his technical discourse. He has several top-notch conferences in his résumé and has published over 150 journal conference articles and book chapters combined. He serves in the advisory committee for several start-ups and forums and does consultancy work for industry on Industrial IoT. He has given over 175 talks at various events and symposiums. He is currently a professor at Galgotias University and teaches students and does research on blockchain and IoT.

Contributors

Rashi Agarwal
Department of Master of Computer
    Application
Galgotias College of Engineering and
    Technology, Greater Noida
Delhi-NCR, India

D. I. George Amalarethinam
Department of Computer Science
Jamal Mohamed College
Tiruchirappalli, India

K. P. Arjun
School of Computing Science and
    Engineering
Galgotias University, Greater Noida
Delhi-NCR, India

Jeevanantham Arumugam
Department of Information Technology
Kongu Engineering College
Erode, India

B. Balamurugan
SCSE
Galgotias University
Greater Noida, Delhi-NCR, India

T. Lucia Agnes Beena
Department of Information Technology
St. Josephs College
Tiruchirappalli, India

Geethu Mary George
Department of Computer Science and
    Engineering
PSG College of Technology, Peelamedu
Coimbatore, India

Dhanalekshmi Gopinathan
Department of Computer Science
Jaypee Institute of Information Technology
Noida, India

Joy Gupta
School of Computing Science and
    Engineering
Galgotias University, Greater Noida
Delhi-NCR, India

A. Ilavendhan
School of Computing Science and
    Engineering
Galgotias University, Greater Noida
Delhi-NCR, India

R. Indrakumari
School of Computing Science and
    Engineering
Galgotias University, Greater Noida
Delhi-NCR, India

L. S. Jayashree
Department of Computer Science and
    Engineering
PSG College of Technology, Peelamedu
Coimbatore, India

S. Karthikeyan
School of Computing Science and
    Engineering
Galgotias University, Greater Noida
Delhi-NCR, India

Supriya Khaitan
School of Computing Science and
    Engineering
Galgotias University, Greater Noida
Delhi-NCR, India

M. Kiruthika
Department of Computer Science and
    Engineering
Jansons Institute of Technology
Coimbatore, India

T. Kokilavani
Computer Science
St. Josephs College
Tiruchirappalli, India

Rajalakshmi Krishnamurthi
Department of Computer Science
Jaypee Institute of Information Technology
Noida, India

T. Krishnaprasath
Department of Computer Science and
    Engineering
Nehru Institute of Engineering and
    Technology (Affiliated with Anna
    University)
Coimbatore, India

D. Rajesh Kumar
School of Computing Science and
    Engineering
Galgotias University, Greater Noida
Delhi-NCR, India

Tapas Kumar
School of Computing Science and
    Engineering
Galgotias University, Greater Noida
Delhi-NCR, India

K. Lalitha
Department of Information Technology
Kongu Engineering College
Perundurai, India

Ch. V. N. U. Bharathi Murthy
School of Information Technology and
    Engineering
Vellore Institute of Technology
Vellore, India

S. Ponmaniraj
School of Computing Science and
    Engineering
Galgotias University, Greater Noida
Delhi-NCR, India

P. Priya Ponnuswamy
Computer Science and Engineering
PSG Institute of Technology and Applied
    Research
Coimbatore, India

C. Poongodi
Department of Information Technology
Kongu Engineering College
Perundurai, India

T. Poongodi
School of Computing Science and
    Engineering
Galgotias University, Greater Noida
Delhi-NCR, India

V. Gokul Rajan
School of Computing Science and
    Engineering
Galgotias University, Greater Noida
Delhi-NCR, India

S. R. Ramya
Department of Computer Science and
    Engineering
PPG Institute of Technology (Affiliated
    with Anna University)
Coimbatore, India

R. Ranjana
Department of Information Technology
Sri Sairam Engineering College
Chennai, India

A. Reyana
Department of Computer Science and
    Engineering
Hindusthan College of Engineering and
    Technology
Coimbatore, India

Sanjay Sharma
School of Computing Science and
    Engineering
Galgotias University, Greater Noida
Delhi-NCR, India

Yogesh Sharma
Computer Science and Engineering
Maharaja Agrasen Institute of Technology
G.G.S.I.P. University
New Delhi, India

T. Sheela
Department of Information Technology
Sri Sairam Engineering College
Chennai, India

M. Lawanya Shri
School of Information Technology and
    Engineering
Vellore Institute of Technology
Vellore, India

Ishita Singh
School of Computing Science and
    Engineering
Galgotias University, Greater Noida
Delhi-NCR, India

P. Sivaprakash
Department of Computer Science and
    Engineering
PPG Institute of Technology (Affiliated
    with Anna University)
Coimbatore, India

Nidhi Snegar
Information Technology
Maharaja Agrasen Institute of Technology
G.G.S.I.P. University
New Delhi, India

T. Subha
Department of Information Technology
Sri Sairam Engineering College
Chennai, India

S. Suganthi
PG and Research Department of Computer
    Science
Cauvery College for Women
Tiruchirapalli, India

R. Sujatha
School of Information Technology and
    Engineering
Vellore Institute of Technology
Vellore, India

D. Sumathi
School of Computer Science and
    Engineering
VIT-AP University
Amaravati, India

1

Blockchain

S. Suganthi
Cauvery College for Women, Tiruchirappalli, Tamil Nadu, India

T. Lucia Agnes Beena
St. Josephs College, Tiruchirappalli, Tamil Nadu, India

D. Sumathi
VIT-AP University, Amaravati, Andhra Pradesh, India

T. Poongodi
Galgotias University, Greater Noida, Delhi-NCR, India

CONTENTS

1.1 Introduction

Blockchain is an emerging innovative technology that has the capability of changing human transactions. It is a form of decentralized Distributed Ledger Technology (DLT) that stores digital data in a secure way, and it is distributed across all nodes connected in a peer-to-peer (P2P) network. A blockchain system is a fusion of several existing technologies such as shared ledger, distributed network, and cryptography, which enhance trust, data security, and transparency, and influences functionality. The distributed database holds a continuously growing list of blocks storing public transactions chained together through mathematical procedures. The blocks are records of transactions that take place among verified, world-wide users that enter a distributed, shared ledger; thus the entered records cannot be changed without the consensus of other users in the group. The decentralized nature, without a centralized server for data storage, enables the network to function even in the event of a node failure. Users can trust the blockchain system without depending on a possibly untrustworthy third party. Cryptographic functions with an immutable and shared ledger system enable technology to transfer data in a reliable and secure manner. Transparency of data and its functioning make verification and tracking of data easy as the data is open to all users in the network. The consensus protocols have a significant effect on the inconsistencies in the confirmation settlement of users and have to handle untrustworthy nodes in the network. Because of this, various improvements have been made on consensus algorithms to make applications function effectively.

Blockchains are initially intended for the purpose of recording transactions in a shared ledger so that data cannot be altered once published. In 2008 blockchain technology, combined with other technologies such as cryptography and other computing concepts, was used to produce electronic cash (cryptocurrency). Bitcoin was the first application to use blockchain-based cryptocurrency [1]. Although blockchains are mainly used in financial services, especially in Bitcoin applications [2], the cutting-edge technology enables blockchain to be incorporated in many sectors, including Internet of Things (IoT), reputation systems, public and social service, privacy, and security [3]. Some of its applications include smart contracts, fraud detection, identity verification [4], supply chain management, online payment, and digital assets. According to the Gartner report, by 2025 blockchain's business value will increase by more than $176 billion, and by 2030 it will grow to $3.1 trillion [5]. Although blockchains have the possibility of being implemented on a huge scale, there are certain inherent issues associated with it. Apart from process inefficiencies and cost-related issues, it has other problems like scalability, security, selfish mining, privacy leakage, and performance issues regarding the consensus algorithm [6]. The basic working mechanism of blockchain is depicted in Figure 1.1.

1.1.1 Evolution of Blockchain

In 1991 research scientists, Stuart Haber and W. Scott, introduced blockchains to store time-stamped documents secured cryptographically. In 1992 with the incorporation of Merkle trees, the system was made more efficient where more than one document could be included in a single block, but it was not used in practice. In 2004 computer scientist and

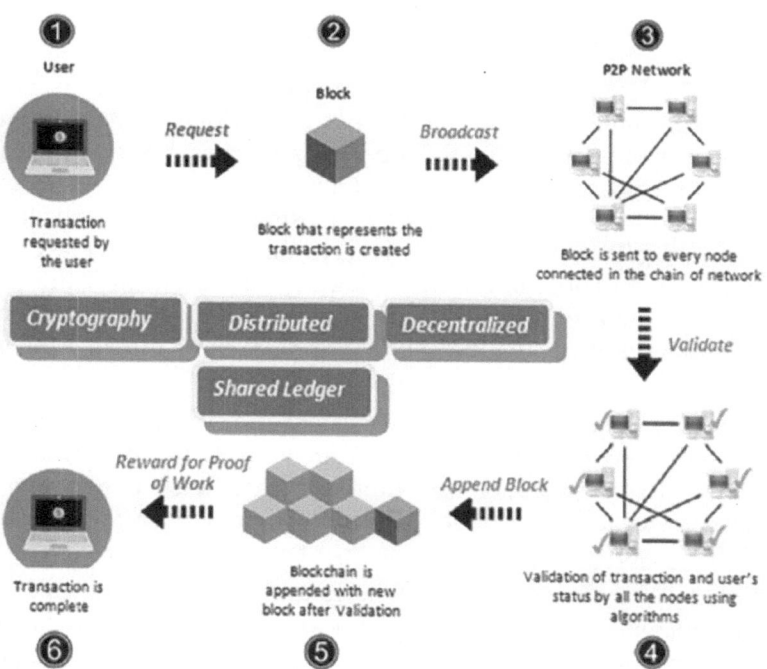

FIGURE 1.1
Working of Blockchain.

cryptographic activist, Hal Finney, introduced a system called Reusable Proof of Work (RPoW), which laid the prototype for the history of cryptocurrencies. In 2008 a model of blockchain was established on white paper and was posted to a cryptographic mailing list under the pseudonym *Satoshi Nakamoto*. This may be a single person or a group of developers working on the unknown blockchains. The system described is a P2P, decentralized cash system called *Bitcoin*. In 2009 the Bitcoin source code was released as open-source software. On January 3, 2009, Satoshi Nakamoto mined the first block (genesis block). On January 12, 2009, Hal Finney received 10 bitcoins from Satoshi Nakamoto and was the first recipient in a Bitcoin transaction. In 2013 Ethereum (Blockchain 2.0) was introduced by a Russian-Canadian programmer, cryptocurrency researcher, and writer, Vilatik Buterin. Emphasizing the need for a scripting language in Bitcoin to build decentralized applications, he started a new blockchain-based computing platform called *Ethereum*, which signified smart contracts that not only record currency but also other assets. Ehereum was implemented on July 30, 2015 and is the second largest public blockchain platform next to Bitcoin. *Ether* is the name of the cryptocurrency used in Ethereum. In December 2015 the Linux Foundation introduced the Hyperledger project with the motive to collaborate across industries by developing blockchains and distributing ledgers with improved reliability and performance. Later in 2018 and onwards many Hyperledger frameworks have been released. Blockchain technology is still growing and evolving with immense potential to be incorporated in a multitude of fields and gain mainstream acceptance.

1.1.2 Characteristics of Blockchain Technology

Blockchain technology has certain built-in characteristics enabling it to function efficiently including the following:

- *Decentralized* – The public blockchains are decentralized and users connected in the blockchain have access to the database, by which they can monitor, modify, and update the data. A trusted third party or a central node is not involved in the validation of transactions thereby reducing the cost and performance issues. Consistency and integrity are ensured by using consensus algorithms; however, consortium blockchains are partially centralized and private blockchains are fully centralized. Though different degrees of decentralization are provided depending on the type of blockchain and the policies employed, no node possesses complete control over the network as blockchain information is maintained by all the nodes.
- *Persistence* – The data in blockchain is persistent and is distributed and maintained by all nodes in the network. The transactions are validated by other nodes before being added to the block, hence it is mostly impossible to delete or alter any data because public blockchains are immutable. The consortium and private blockchains can be altered if the majority of the nodes want to alter.
- *Transparency* – The data is open to all users in the network and each record is time stamped, which makes verification and tracking very easy and transparent.
- *Provenance* – The origin of every transaction inside the blockchain ledger can be easily tracked.
- *Anonymity* – Real identity of the users is not exposed as they participate with the generated address in the network. Also in some systems identity is exposed and

the degree of exposure depends on the type of the system and policies implemented with the blockchains.

- *Autonomy* – The nodes on the blockchain network are independent and other nodes cannot intervene in its functioning.

1.2 Components of Blockchain

Components that play a significant role in the blockchain process include cryptographic hashing functions, asymmetric keys, shared ledger, transactions, blocks, and the consensus protocols. A brief description of these components follows:

1.2.1 Cryptographic Hash Functions

Cryptographic hash functions are different from other hash functions in that they are more suitable for applications involving cryptography, security, or privacy. They have many applications including encryption, password security maintenance, digital signatures used in e-commerce protocols, and Bitcoin generation. It is a mathematical function that takes any input such as a file, text, or an image, and produces a single fixed-size output for that particular input. Thus, the set of mathematical transformations applied to the input data gives an output bit string, which is compressed and called a *digest, hash, tag*, or *hash value*. This output value is uniquely produced for that particular input and any modifications to the input data will produce a completely different output digest value. The property that no two different inputs will hash to the same output is called *collision resistance*, and the digest output generated is random and irrelevant to the input data. Collision resistance makes it is very hard to predict relevant information about the input data using the digest value. In blockchain, deriving addresses, creating unique identifiers, digital signing, authenticating block data, and block headers are done through hashing functions. Also, cryptographic hash functions should be computationally efficient and compute outputs quickly.

1.2.2 Asymmetric-Key Cryptography

Asymmetric-key cryptography is an encryption mechanism used in blockchains that makes use of asymmetric keys and cryptographic hash functions. It is asymmetric in the sense that each user possesses two different keys called the *private key* and *public key* for the encryption and decryption processes involved in blockchains. This coupled with cryptographic hash functions are used in digital signatures for signing transactions and in untrusting environments. Digital signatures are a digital alternative for handwritten signatures which provides data integrity, data origin authentication, and non-repudiation. Digital signatures use Elliptical Curve Digital Signature Algorithm (ECDSA) [7]. Digital signing consists of two phases: the *signing phase* and the *verification phase*. In a blockchain transaction when a user wants to start a transaction, the private key is kept secret with the user and copies of the authentic public key are known by all users on the network. Before a message is transmitted mathematical transformations are applied to the message and the private key with cryptographic hashing functions are applied to the message, and the resultant message digest is combined to provide the digital signature, which is a sequence of numbers. In the verification phase the user receiving the message decrypts that message using the public key provided to them, and verifies the message by comparing the hash value.

1.2.3 Transactions

Transaction takes place when two different parties interact by transfer of data across different nodes. Transaction takes place when data are transmitted over the network, which involves information such as the sender's identifier, digital signature, public key, transaction inputs, and outputs. The digital data to be transferred are called *inputs* and the receiver account that gets the digital asset is called *outputs*. It also contains details of the receiver's identifier and how much data should be transferred. The transactions received over a particular period of time are combined into a transaction block. The size of the transaction and the block determine the number of transactions stored in the block. The transactions are hashed in pairs to get a single digest value. When a user requests a transaction all the nodes in the network receive this message along with other transaction messages. Transactions are verified with the digest value by other users in the network and recorded in the ledger book after validation. The transaction will have a reference to the origin or source of the data, which can be a reference to past events or the origin event in the case of new transactions. Transactions are not encrypted, so users can publicly view all technical details.

1.2.4 Ledgers

Blockchain uses a distributed ledger database that contains a collection of transactions shared and synchronized across different locations among multiple users. Cryptography and digital signatures accurately store transactions and contracts in a decentralized way, so validation and verification of transactions by a central authority is eliminated. Blockchains are resistant to cyberattacks because the records become unalterable once they are stored. Also due to its distributing nature, copies of records are spread across many computers, so modifying a single record requires modification to all other copies of the record, which is impossible. The flow of information is enhanced in ledgers, thereby making an audit trail easy for accountants to follow.

1.2.5 Blocks

Blocks are records holding unalterable digital information related to transactions that together form a blockchain. They are like ledger pages containing a collection of zero or more transactions. A block is created by a publishing node, and transactions are added to the block.

Logically a blockchain is a chain of specific digital information stored in the form of blocks, and any new block can be appended to the chain of blocks. Each block contains a *block header* containing metadata about the block and *block data* containing the set of transactions and other related data. Each block header, except the very first block, contains a link to the immediately previous block's header. The link is usually a reference, which is the hash value of the previous block or the parent block. Thus, a change of data in a block would change its hash value and reflect in the subsequent blocks. The very first block of a blockchain is called the *genesis block* and it does not have a parent block. Once a block reaches a certain number of approved transactions, a new block is formed. A block usually consists of a block header and block data; this may vary depending on blockchain implementation.

Block header consists of:

- *Block version* – indicates which set of block validation rules to follow.
- *Timestamp* – contains current approval time for a particular block as seconds in universal time since January 1, 1970.
- *Parent block hash* – 256-bit hash value pointing to the previous block.
- *Merkle tree root hash* – the hash value of all the transactions in the block.
- *nBits* – target threshold, which is the mining difficulty or block-creation difficulty of a valid block hash.
- *Nonce* – a four-byte field usually starting with zero and increasing for every hash calculation. It changes the hash output of the block contents.

Block data consists of:

- *Transaction counter* – maintains the total number of successfully completed transactions.
- *Transaction data* – depends on the purpose of transactions performed and the data stored in the fields of business data, healthcare data, or data related to Bitcoins.

1.2.5.1 Merkle Tree

The Merkle binary tree is primarily a data structure used to store the hash value of different blocks in a large data set supporting the efficient verification process of the complete data available in the set. In every block, all transactions are summarized with the help of a digital fingerprint. Participants can validate transactions using the Merkle tree whether it belongs to a particular block or not. The hierarchical structure in the Merkle tree is made by hashing pairs of children nodes, and the same process is continued until the root node known as the *Merkle root* is obtained. The hashing process is performed between pairs using a bottom-up approach from each transaction. The leaf node contains the hash value of each transactional data, and the non-leaf node contains the hash value of the previous hash. The root node summarizes the complete information about transactional data ensuring data integrity. If any details in the transactional data are altered, it disturbs the root node because the blocks are interlinked. If any transaction is altered in the hashing pairs, the root node is also modified. If any attacker or hacker disturbs any transaction in a block, it is much easier to track or identify unauthorized access by recomputing the hash value of the blocks.

1.2.6 Consensus Algorithms

Consensus is the agreement among groups of nodes about their data, and consensuses play a major role in defining the efficiency and scalability of the system being implemented. In blockchains the nodes are decentralized and there is no single node to check the consistency of the ledgers; some protocols are needed to ensure this. Also, in a group of distributed nodes, reaching an agreement of who publishes the next block is difficult when untrustworthy nodes may also exist. This situation is similar to the classic Byzantine Generals (BG) Problem. There are different approaches for reaching a consensus mentioned next.

1.2.6.1 Proof of Work (PoW)

Proof of work is a consensus algorithm in which the main strategy is to select a node for publishing a block of transaction from a group of competing distributed nodes. All the nodes compete by performing lots of calculations, which proves that the node is legitimate and consumes a lot of work done on the part of users. The calculations are done by frequently calculating the block header's hash value by changing the nonce. The node that has computed a hash value less than or equal to certain a target value is selected. The selected node can publish the block and is rewarded for his work. All other nodes validate the selected block, and once validated the transaction block is added to the chain.

1.2.6.2 Proof of Stake (PoS)

This is a type of consensus algorithm that considers the amount of stake or cryptocurrencies held by an individual user node. The users stake the cryptocurrencies in order to bid on a proposal to publish the block. The person with more stake or cryptocurrency is considered more legitimate and given the chance to publish the block. All other nodes validate the transactions. When the number of attestations from other nodes reaches a target value, the block is appended to the block chain. In this procedure, the nodes that publish the block, as well the nodes which validate the transactions, receive rewards. The major drawback is that the person holding more stake overpowers other users, but this can be altered by changing the strategy formula through methods such as combining the stake size with age of the users.

1.2.6.3 Practical Byzantine Fault Tolerance (PBFT)

Byzantine fault tolerance is the ability of a distributed network to function and reach a consensus, even when the network nodes fail to respond, or they respond incorrectly. The term is derived from the *Benzantine Generals Problem*. It can handle up to one third of malicious user nodes in a group of nodes. The node for publishing is selected in a round. Based on certain rules a primary node is selected in each round, which carries out the transactions and all other nodes are secondary nodes. The process has three phases: *pre-prepared*, *prepared*, and *commit*. If a node gains two thirds of the votes from all other nodes in each phase, it can pass through the successive phases.

1.2.6.4 Delegated Proof of Stake (DPOS)

In PoS the block producers are selected depending on the stake they hold. But DPoS differ from PoS in that the delegates or the representatives are elected by the stakeholders for block creation and validation. In DPoS, the system has only a few nodes to validate resulting in the easy confirmation of blocks. The nodes are rewarded for their work and given the status of publishing node, and they cannot act maliciously as they can be voted out.

1.2.6.5 Round Robin Consensus Model

This is used in permissioned blockchains and in this consensus model the blocks are published by the nodes only when they get their turns. A node has to wait for several block creation cycles until they get a turn. If a node is not available for publishing within their time limit, other nodes will be allowed to publish.

1.2.6.6 Proof of Authority (Identity) Model

This consensus model is used in permissioned systems where a publishing user identity is verified and included in the blockchain. The reputation of the user can be built on behavior and determine the chance to publish a block.

1.2.6.7 Proof of Elapsed Time Consensus Model (PoET)

In this model a random wait time is allotted by software when a request is made for publishing a block. The node waits for that particular elapsed time and then starts publishing a block. A signed certificate is issued with respect to the randomly allotted time and is published by the user along with their block. This process is very transparent and makes other users aware of when a malicious user is idle for a period of elapsed time and starts publishing his blocks earlier.

1.3 Types of Blockchain

The researchers [8] classified blockchain based on the nature of data accessibility, the need of authorization to participate in blockchain and core functionality, and smart contract support. The blockchain categorization is listed in Table 1.1.

1.3.1 Public Blockchain

Public blockchains are open source. In public blockchain any users, developers, miners, or community members are allowed to participate in the process. All transactions that take place on public blockchains are fully transparent, meaning that anyone can check the transaction and verify it. Transactions are documented as blocks and connect together to form a chain. Each new block must be time stamped, and all computers connected to the network, known as nodes, must validate the block before it is written into the blockchain. A public blockchain is distributed, decentralized, and immutable. Bitcoin, Ethereum, Monero, Zcash, Steemit, Dash, Litecoin, and Stellar are examples of public blockchain [9]. The public blockchain is depicted in Figure 1.2.

TABLE 1.1

Types of Blockchain

Area of Concern	Blockchain Types
Nature of data accessibility	• Public blockchain • Private blockchain • Community/federated • Consortium blockchain • Hybrid blockchain
Authorization to participate in blockchain	• Permissionless blockchain • Permissioned blockchain • Hybrid blockchain
Core functionality and smart contract support	• Stateless blockchain • Stateful blockchain

FIGURE 1.2
Public Blockchain.

In a public blockchain, the following activities can be done by any user without the permission of a centralized entity:

- After downloading the code from the Internet, users can run a full node using their local device and validate transactions in the network.
- By installing an application on their device, users can mine a block of transaction, write data to the blockchain in consequence, participate in the consensus process, and receive network tokens in the process.
- Users can view all transactions that occurred on the blockchain using public block explorer software or perform chain analysis on the data stored on a full node.

1.3.2 Private Blockchain

This type of blockchain is suitable for organizations. The participants are permitted to read or write in the blockchain only by a single authority who can give selective permission to the participants based on their role in the organization. The information in the private blockchain is secured using cryptography. Compared with the public blockchains, private blockchains are rapid, competent, and more commercial. The Figure 1.3 shows the network of private blockchain.

Ripple (XRP) and Hyperledger are examples of private blockchain. Ripple is a nonproprietary protocol that uses RippleNet to link up payment providers and banks for easy money transfer overseas. Hyperledger is a non-proprietary project involved in associating world leaders in banking, supply chains, finance, and technology [10].

1.3.3 Consortium Blockchain

Consortium blockchain is a special type of private blockchain referred to as a *community/ federated blockchain*. In this type, the authority is not a single entity, rather it is a consortium formed by a group of enterprises. The enterprises involved in this blockchain are

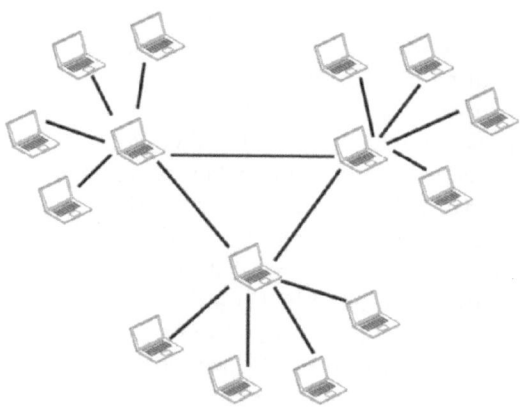

FIGURE 1.3
Private Blockchain.

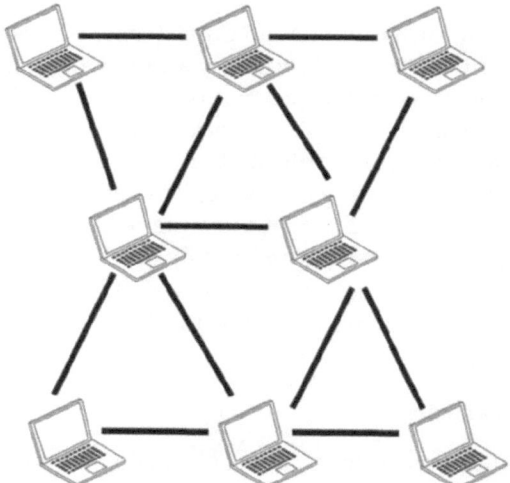

FIGURE 1.4
Consortium Blockchain.

benefitted through some characteristics of their industry [11]. Examples of consortium blockchains are Corda and Quorum [12]. Figure 1.4 represents the network of consortium blockchain.

Corda is a non-proprietary blockchain venture dedicated to collaboration of companies. The companies involved in Corda exploit blockchain network for smart contract privacy and inter-business operation. Quorum is a distributed ledger policy established on Ethereum. Quorum, with the permissioned realization of Ethereum, is able to facilitate the smart contract privacy and inter-business operation for its industry collaborators.

1.3.4 Hybrid Blockchain

The hybrid blockchain is the combination of the public and private blockchain for the better business operations. The company does not need to be concerned about the information leak as the business operations are performed under a controlled environment. Thus, the hybrid blockchain facilitates the organizations having good communication with their stakeholders by incorporating both public and private blockchains. The hybrid blockchain participants can decide which members can take part in the blockchain and which transactions are to be made public. Due to the immutability characteristics of the blockchain, the security of the transaction is preserved from hackers to a certain extent. The transaction cost is low in hybrid blockchain as information has to be verified only by a few powerful nodes.

Hybrid blockchains are suitable for Internet of Things (IoT) applications. The devices involved in the IoT network can be sheltered with private blockchain, and sharing of data can be made possible with the public blockchain. Another important application where hybrid blockchains can be applied is finance. XinFin, a global trade and finance organization, applies Ethereum for the public module and Quorum for the private module in their accomplishment [13] and utilized Proof of Stake (DPOS) as a consensus mechanism. Banking, supply chain, governments, and enterprise services are the other sectors where hybrid blockchain can be employed.

1.3.5 Permissionless Blockchain

In decentralized ledger platforms anyone can publish new blocks because access rights are not controlled by any authority [14]. Since it is a non-proprietary software anyone can download and use it. All users in the permissionless blockchain network can read and write into the ledger. It is also termed as *public blockchain* as it permits anyone to link in the network and make a transaction. The information on this blockchain is openly accessible and the ledger copies are available worldwide. This enables hackers to introduce new blocks in such a way that they can threaten the system. To avoid the intruders, consensus systems such as *proof of work* and *proof of risk* can be used. As it is open to all, there is contradiction between the characteristics, speed, and scalability. The notable examples of permissionless blockchain are Ethereum, Bitcoin, Dash, Monero, and Litecoin.

1.3.6 Permissioned Blockchain

Permissioned blockchains, also termed *private blockchain*, are thought of as a closed environment where anyone who wants to publish a block, read, and validate a block needs to get the approval of a centralized authority. This type of blockchain is appropriate for banks, companies, and enterprises where the permitted members and their access rights can be pre-defined for smooth commercial operations.

Permissioned blockchain are also applicable at the community level. The enterprises that have the common understanding can encourage their business partners to document their transactions on a public distributed ledger and apply consensus modules to maintain trust among them. This transparency may assist them in making better corporate decisions. Periodic and explicit reviews can be made easily for better perception among business associates. Perfect examples of permissioned blockchain are Corda, Ripple, Hyberledger Fabric, and Quorum.

1.3.7 Stateless Blockchain

The value inside a block at any moment is called a *state*. As the transactions are evolving and processed, the virtual machine's state changes. To optimize the transactions and verify the transaction using computing hashes in a sequential processing, stateless blockchain system is applied. Bitcoin, IOTA, Ripple, NEM, MultiChain, BigChainDB, and OpenChain are examples of stateless blockchains. Statelessness leads to the formation of light nodes, which only hold the sequence of headers and do not execute any transactions or associated states. A stateless node carries zero information regarding the state when it arrives online. These new nodes require lower disk, I/O, and memory usage.

Stateless clients and stateless miners are resolved by sharding in Ethereum 2.0 [15]. Since all the nodes are stateless, it benefits the blockchain by faster processing and rises the scalability. Therefore, Ethereum blockchains are built in a linear fashion. One disadvantage is the traffic jams that may lead to inefficient data processing. To avoid this, sharding is applied as the Ethereum system is divided into clusters termed as *shards*. Each shard holds an independent state. Transactions are allotted to various shards in the network so that processing is performed in a parallel manner improving the efficiency. The technique of sharding also increases the security of the blockchain by rearranging clients among shards regularly.

1.3.8 Stateful Blockchain

This type of blockchain offers smart contract and transaction processing facilities. Stateful blockchains can optimize and protect the logic states of a multifaceted company. Ethereum also has a stateful blockchain system where each node keeps a replica of the blockchain. Using this replica, users are able to validate their current version of the state and that it began at a commonly agreed upon genesis block. Stateful clients will have higher load on I/O, memory, and disk while using less bandwidth. Some of the examples of stateful blockchain are Corda, Hyperledger Fabric, Kadena, Tezos, Sawlooth, Hydrachain, and Quorum.

1.4 Blockchain Use Cases

1.4.1 Blockchain in Supply Chain Management

Supply chain management involves certain activities like the cohesive planning, flow of information from one department to another, material flow, financial capital flow, and the management of goods. Cooperation among various stakeholders in various departments is complex. With the deployment of blockchain, the supply chain shows tremendous progress in transparency and exhibits a great impact from storage management to the delivery process. Sharing information among various stakeholders is easier due to data interoperability. Due to the absolute behavior of blockchain, tampering is avoided, and as a result of this, trust is maintained between parties. Next, many areas where blockchain could be implemented will be discussed.

1.4.1.1 Provenance Tracking

Tracking of records is a difficult task for big organizations. The information about the product could be accessed through the sensors deployed in products as well as RFID tags. Blockchain aids in tracing the product from the origin to the current time, which helps to detect frauds in supply chain management. Through the tracking process, handling a crisis could be done properly with a detailed analysis on products that are purchased through the location of consumers and retailers. As soon as the product is identified, verification of audit logs could be done easily once reason for the contamination is identified. For instance, Walmart's blockchain pilot in China deployed blockchain where tracing of a package of mangoes from the farm to the store is done within a few seconds. A special feature is that the information is shared among all stakeholders promptly with self-reliance across the trusted network [16]. The largest retailer in Europe, Carrefour, has deployed this block-chain technology in tracing the production of unconfined chicken in the Auvergne region of central France. Smartphones are used to scan a code available on the package to get information about details of the feed given to the chicken, how the chicken grew, and meat processing. In these two instances, it is observed that the deployment of the blockchain in various phases of supply chain reduces failure at certain points, enhances trust, and increases security and transparency. Implementation of DLT might augment trade volume by 15%. Also, ethical production and ingestion of commodity is expanded through it.

1.4.1.2 Transparency

Open access to data is public in addition to gaining access to certifications and claims. Authentication could be checked by third parties as the registration is done on the Ethereum blockchain.

One example of blockchain-based e-commerce market is Coupit, which checks the status of sellers and buyers. A claim is listed as soon as the transaction is recognized. When the transaction process is initiated, information is hashed, and the blockchain is updated so that the required information is made accessible to approved users that participate in transactions.

1.4.2 Blockchain in Logistics

Implementation of blockchain in logistics helps:

- Automation being done to produce cost savings besides the generation of secured and error-prone processes.
- Data exchange being augmented among stakeholders so that the prediction of logistics process and transparency could be established.
- Amalgamation of blockchain and IoT provides more perceptions about the prevention measures that are handled against counterfeit and transportation conditions.

1.4.2.1 DexFreight

DexFreight is a freight exchange based on blockchain that is communicated between shippers and carriers so that negotiations can be done directly. An open platform and communication with partners are also provided to users. Information associated with a trucking company is being stored so that the onboarding procedure for subcontractors is done rapidly. Therefore, shippers could focus on negotiations associated with freight tasks. As stated in a freightwaves.com article from 2018, the complete process of contracting freight transportation is done based on the blockchain [17]. In another freightwaves.com article from 2019, the authors indicated that combined work of DexFreight and CargoX provided documents for contracting and the complete original transportation process was moved under one platform [18].

1.4.2.2 GS1

Digitizations of pallet notes are done to move the freight. A pallet-sharing program facilitates the movement of freight in a flexible manner where information about the owner, flow of the pallets, and state of the pallets is required. Blockchain is deployed for the information storage and transfer about the pallets among various stakeholders. An article published by gs1-germany.de states that digitization of this procedure is done [19]. Once the trial period is completed, the project is found to be a success and the system is developed [20].

1.4.3 Blockchain in Banking

Apart from efficiency, transparency, and security, speed is also considered as a significant benefit that makes blockchain an important factor for deploying in banking. There are several uses that could be carried out in the banking field described here:

1.4.3.1 Digital Identity Verification

Identity registration must be done on the blockchain one time. Storing this information on a blockchain assures security. Through this enhanced verification process, customers and companies experience great benefits.

1.4.3.2 Auditing and Accounting

Conventional double-entry bookkeeping systems could be streamlined, and simplification of compliance could be done with the blockchain. Entries in the joint register would be done in a distributed manner creating transparency and security. Verification of all transactions would be done similar to a digital notary.

1.4.3.3 Faster Payments

Deployment of blockchain in banking sectors diminishes the necessity for the verification task of third parties, which quickens processing time for conventional bank transfer processes. A decentralized channel was launched for payments that facilitating bank sectors used to make state-of-art technologies so that payments could be made faster and processing fees could be lowered. Hence, a new level of service was launched through the introduction of new products and endeavoring pioneer start-ups.

1.4.3.4 Clearance and Settlement Systems

Conventional centralized SWIFT protocol takes care of only the payment process. Intermediaries are involved in the process of money transfer, which results in more expenses and time consumed. By using blockchain there is no need for depending on the regulatory bodies. Also tracking of public and transparent transactions is done easily.

1.4.3.5 Asset Management

Decentralization of digital assets is constructed to facilitate the financial market. Rights of an asset can be transferred with the help of cryptographic tokens that represent off-chain assets. Pure digital assets are obtained with currencies such as Ethereum and Bitcoin. Removal of an intermediate person diminishes the asset exchange fees and therefore hastens the process.

1.4.4 Blockchain in Education

Implementation of blockchain is still in its primary phase. Validation and sharing of academic certificates are made possible with blockchain [21]. Various fields such as certificates management, learning outcome management, transfer of fees and credits, and security of collaborative environments are discussed here:

1.4.4.1 Certificates Management

This deals with various forms such as transcripts, academic credentials, accomplishment records, and certificates of students. Blockchain is used to provide digital certificates to students. In an article about cyberschool published in *Pedagogy, Culture & Society*, school is used as a certification agent who compensates for the blockchain certification platform [22]. Authorization by education providers along with privacy preserving is done for certificates that are provided to students. Thus, it becomes flexible for students to share certificates whenever it is required. Verification and issuing of transcripts are made possible with the development of novel blockchain-based education records based on the principle of decentralization [23]. Individuals would be able to access the data records; however, accessing and modifying the data stored in the system is given only to the proficient institutions under some rules with limited constraints.

1.4.4.2 Learning Outcome Management

Certain blockchain applications could be developed so that the enrichment of the learning objectives and the fulfillment of competencies could be obtained. Qualitative and quantitative metrics are used for the evaluation of students' performance and done through the blockchain. Multilearning activities are made so that the performance of students can be assessed [24]. A learning environment based on the blockchain has been proposed with apt, reliable support, and significant feedback [25]. The objective of the application is to boost critical thinking, problem solving, and the learning process through skills by means of proper communication and improved cooperation.

1.4.4.3 Collaborative Environment

The ubiquitous learning (u-learning) system is based on blockchain technology [26] so that collaborative learning environments are provided for students at any time regardless of their locations. This system is supported by an interactive multimedia system so that students and teachers may communicate effectively and efficiently. Through the development of blockchain as a School Information Hub (SIH) [27], analysis, collection, and reporting of data with respect to the school systems aids in the decision-making process.

1.4.4.4 Transfer of Fees and Credits

Blockchain has been implemented in various universities, institutions, or organizations so that transfer processes of fees or credential records are done with an effective trust and high security. Through this technology, the need for the intermediary could be removed as the system facilitates the process of transfer through tokens which could be in the form of courses, diplomas, and certificates [28]. Secured transfer process could be handled in each educational institution that has its personal EduCTX address.

Deployment of blockchain in the education sector results in various benefits such as transparency, sharing of data, and improved trust. A detailed review has been done in the journal, *Applied Sciences*, that focuses on various advantages of blockchain in education, issues due to the establishment of blockchain, and various blockchain-based educational applications [29].

1.4.5 Environmental Awareness and Waste Management

Blockchain has been initiated in waste management activities, such as in Canada where plastic waste is decreased through the global recycling business. Rewards in the form of blockchain-secured digital tokens are provided to individuals who bring plastic waste to recycling centres. With the help of these tokens, purchasing items like phone charging units and food in stores are done through the plastic bank app [30]. Recycling and sorting waste produced by grocery chains is observed at swachhcoin.com [31].

1.5 Challenges and Opportunities

Though blockchains have a vast scope of being implemented in many fields, there are many related issues that should be seriously considered.

- *Security and Privacy of the Data*
  Blockchain uses a decentralized system in which data is shared among different services and nodes and so there is a potential chance of data leakage. Everyone on the chain has access to the data against the centralized system in which authorization is provided by an intermediate trusted third party. The data requiring security includes corporate financial data, personally identifiable information, daily individual activities, and medical records [32]. Blockchain relies heavily on cryptographic algorithms, smart contracts, and software which may also have flaws and loopholes making many organizations and industries reluctant to apply blockchain technology.

- *Storage*
  Blockchains have limited on-chain data storage with decentralization and hashing architecture [33]. As people are people becoming aware of technology, the amount of data generated is voluminous with the increasing number of users, so blockchain applications should be designed taking scalability issues into consideration.

- *Standardization*
  Blockchain is used in different infrastructure and applications; a high level of standardization is required. Implementation of predefined universal standards for size, format, and nature of data is essential. As a relatively new technology the potential benefits, as well as related issues, are unknown and so it faces the problem of legal regulations acceptance from many countries [4].

- *Scalability*
  Blockchain technology should be able to handle an increasing number of users and devices such as sensors, smart devices, or IoT which will be more prevalent in the near future. In blockchain technology the chain is growing at a rate of 1MB per block every 10 minutes in Bitcoin and there are also copies of data stored in the nodes [34].

The transactions have to be stored and maintained for validation. Also, the number of transactions processed per second is less and would be a major challenge when the number of users increases. The computational overhead required is directly proportional to the increasing number of users and devices, which in turn leads to performance and synchronization issues.

- *Interoperability*
  Interoperability is a major issue as most blockchain systems are not designed to operate with other systems based on blockchain. The sharing of data among various commuting providers and services, which is very vital in healthcare systems for example, is a major issue in blockchain. Blockchain systems should be designed to be interoperable with different systems.

- *Key Management*
  In blockchain technology the data is distributed and all users in the block have access to the data. So, the cryptographic process that uses private and public keys is used for the encryption and decryption of data. Currently blockchain uses one key for all blocks and any leakage of this key may leave data in the whole block vulnerable [4].

- *Blockchain Vulnerabilities*
  The inherent blockchain framework is vulnerable to many malicious attacks such as a 51% attack (in which a user dominates the other users with most of the computational resources), and double spending attack (in which a user spends the same cryptocurrency for more than one transaction).

- *Regulations and Governance*
  The advent of any new technology poses serious problems in framing regulations for the system to maintain the integrity and safety of the system. Blockchains function as part of a government infrastructure that needs proper rules, and governance should be interpreted and implemented.

- *Social Challenges*
  Blockchain is a new and evolving technology, so societal acceptance and shifting from traditional technology unaware of its pros and cons is a big challenge.

- *Lack of Human Resource*
  Blockchain technology is considered relatively new and so there is a lack of experts and developers who could manage and sort out related issues. The shortage of trained and skilled professionals makes organizations hesitant to adopt the new technology.

- *Accountability*
  The transparent, sharable, and verifiable nature of data in blockchain plays an effective role in the accountability of data. It provides a multilayered data protection mechanism through decentralization, consensus, and hashing algorithms. Once data are stored, it is immutable without consent from other users on the network, and it creates trust and accountability from all parties involved.

- *Accuracy*
  In blockchains data can be shared and verified by all parties connected in the chain and updated immediately. The data are provided in a reliable and timely manner ensuring the accuracy of data.

- *Cost Efficient*
  The users can make smart financial decisions with the help of accurate and documented information provided through blockchains. Also, the operation cost of data analysis and distribution spent on third party service providers is eliminated.

- *Anti-Counterfeiting Actions*
  Fraudulent actions can be accidental or malicious and offenders can be punished accordingly. Blockchains hold transactions that are time-stamped and stored in a distributed decentralized way thereby ensuring stakeholders of a particular blockchain network to store, verify, and validate the authenticity of information from its origin. This inherent design makes fraudulent actions transparent to all users on the network [35].

- *Improving Research and Development*
  Anonymity of the data in blockchain plays an effective role in conducting research. Also, at present there are fake data that affect research quality. Blockchain produces a transparency in tracking these data from the real generators to the end users who analyze the data. Blockchain technology coupled with other technologies such as artificial intelligence and machine learning can take research in many fields to enormous heights.

1.6 Conclusion

Blockchain is expected to revolutionize many sectors with its potential characteristics ensuring secure transactions. Beyond the technology being adapted in many fields there are problems associated with storage, scalability, standardization, interoperability, and vulnerabilities related to the consensus algorithm, which needs further research. Also as an emerging technology, more efforts are needed to ensure its privacy and security. In this chapter an overview of blockchain is presented with its significant characteristics and components. The types of blockchain are analyzed in detail depending on data accessibility, authorization, and core functionality. Later the use cases of blockchain are elaborated with illustrations to help the readers understand the potentiality of blockchain in various fields. Finally, challenges and opportunities of blockchain are evaluated in depth, giving a glimpse into the future of technology.

References

[1] Yaga, D., P. Mell, N. Roby, and K. Scarfone. 2018. "Blockchain Technology Overview." *Natl Instit Stand Technol*. doi:10.6028/NIST.IR.8202.

[2] Nofer, M., P. Gomber, O. Hinz, and D. Schiereck. 2017. "Blockchain." *Business Informat Syst Eng* 59 (3): 183–187. doi:10.1007/s12599-017-0467-3.

[3] Zheng, Z., S. Xie, H.-N. Dai, X. Chen, and H. Wang. 2014. "Blockchain Challenges and Opportunities: A Survey." *Int J Web Grid Services* 14 (4): 352.

[4] McGhin, T., K.-K. R. Choo, C. Z. Liu, and D. He. 2019. "Blockchain in Healthcare Applications: Research Challenges and Opportunities." *J Network Comp Appl* 135: 62–75.

[5] https://www.gartner.com/en/newsroom/press-releases/2019-07-03-gartner-predicts-90--of-current-enterprise-blockchain.

[6] Zheng, Z., S. Xie, H. Dai, X. Chen, and H. Wang. 2017. "*An Overview of Blockchain Technology: Architecture, Consensus, and Future Trends.*" Paper presented at *IEEE 6th International Congress on Big Data*, Honolulu, Hawaii, US, June 25–30.

[7] Johnson, D., A. Menezes, and S. Vanstone. 2001. "The Elliptic Curve Digital Signature Algorithm (ECDSA)." *IJIS* 1: 36–63. doi:10.1007/s102070100002.

[8] Shrivas, M. K., and T. Yeboah. 2018. "The Disruptive Blockchain: Types, Platforms and Applications."

[9] https://blockchainhub.net/blockchains-and-distributed-ledger-technologies-in-general/. Accessed June 1 2020.

[10] https://hedgetrade.com/what-is-a-private-blockchain/. Accessed June 1 2020.

[11] https://dragonchain.com/blog/differences-between-public-private-blockchains/. Accessed June 2 2020.

[12] https://www.mycryptopedia.com/consortium-blockchain-explained. Accessed June 2 2020.

[13] https://101blockchains.com/hybrid-blockchain. Accessed June 3 2020.

[14] Yaga, D., P. Mell, N. Roby, and K. Scarfone. 2019. "Blockchain Technology Overview." arXiv:1906.11078.

[15] https://docs.ethhub.io/ethereum-roadmap/ethereum-2.0/stateless-clients/. Accessed June 3 2020.

[16] Forbes. 2018. "3 Innovative Ways Blockchain Will Build Trust in the Food Industry." https://www.forbes.com/sites/samantharadocchia/2018/04/26/3-innovative-ways-blockchain-will-build-trust-in-the-food-industry/#839519f2afc8.

[17] Prevost, C. 2018. "DexFreight Completes First Truckload Shipment Using Blockchain." Accessed March 25 2019. https://www.freightwaves.com/news/blockchain/technology/dexfreight-completes-first-truckload-shipment-using-blockchain.

[18] Rajamanickam, V. 2019. "CargoX's Blockchain-Based Smart Bill of Lading Solution Is Now on DexFreight's Platform." March. Accessed May 13 2019. https://www.freightwaves.com/news/blockchain/cargoxs-blockchain-based-smart-bill-of-ladingsolution-is-now-on-dexfreights-platform.

[19] Uhde, T. 2018. "Blockchain-Technologie und Architekturim Pilot-Projekt." December. Accessed March 25 2019. https://www.gs1-germany.de/innovation/blockchainblog/das-passende-system-blockchain-technologie-und-architek/.

[20] Nallinger, C. 2018. "Hat der Palettenscheinausgedient?" December 17. Accessed April 3 2019. https://www.eurotransport.de/artikel/palettentausch-mithilfe-der-blockchainhat-der-palettenschein-ausgedient-10634469.html.

[21] Chen, G. B. Xu, M. Lu, and N.-S. Chen. 2018. "Exploring Blockchain Technology and Its Potential Applications for Education." *Smart Learn Environ* (5)1.

[22] Nespor, J. 2018. "Cyber Schooling and the Accumulation of School Time." *Pedag Cult Soc*: 1–17.

[23] Han, M., Z. Li, J. S. He, D. Wu, Y. Xie, and A. Baba. 2018. "*A Novel Blockchain-Based Education Records Verification Solution.*" In *Proceedings of the 19th Annual SIG Conference on Information Technology Education*, Fort Lauderdale, FL, US, October 3–6.

[24] Farah, J. C., A. Vozniuk, M. J. Rodríguez-Triana, and D. Gillet. 2018. "*A Blueprint for a Blockchain-Based Architecture to Power a Distributed Network of Tamper-Evident Learning Trace Repositories.*" In *Proceedings of the IEEE 18th International Conference on Advanced Learning Technologies (ICALT)*, Mumbai, India, July 9–13.

[25] Williams, P. 2018. "Does Competency-Based Education with Blockchain Signal a New Mission for Universities." *J High Educ Policy Manag* 41: 104–117.

[26] Bdiwi, R., C. De Runz, S. Faiz, and A. A. Cherif. 2018. "*A Blockchain Based Decentralized Platform for Ubiquitous Learning Environment.*" In *Proceedings of the 2018 IEEE 18th International Conference on Advanced Learning Technologies (ICALT)*, Mumbai, India, July 9–13.

[27] Bore, N., S. Karumba., J. Mutahi, S. S. Darnell, C. Wayua, and K. Weldemariam. 2017. "*Towards Blockchain-Enabled School Information Hub.*" In *Proceedings of the 9th International Conference on Information and Communication Technologies and Development*, Lahore, Pakistan, November 16–19.

[28] Hölbl, M., A. Kamisali'c, M. Turkanovi'c, M. Kompara, B. Podgorelec, and M. Heri'cko. 2018. "*EduCTX: An Ecosystem for Managing Digital Micro-Credentials.*" In *Proceedings of the 28th EAEEIE Annual Conference (EAEEIE)*. Hafnarfjordur, Iceland, September 26–28.

[29] Alammary, A., S. Alhazmi, M. Almasri, and S. Gillani. 2019. "Blockchain-Based Applications in Education: A Systematic Review." *Appl Sci.* 9: 2400. doi:10.3390/app9122400.

[30] Steenmans, K., and P. Taylor. 2018. "A Rubbish Idea: How Blockchains Could Tackle the World's Waste Problem." https://theconversation.com/a-rubbish-idea-how-blockchains-couldtackle-the-worlds-waste-problem-94457.

[31] Swachhcoin. 2018. Decentralized Waste Management System. https://swachhcoin.com.

[32] Phan The Duy, Do Thi Thu Hien, Do Hoang Hien, Van-Hau Pham. 2018. "A Survey on Opportunities and Challenges of Blockchain Technology Adoption for Revolutionary Innovation." *Assoc Comp Mach.* December. doi:10.1145/3287921.3287978.

[33] Onik, Md. M. H., S. Aich, J. Yang, C.-S. Kim, and H.-C. Kim. 2019. "Blockchain in Healthcare Challenges and Solutions." *Big Data Anal Intell Healthc Manag.* doi:10.1016/B978-0-12-818146-1.00008-8.

[34] Reyna, A., C. Martín, J. Chen, E. Soler, and M. Díaz. 2018. *On Blockchain and its Integration with IoT Challenges and Opportunities, Future Generation Computer Systems.* Vol. 88. Elsevier, pp. 173–190.

[35] Poongodi, T., R. Sujatha, D. Sumathi, P. Suresh, and B. Balamurugan. 2020. *Blockchain in Social Networking, Cryptocurrencies and Blockchain Technology Applications.* John Wiley & Sons, Inc, pp. 55–76.

2

Internet of Things

S. Karthikeyan and B. Balamurugan

Galgotias University, Greater Noida, Delhi-NCR, India

CONTENTS

2.1 Overview of Internet of Things

2.1.1 Introduction

There are many applications in Internet of Things(IoT) that are not limited to smart homes, smart retail, intelligent transportation systems, agriculture, and so on. One of the most vital application is the role of IoT is healthcare. As it is used to save human lives in certain situation where people are not able to receive help, an IoT device can be handy and used in healthcare such as with wearable, real time location services, hand hygiene compliance, and remote monitoring.

2.1.2 History of IoT

IoT was initiated by Kevin Ashton to reduce human workload in 1999. Initially Radio-frequency identification (RFID) was the basic requirement for implementing IoT. RFID can be attached to any physical object for tracking location. Then sensors, actuators, and other standards evolved with IoT [1].

2.1.3 Characteristics of IoT

The major characteristics of IoT are listed as follows.

2.1.3.1 Interconnectivity

In IoT any type of devices can be connected and communicate depending on the needs of the users [1].

2.1.3.2 Heterogeneous Devices

Sensors, actuators, gateways, wireless networks mobiles, personal computers, and many others can be used in IoT environment making it a heterogeneous platform [1].

2.1.3.3 Dynamic Nature

The major benefit of the IoT environment is that devices can perform based on changes given by the user making the IoT environment dynamic in nature [1].

2.1.3.4 Ensuring Safety

IoT ensures protection for individual users in different geographical area and data security is done by using encrypting message transmission.

2.1.3.5 Intelligence

As IoT is combination of applications and embedded devices, it can make a decision automatically in certain environments depending upon the situation.

2.1.3.6 Sensing

It does the input operation by observing the environment and sending a notification in case of any abnormalities in the environment [1].

2.1.3.7 Energy Efficiency

As most IoT devices work using batteries, it is not necessary to recharge it often, and consumption of energy is minimal comparing to the other methods.

2.1.3.8 Cost

IoT is affordable and replaces regularly-paid workers by performing the same routine work.

2.1.4 Advantages of Internet of Things

There are various advantages of IoT in all sectors including healthcare, industry, education, agriculture, and others. Below are a few advantages to implementing IoT [1]:

- Affordable
- Easy user interface
- Small kit size
- Accuracy of the data
- Tracking of objects
- Performance
- Reduces the workload
- Productivity

- Automation
- Saves time
- Improved quality of life

2.1.5 IoT Challenges

The following are the major challenges associated with implementing IoT [1].

2.1.5.1 Data Scalability

The amount of data will not be static in nature in IoT; it requires increasing storage space as well as other resources based on the environment [1].

2.1.5.2 Huge Data Storage

Because the sensors record data continuously, a massive amount of data is stored in the IoT cloud, meanwhile the old data will change to big data after a certain number of years [2].

2.1.5.3 Bandwidth Issues

As the name implies, IoT works on the internet; if the internet is not reliable and robust, there will be issues such as a delay or a failure in data transmission.

2.1.5.4 Analytics of Data

The IoT sensors process 'N' number of data, such as structured, unstructured, and semi-structured data which makes it challenging to make any business decisions with available data analytics method.

2.1.5.5 Privacy and Security

As each and every recorded IoT data are stored in the IoT cloud, it becomes easy for an attacker to hack the centralized cloud in order to access data.

2.2 IoT Architecture

Sensors, actuators, or micron rollers are the first element or component in the environment that records the data, and then the recorded data will be transferred to the cloud server by using IoT gateway, where the user will get the notifications or alerts through a user interface such as mobile or web. Data analytics will be done in a big data component when any insights need to be done [3].

2.2.1 IoT Devices

The major components of IoT include:

- Sensors
- Actuators

- IoT gateways
- Cloud and
- User interface

Figure 2.1 displays the components of IoT and the way data are recorded, stored, analysed, and retrieved in the environment.

2.2.1.1 Sensors

A sensor detects physical changes in the environment and sends information to the corresponding destination. It senses only the specific condition of the environment [4]. Sensor types are described in Figure 2.2. Both the sensors are compared in Tables 2.1 and 2.2.

2.2.1.2 Actuators

An actuator is a machine's component that controls a mechanism in the system. It needs energy and signal control; after receiving control, it converts energy into mechanical motion [5].

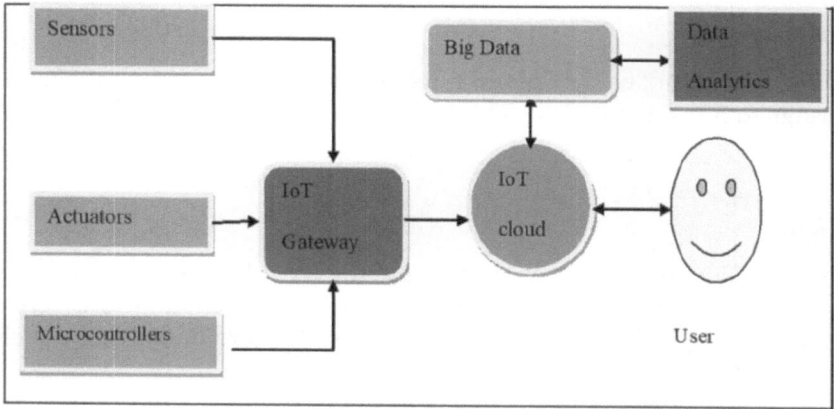

FIGURE 2.1
IoT Architecture.

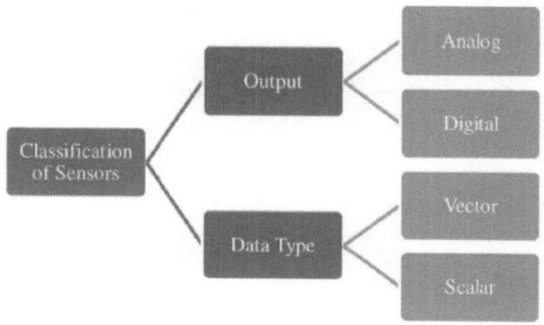

FIGURE 2.2
Sensors Classification.

TABLE 2.1

Comparison of Analog and Digital

S.No	Analog	Digital
1	Continuous values	Non-continuous values
2	Output can be any numeric value	Output can be zero or one only
3	It measures speed, temperature, strain, pressure, and displacement	It measures chemical and liquids

TABLE 2.2

Comparison of Scalar and Vector

S.No	Scalar	Vector
1	It measures by using magnitude of quantity	It measures by using magnitude, direction, and orientation of quantity
2	Colour sensor, pressure sensor, temperature sensor, and strain sensor	Acceleration sensor, image sensor, sound sensor, and velocity sensor

2.2.1.3 IoT Gateways

The recorded data cannot be transferred directly from the environment to the IoT cloud– it needs a bridge to transfer data and acts as message carrier for transmitting data as shown in Figure 2.3.

2.2.1.4 Cloud

The cloud is a major component of IoT because the recorded sensor data gets stored in a cloud server using the internet through the IoT gateway transmission. It is very difficult to own a data centre with a building and other assets to store data; here, the cloud helps

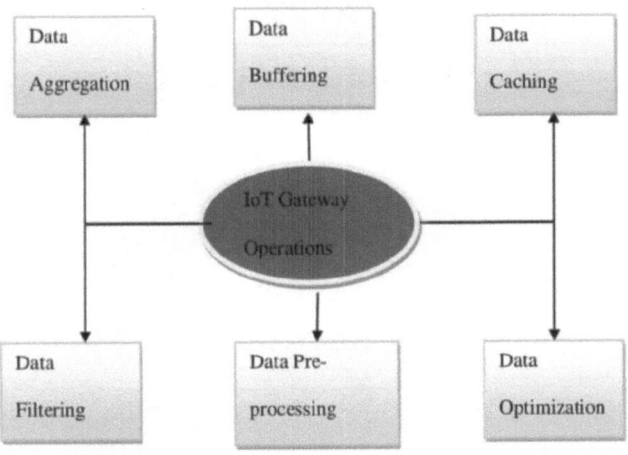

FIGURE 2.3
IoT Gateway Operations.

people to rent servers based on requirements and a pay-as-you-go model, which is biggest advantage of cloud computing.

The major benefits of IoT include:

- Rapid elasticity
- Utility computing
- Resource pooling
- Affordable computing
- High availability
- Disaster recovery
- Scalability

2.2.1.5 Data Analytics

Data analytics is the process in which data is analyzed irrespective of size and structure, and the final decision will be based on insights. In healthcare, a patient can be a diagnosed based on a past record of similar patient's data. Many businesses improve their productivity and revenue based on past IoT data.

2.2.1.6 User Interface (UI)

The user interface is the place where all IoT data and alerts will be received by the user; the interface can be any mode such as mobile application, web application, and any other medium where a person can receive notifications about the IoT environment. The following constraints are to be validated in creating a UI for IoT environment.

- Connectivity
- Physical UI
- Accuracy
- Security
- User-friendly design

2.2.2 IoT Protocols

Protocols are a set of rules that governs communication among IoT devices such as sensors, actuators, gateways, hubs, mobiles, personal computers, and other gadgets that communicate using messages to a user. This communication among devices follows certain protocols, which instruct the IoT environment to reach the exact location and identity of the user.

2.2.2.1 Connectivity

Low-Power Wireless Personal Area Networks over IPv6 (6LowPAN)
Low-power wireless personal area networks over IPv6 permits tiny devices with restricted processing potential to transfer data by using wireless networks with an internet protocol. It even establishes connections between low-power devices and the internet [6].

Routing Protocol for Low Power and Lossy Networks (RPL)
RPL is a proactive protocol vulnerable to loss of packets. It consumes less power in data transmission such as multi-hop, many to one, and one-to-one messages.

2.2.2.2 Identification

uCode
uCode is the unique identification number assigned to individual objects. It can be applied to both physical content as well as invisible digital data content [6].

Electronic Product Code (EPC)
EPC enables the use of RFID for creating a smart industry with a worldwide network. Header, EPC manager, object classification, and serial number are the four categories in EPC [6].

Uniform Resource Identifier (URI)
URI is an identification resource on the internet that contains a sequence of characters making communication easy between a user and the World Wide Web as shown in Figure 2.4.

2.2.2.3 Data Protocols

Constrained Application Protocol (CoAP)
A constrained application protocol performs communication between machine-to-machine (M2M) applications like smart traffic and automatic navigation. It works on the principle of the request-response model between endpoints.

Message Queue Telemetry Transport
Message queue telemetry transport is an ISO standard that does the operations of publishing and subscribing with the TCP/IP protocol. It aims to establish connectivity between applications in one end and networks on the other end as shown in Figure 2.5.

Advanced Message Queuing Protocol (AMQP)
Advanced message queuing protocol transfers business messages between applications, which connect between systems and organization processes.

Benefits of AMQP
- Interoperability
- Security
- Reliability
- Routing
- Queuing

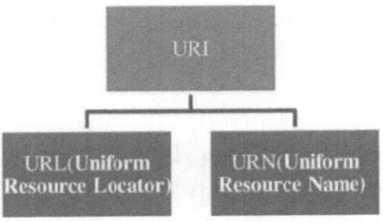

FIGURE 2.4
URI Classification.

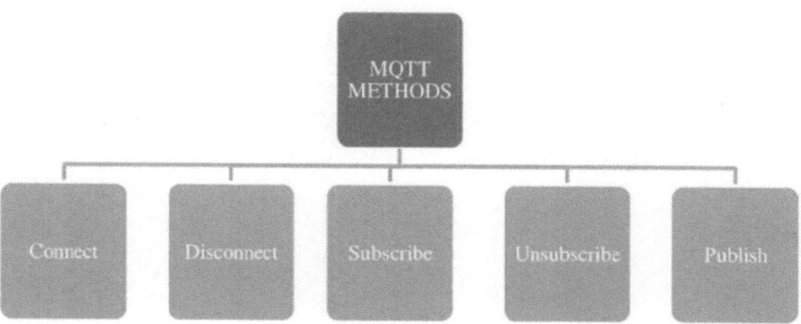

FIGURE 2.5
MQTT Methods.

Simple Sensor Interface (SSI)
    Simple sensor interface is designed for data transfer between computers or user terminals and smart sensors.

2.2.2.4 Communication Protocols

The following communication protocols have immediate importance to consumer and industrial IoTs:

IEEE 802.15.4
    This is the most familiar and reliable standard for a low data-rate wireless personal area network (LR-WPAN). Star and mesh topologies are defined in the framework and two networks are enabled called beacon and non-beacon.

ZigBee
    These new additional layers promote authentication with valid credentials, encryption, routing, and data forwarding [2].

Radio-Frequency Identification (RFID)
    Radio-frequency identification is where data are stored digitally in RFID tags, and is composed of an integrated circuit and an antenna. The major applications include tracking individuals as well as objects, monitors access to required areas

Near Field Communication (NFC)
    Near field communication does the communication between devices over a distance of four centimeters or less. It enables short-range wireless transmission like switching off a light, doing transactions, and transferring digital content for making life easier for individuals.

There are two categories of NFC:

    Active (smartphone) and

    Passive (NFC tags)

    Bluetooth

    Bluetooth enables short-range communication between devices by ad-hoc technology intended for replacing cable shown in Figure 2.6.

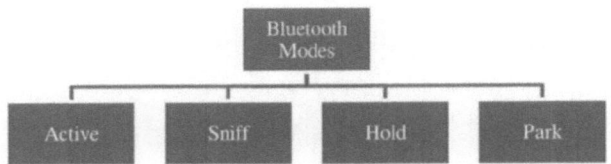

FIGURE 2.6
Bluetooth Models.

Wireless Highway Addressable Remote Transducer Protocol (Wireless HART)
   Wireless HART is the medium for transferring many radio waves, WLAN, Bluetooth, and ZigBee technologies. A mesh network has been used where all stations form a network. It works only on a 2.4 GHz ISM band and the major objectives are forming networked smart field devices.

2.2.3 IoT Applications

Every emerging technology will have a unique application in a real-time environment such as cloud computing, fog computing, artificial intelligence; here IoT came into picture to sort out problems faced in society. IoT makes it is easy to inculcate any real time problems, and it enables the user or the individual to reduce the work of safeguarding entities.

The applications of IoT are classified into categories such as:

2.2.3.1 Consumer Applications

- Smart homes
- Wearable
- Connected cars
- Assets tracking

2.2.3.2 Educational Applications

- Attendance monitoring system
- Safety features to students
- Anytime and anywhere learning
- Boards into IoT enabled boards

2.2.3.3 Industrial Applications

- Digital factory
- Management
- Production monitoring
- Safety and security
- Quality control

2.2.3.4 *Agricultural Applications*

- Precision farming
- Smart greenhouse
- Drone monitoring
- Crop yield analysis

2.3 Relevance of Blockchain

2.3.1 Blockchain Technology

Blockchain technology arranges a block of chain where block represents the digital information and chain represents the database server that holds the data. The major difference is the data holding digital information is decentralized in nature [7,8]. It is also called DLT based on P2P where data will be secured by cryptographic functions, and it is transparent and cannot be changed [9].

$$Block + Chain = Digital\ Data + Database$$

2.3.2 Blockchain Component

The blockcahin technology consist of four major components are as follows [10]:

- Node Application
  It is the application specific to the needs of the user connected to the internet.
- Shared Ledger
  It is available inside the node application. Once the application is active, shared ledger can be seen [11].
- Consensus Algorithm
  It also belongs to the node application, which defines the constraint of the ecosystem.
- Virtual Machine
  It is last logical component of blockchain in node application. A virtual machine is the abstract machine of the real physical system.

2.3.3 Advantages of Blockchain

The benefits of blockchain are numerous around the globe; here are few major ones to be pointed out in various sectors [12].

- Greater transparency
- Process integrity
- Security
- Logistics

- Affordability
- Decentralization
- Stability

2.3.4 Applications of Blockchain

There are many applications oriented with blockchain in different sectors such as banking, healthcare, logistics, and security [13]:

- Personal identification
- Supply chains monitoring
- Data sharing
- Claim processing of insurance
- Digital voting
- Blockchain IoT
- Music royalties tracking
- Food safety
- Immutable data backup
- Medical recordkeeping
- Weapons tracking
- Equity trading
- Cross-border payments

Blockchain integration into IoT leads to improvement in scalability, reliability and privacy.

Blockchain provides transparency in data and the decentralized server helps to store the data securely [14]. It is a promising innovation that supports data integrity, confidentiality, and availability [15].

Table 2.3 describes the major differences between the IoT and blockchain [16].

TABLE 2.3

Comparison between IoT and Blockchain

S.No	Performance Factors	IoT	Blockchain
1	Data storage	Centralized single cloud	Decentralized database
2	Security	One of the challenging factors	It provides better security
3	Internet bandwidth	It consumes high bandwidth	It consumes less bandwidth
4	Applications	Cryptocurrency	Smart home
5	Devices scalability	More number of devices can be added	Performance may go down if heterogeneous devices used

2.4 IoT Security Framework

Security is one of the most challenging factors in storing data in the IoT cloud. Here IoT is protected with three different layers called the application, network, and perception layers [17].

The major issue in providing security to IoT devices is due to the storage of data in the IoT cloud because when the data gets stored in central cloud, it is vulnerable to any attacks by an intruder [18].

Here blockchain nodes will be used to communicate between network layer and the user interface. Blockchain and IoT are integrated, which can be seen in Figure 2.7 [19].

2.4.1 IoT Layers

There are three different layers in IoT data transmission from the environment to the user interface listed next [20].

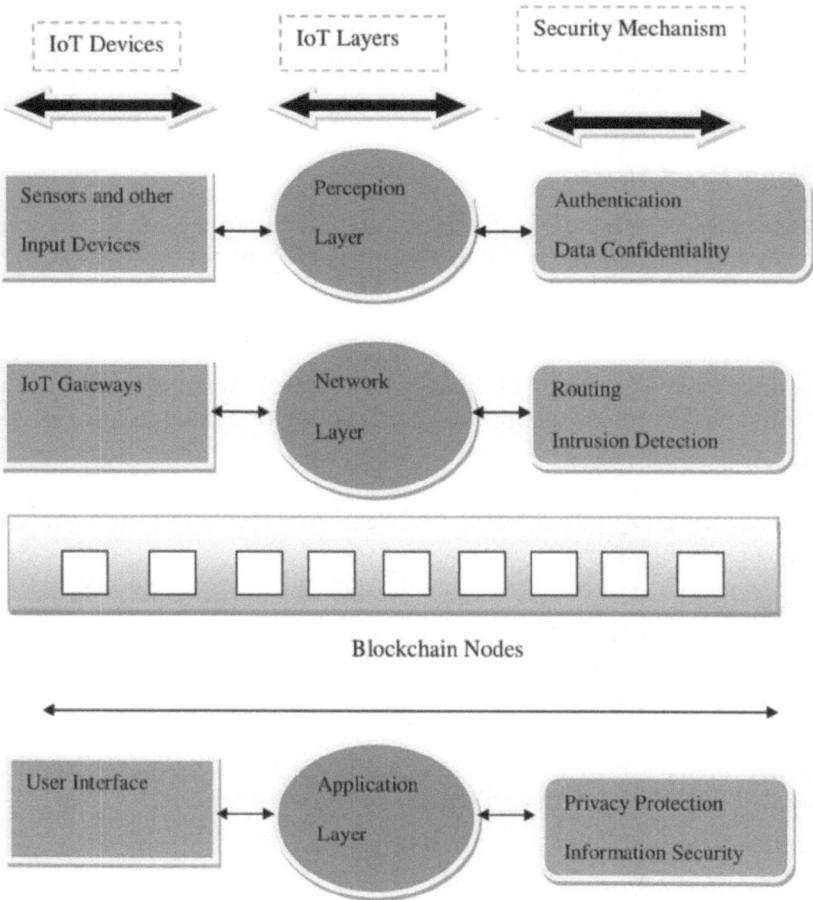

FIGURE 2.7
IoT Security Gateway with Blockchain.

2.4.1.1 Application Layer

The application layer is responsible for retrieving stored data from the IoT cloud; here the blockchain nodes act as a carrier in the transmission of the data.

2.4.1.2 Network Layer

The recorded data will be sent to the IoT cloud through the IoT gateway by an internet service provider; the most common user protocol is UDP.

2.4.1.3 Perception Layer

This is the layer where IoT data recording starts from the environment such as a home, office, or any other places where sensors are fixed.

2.4.2 Security Attacks in IoT

There are 'N' number of attacks that are happening in and around the globe in the IoT environment [21].

Here in this chapter, attacks are classified into two types called identity and non-identity in Figure 2.8. Then the attacks are further divided based on the mode [22]called identity based and non-identity based attacks.

2.4.2.1 Identity-Based Attacks

The intruder causes these attacks by giving a false identity to the server or the environment in order to get data. The sender or the receiver may or may not be aware of the user is real or not.

Brute Force Attack
> These attacks break security by making brute force attack, which does the permutation of the password ora dictionary attack, falling under same category [23].

Spoofing
> Spoofing can be done in two ways called IP and MAC spoofing. In IP spoofing intruders will use the IP address of an IoT users mobile [24]. Similarly MAC spoofing will be done by attackers.

Key Logger Attack
> This application will monitor words that are typed by the IoT user in the mobile application, and then it will be used by an intruder where an intruder can enter the IoT server without breaking the security.

Active Attacks
> All active attacks such masquerade, modification of messages, repudiation, and replay will fall in identity-based attacks by the intruders.

2.4.2.2 Non-Identity Based Attacks

The major objective of the intruder is to deny access to the IoT user or monitor internet traffic. The intruder will cause these attacks by not giving a false identity to the server or

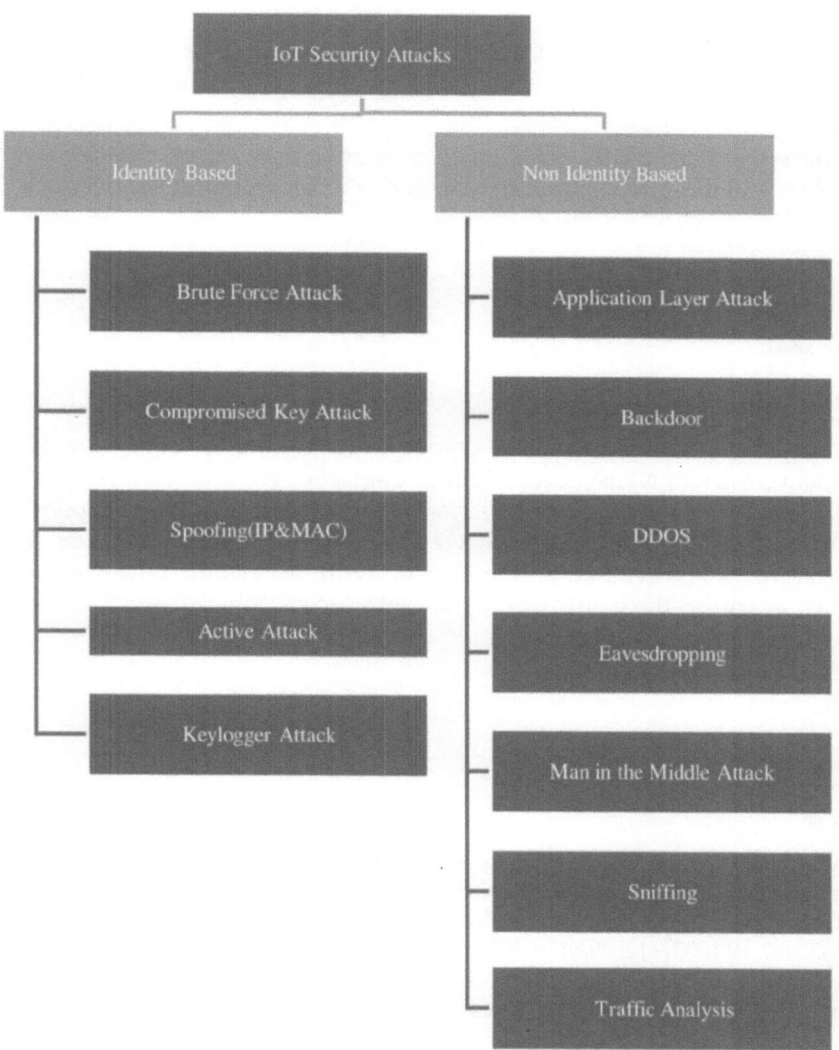

FIGURE 2.8
IoT Security Attacks.

the environment in order to get data. The sender or the receiver may or may not be aware of the attacks as the intruder's objective is to steal data or block traffic.

Application Layer Attack
This attack will intrude the IoT application in the mobile or web environment, thereby stopping access to the IoT user.

Distributed Denial of Service (DDOS)
Here the IoT user will not be able to login to the IoT environment as the intruders will add more traffic to the current environment and block the IoT user from entering the system.

Eavesdropping
It is the process of listening to the interaction between the IoT users and getting data without the knowledge of users.

Man in the Middle Attacks
Interception and decryption are the two modes of man in the middle attacks, which involve physical proximity and the latter involves suspicious software.

Sniffing
Intruders use both hardware as well as software to attack the IoT environment; intruders will monitor the traffic by fixing a packet sniffer in the network. There are two modes called active and passive sniffing.

2.5 Conclusion

As the entire globe is evolving almost everyday with new changes, IoT enhances quality of life by reducing work in all aspects including a smart home, smart transportation, and so on. In this chapter, IoT environment such as IoT protocols and IoT components such as sensors and gateways are described with real-time applications of IoT. One of the major challenges, security, is discussed and addressed with blockchain, which is the decentralized network, enhances the security of IoT. Attacks such as identity and non-identity based attacks are discussed with classification.

References

[1] Williams, P. 2018. "Does Competency-Based Education with Blockchain Signal a New Mission for Universities." *J High Educ Policy Manag* (41): 104–117.

[2] Nofer, M., P. Gomber, O. Hinz, and D. Schiereck. 2017. "Blockchain." *Business Inf Syst Eng* 59 (3): 183–187. doi: 10.1007/s12599-017-0467-3.

[3] Rajamanickam, V. 2019. "CargoX's Blockchain-Based Smart Bill of Lading Solution Is Now on DexFreight's Platform." March. Accessed May 13 2019. https://www.freightwaves.com/news/blockchain/cargoxs-blockchain-based-smart-bill-of-ladingsolution-is-now-on-dexfreights-platform.

[4] Uhde, T. 2018. "Blockchain-Technologie und Architekturim Pilot-Projekt." December. Accessed March 25 2019. https://www.gs1-germany.de/innovation/blockchainblog/das-passende-system-blockchain-technologie-und-architek/.

[5] Nallinger, C. 2018. "Hat der Palettenscheinausgedient?" December 17. Accessed April 3 2019. https://www.eurotransport.de/artikel/palettentausch-mithilfe-der-blockchainhat-der-palettenschein-ausgedient-10634469.html.

[6] Prevost, C. 2018. "DexFreight Completes First Truckload Shipment Using Blockchain." Accessed March 25 2019. https://www.freightwaves.com/news/blockchain/technology/dexfreight-completes-first-truckload-shipment-using-blockchain.

[7] Yaga, D., P. Mell, N. Roby, and K. Scarfone. 2018. "Blockchain Technology Overview." *Natl Instit Stand Technol*. doi: 10.6028/NIST.IR.8202.

[8] Zheng, Z., S. Xie, H. Dai, X. Chen, and H. Wang. 2017. *"An Overview of Blockchain Technology: Architecture, Consensus, and Future Trends."* Paper presented at *IEEE 6th International Congress on Big Data*, Honolulu, Hawaii, US, June 25–30.

[9] Zheng, Z., S. Xie, H.-N. Dai, X. Chen, and H. Wang. 2014. "Blockchain Challenges and Opportunities: A Survey." *Int J Web and Grid Services* 14 (4).

[10] Johnson, D., A. Menezes, and S. Vanstone. 2001. "The Elliptic Curve Digital Signature Algorithm (ECDSA)." *IJIS* 1: 36–63. doi: 10.1007/s102070100002.

[11] https://hedgetrade.com/what-is-a-private-blockchain/. Accessed June 1 2020.

[12] McGhin, T., K.-K. R. Choo, C. Z. Liu, and D. He. 2019. "Blockchain in Healthcare Applications: Research Challenges and Opportunities." *J Netw Comp Appl* 135: 62–75.

[13] https://docs.ethhub.io/ethereum-roadmap/ethereum-2.0/stateless-clients/. Accessed June 3 2020.

[14] Yaga, D., P. Mell, N. Roby, and K. Scarfone. 2019. "Blockchain Technology Overview." arXiv:1906.11078.

[15] Forbes. 2018. "3 Innovative Ways Blockchain Will Build Trust in the Food Industry." "https://www.forbes.com/sites/samantharadocchia/2018/04/26/3-innovative-ways-blockchain-will-build-trust-in-the-food-industry/#839519f2afc8.

[16] Shrivas, M. K., and T. Yeboah. 2018. "The Disruptive Blockchain: Types, Platforms and Applications."

[17] https://www.mycryptopedia.com/consortium-blockchain-explained. Accessed 2 June2020.

[18] https://101blockchains.com/hybrid-blockchain. Accessed June 3 2020.

[19] https://dragonchain.com/blog/differences-between-public-private-blockchains/. Accessed June 2 2020.

[20] https://blockchainhub.net/blockchains-and-distributed-ledger-technologies-in-general/. Accessed June 1 2020.

[21] Chen, G. B. Xu, M. Lu, and N.-S. Chen. 2018. "Exploring BlockchainTechnology and Its Potential Applications for Education." *Smart Learn Environ* (5): 1.

[22] Nespor, J. 2018. "Cyber Schooling and the Accumulation of School Time." *Pedag Cult Soc*, 1–17.

[23] Han, M., Z. Li, J. S. He, D. Wu, Y. Xie, and A. Baba. 2018. *"A Novel Blockchain-Based Education Records Verification Solution."* In *Proceedings of the 19th Annual SIG Conference on Information Technology Education,* Fort Lauderdale, FL, US, October 3–6.

[24] Farah, J. C., A. Vozniuk, M. J. Rodríguez-Triana, and D. Gillet. 2018. *"A Blueprint for a Blockchain-Based Architecture to Power a Distributed Network of Tamper-Evident Learning Trace Repositories."* In *Proceedings of the IEEE 18th International Conference on Advanced Learning Technologies (ICALT),* Mumbai, India, July 9–13.

3

Artificial Intelligence

Ishita Singh, Joy Gupta, and K. P. Arjun

Galgotias University, Greater Noida, Delhi-NCR, India

CONTENTS

3.1 Introduction

AI is a type of approach to make a machine determine how smartly a human brain can think or perform any task. It is the study [1] of how humans think, learn, work, and decide when it comes to problem solving. Furthermore, the whole study gives a software system as an output, which is very intelligent. AI aims to enrich computer functions that are associated with human knowledge, for instance, reasoning, problem-solving, and learning. There are various long-lasting objectives within the intelligence sector. The objectives of AI analysis embrace knowledge representation, reasoning, planning, natural language handling, understanding, learning, training, realization, and skill to change and edit objects. The standard AI views embrace analytical methods, arithmetical and statistical intelligence, and customary scrambling AI. The field of computer science enhances AI within each and every field.

We see AI [2] as a technology that helps different people in different parts of the world attain and archive a lot by collaborating with sensible software. Consider AI as putting a human face on technology. AI is automation that may acquire knowledge from a huge amount of accessible data from any part of the modern world, it understands our kind of language and responds in that kind, and interprets the world the way that we do. During this chapter, we will explore how this kind of technology works, how AI can revolutionize the way computers and humans work together to form a far better, safer world. With the help of current advances and cameras in AI, different actions are taken to enhance the well-being and safety of people at the right time and right place. As our world becomes progressively digitized, so will our ability to attach to technology. For instance, when a threatening liquid flows in any factory, a camera acknowledges it and quickly shares it with those who want it foremost, sanctioning them to safeguard different workers from a risk that may affect them by coming in contact with the liquid. Digitalization in the world converges to help make each and every person more secure, productive, and safe.

The scope [3] of AI is broad and extensive. We consider the broadly common booming research areas in the domain of AI represented in Figure 3.1. There is a huge collection of implementations where AI is busy supporting the common man in their daily lives. One of the most potent and exciting technologies, which comes in the real-world application, is machine learning. Machine learning (ML) techniques drastically change a computer application, and it simulates human-decision making using neural networks. ML is a domain of AI where we bring AI into the equation by learning the input data. The process of making machines learn through provided data is nothing but machine learning. Devices that are

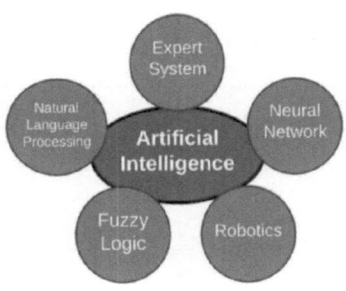

FIGURE 3.1
Artificial Intelligence Domains.

trained with a massive volume of data perform the task more accurately, and it can predict a result more precisely.

With the help of ML [4], we can develop intelligent systems that make decisions autonomously. The algorithms learn from the past occurrence of data through statistical analysis and pattern matching. Then the algorithms provide researchers with accurate results based on learned data. Data is the spine of machine learning algorithms. With the help of historical data, we can produce more data by training these machine learning algorithms. The whole of ML combines computer science, mathematics, and statistics.

3.2 Intellectual Overview of AI

3.2.1 Problem Solving by Searching Algorithms

In this section, we will discuss footprints that should be followed in the field of designing software to resolve a particular issue [5]. First, we have to define an issue clearly and define operators like the starting node and the goal node(s). Then we have to examine the issue and identify where it falls with respect to problems. After this, we must analyze and represent the knowledge required by the task. Then finally, we must choose one or more techniques for solving problems and applying those techniques to the issues. We will talk about ways that knowledge about task domains can be encoded in problem-solving programs and the techniques for merging problem-solving techniques with knowledge to solve several important classes of problems.

3.2.1.1 Distinct Problem and Solutions

An issue can be analyzed by the following elements:

- The *primary node* where agents start the work.
- An illustration of all possible *actions* available to the agents.
- An account of each and every operation, which is known as the *transition models*.
- The *goal node*, which differs from the goal node from other nodes.
- A *path cost* operation that allocates a numeric cost to each path.

3.2.1.2 Searching for Solutions

The initial step is to finalize the goal node. We do this by applying legal actions to the current state, therefore generating a new set of states. In this case, Figure 3.2(a) shows a parent node (Shiv), which has three child nodes (Manish, Sanjay, Rajesh). Here the parent node leads to three branches, which are known as the children of Shiv. These three child nodes likewise are known as a leaf or terminal node (terminal nodes don't have children in the tree) only in part of Figure 3.2(a). As we move ahead, we will find new leaf nodes.

Suppose we choose Manish, then we will see if it is the goal node (it is not) and then expand it to Tanish and Anish from Figures 3.2(b) and 3.2(c). We can choose any two of them or go back to Sanjay or Rajesh. Now here, Tanish and Anish are the leaf nodes. We will stop expanding the nodes when there are no more nodes to expand, or a solution is found.

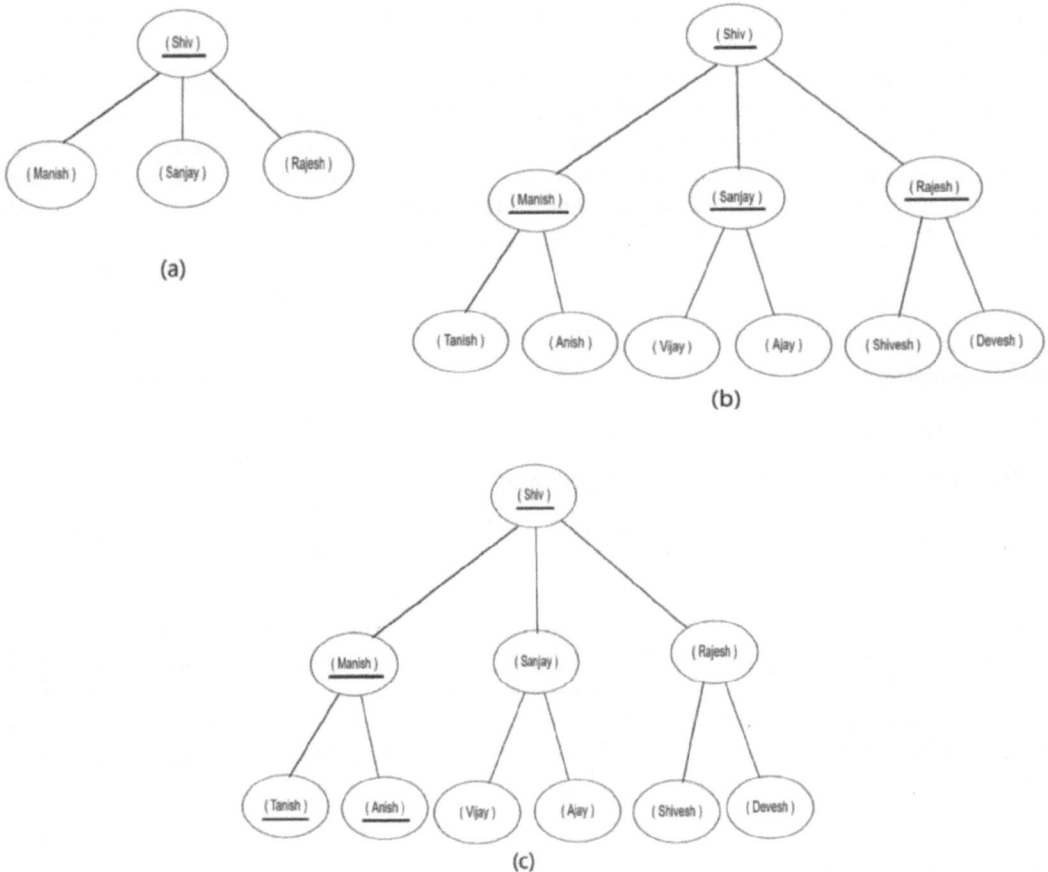

FIGURE 3.2
Example of Searching Solution.

Search algorithms share these basic structures; they initially differ accordingly based on the next chosen node to expand the search strategy.

Figure 3.2(c) shows the shortest path to travel from Shiv to Anish in the form of a partial search tree. The nodes which are underlined are the expanded ones.

3.2.1.3 Uninformed Search

The name of an uninformed search [6] is also known as 'blind search' because they do not carry any extra knowledge about the nodes. They have little information about the steps involved in traversing the tree. The uninformed search types are illustrated in Figure 3.3.

3.2.1.3.1 Breadth-First Search Algorithm

Breadth-first search (BFS) is used to traverse trees and graphs [7]. It is the most common search algorithm. This uses its algorithm to search breadth wise in trees or graphs. This algorithm begins the process of searching from the root node of the tree and then spreads

FIGURE 3.3
Uniformed Search Strategies.

to the next node on the ongoing level. Then after completing its search on the first level, it moves to nodes of the next level. It is implemented using the FIFO queue data structure. Figure 3.4 shows the work of breadth-first search with an example.

The advantages of breath-first searching are, if any solution exists, BFS will provide a solution, and if the dilemma contains more than one solution, then BFS will give the exact solution. The disadvantages are that the algorithm demands a tremendous amount of storage as it needs to save every level or node it traverses. Another disadvantage is that it turns into a time taking algorithm when the root node is distant.

Algorithm for Breadth-First Search

Step 1: Initially set FLAG = 1 for marked as ready state for each node in G.
Step 2: Insert the initial node A to QUEUE and change its status of FLAG = 1 to FLAG = 2 marked as a waiting state.
Step 3: Repeat Steps 4 and 5 until QUEUE is empty.
Step 4: Delete a node N from QUEUE. Process it and set its FLAG = 3 marked as a processed state.
Step 5: Insert all the neighbors of N to QUEUE which are in the ready state (FLAG = 1) and change their FLAG = 2 (waiting state).
[END OF LOOP]
Step 6: EXIT

3.2.1.3.2 Depth-First Search

This type of search is very common to implement with a recursive function that keeps calling itself for traversing the trees or graphs in a data structure [8]. This is because the

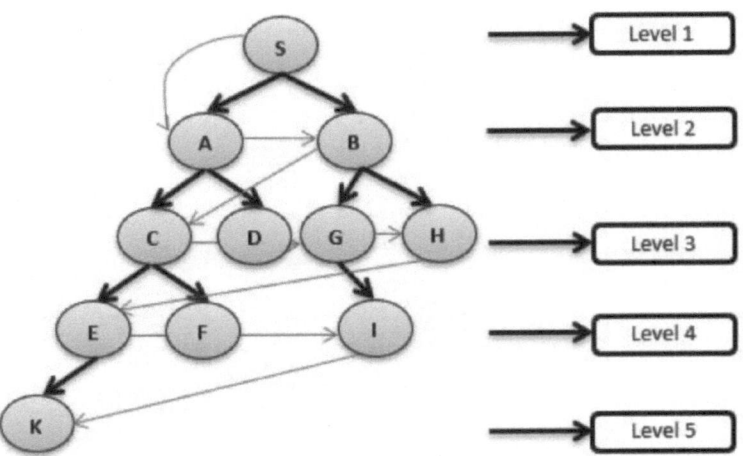

FIGURE 3.4
Example of Breath-First Search.

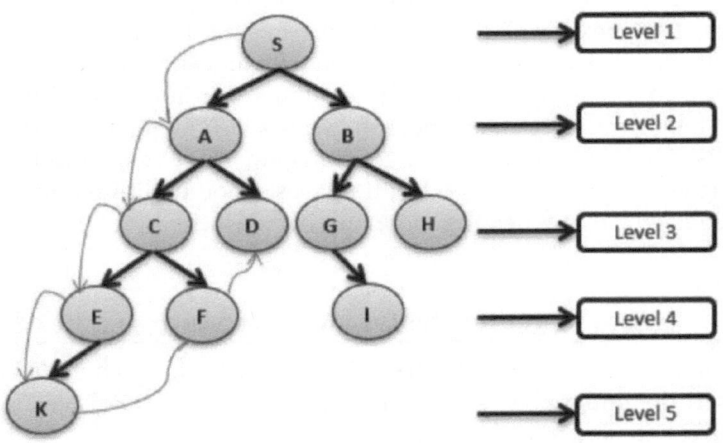

FIGURE 3.5
Example of Depth-First Search.

depth-first search algorithms begin the search from the starting node and keep chasing every path to its most significant last node, then it goes to the next path. It is implemented using the LIFO stack data structure. Figure 3.5 shows the working of depth-first search with example.

The advantages of breadth-first search are that it uses very little storage as compared to breadth-first search, and the time taking of the algorithm as it does not take much time to reach to the goal node. The disadvantages are there is no guarantee of finding solutions as many states keep repeating and the algorithm may go through the infinite loop as it goes deep down searching sometimes.

Algorithm for Depth-First Search

Step 1: Initially set FLAG= 1 for marked as ready state for each node in G.
Step 2: Starting node A is Push on the STACK and change its FLAG = 2 marked as a waiting state.
Step 3: Repeat Steps 4 and 5 until STACK is empty.
Step 4: Top node N is Pop from the STACK. Process it and set its FLAG = 3 marked as a processed state.
Step 5: Push on the STACK all the directed connect node of N that are in the ready state (whose FLAG = 1) and set their FLAG = 2 marked as waiting state.
[END OF LOOP]
Step 6: EXIT

3.2.1.3.3 Depth-Limited Search

This algorithm is the very same as the depth-first search algorithm, just with an addition of a prearranged checkpoint. A depth-first search has some desirable properties like space complexity. In any case, if an off-base branch extended with no arrangement on it, at that point it may not end. So they present a depth-limited search on branches to be extended and don't grow a branch below this depth. Depth limit search is generally useful on the chance that you know the most extreme depth of the solution. Figure 3.6 shows the working of a depth-limited search algorithm with an example.

The advantage of a depth-limited search is its memory efficiency. The disadvantage is it is not recommended to be used when there is more than one solution to any problem.

3.2.1.3.4 Uniform Cost Search

This search algorithm is used for the traversal of weighted graphs or trees [9]. It is done with the help of the priority queue. The algorithm is identical to the breadth-first search when the cost of each path is the same as all edges. The prime goal of this algorithm is to find the goal node. Figure 3.7 shows the working of a uniform cost search algorithm with an example.

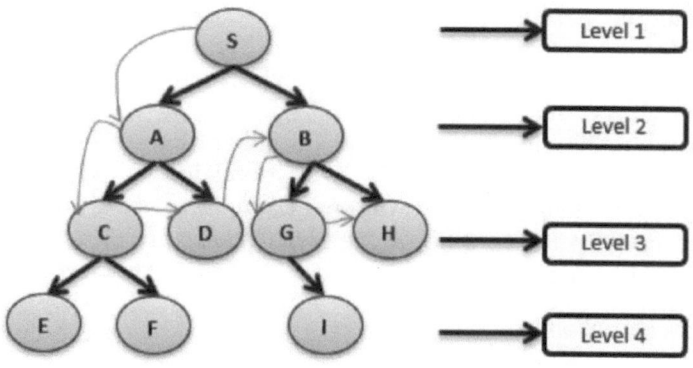

FIGURE 3.6
Example of Depth-Limit Search Algorithm.

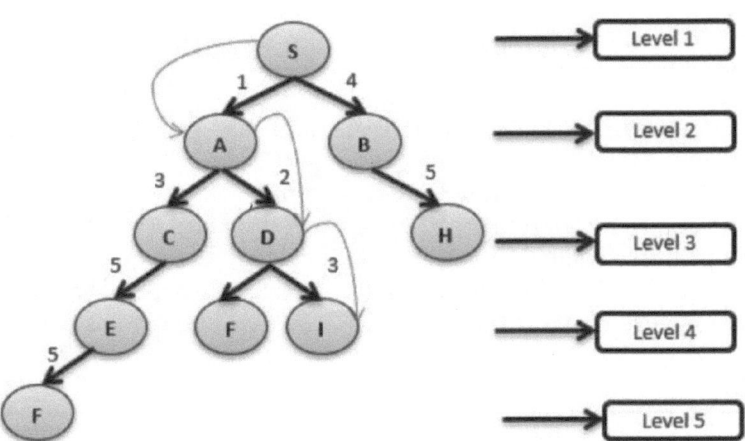

FIGURE 3.7
Uniform Cost Search Algorithm.

The advantage of a uniform cost search is that it is most favorable as it will choose the path with the lowest cost. The disadvantage is that this algorithm is not concerned about the number of steps involved in searching, but is concerned about the path cost. Due to this, it may end up stuck in an infinite loop.

3.2.1.3.5 Bidirectional Search

The purpose of a two-way ransacking [10] is to decrease search time via looking frontward toward the beginning and vice versa towards the objective at the same time. If the two activities that outskirt converge, the rule might recreate a simple way that expands from the start via the metacentre to the final point. Another issue to notice regarding two-way search is guaranteeing that two points meet at the same place. For instance, a depth-first bidirectional search won't function well because of its different functionality regarding frontier point. A bidirectional breadth-first search is sure to meet needs.

A blend of breadth- and depth-first searches in two different directions would ensure the necessary crossing point of search strategies, yet the reason for finding out the direction is difficult. Selection choice relies on the expense of upholding both search strategies meeting one of its conditions. Figure 3.8 shows the working of a bidirectional searching algorithm with an example.

3.2.1.4 Informed Search

An informed search makes an effort to ease the number of searches that may be done by taking smart and sharp options for each and every node designated for the extension [11]. It is also known as a system that has additional knowledge about the estimated interval from an ongoing node to the goal node. Mainly, all these are done by using a function known as the heuristic function.

3.2.1.4.1 Heuristic Function

This function is accustomed to an informed search [12]. It estimates the assuring path between a state and the goal state. It evaluates how close the agent is from the goal by taking the ongoing node as an input. This search algorithm may not provide the finest

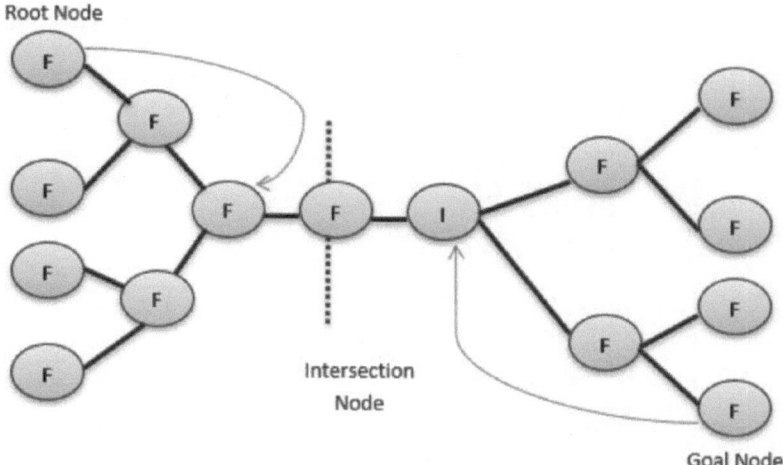

FIGURE 3.8
Bidirectional Searching Algorithm.

explanation, but in a single search, it gives a guarantee to find a favorable explanation. It is expressed ash(x), and the optimized path cost is calculated by this function between any two nodes. Now we will also see while solving examples on the heuristic function that its value can never be negative.

$$h(x) \leq h*(x) \tag{3.1}$$

Where $h(x)$ is the cost of heuristic, and $h*(n)$ is the cost that is being estimated. The cost of estimation should always be greater, or it can be equal to the cost of heuristic.

3.2.1.5 Pure Heuristic Search

A pure heuristic search is the most accessible form of informed search algorithms. It expands nodes on the basis of the heuristic value $h(x)$. It maintains two lists: the 'OPEN' list and the 'CLOSED' list. In the OPEN list, it places nodes that have not yet expanded; in the CLOSED list, it places nodes that have already expanded. Figure 3.9 shows the types of pure heuristic search algorithms.

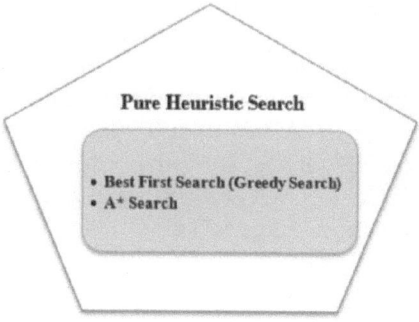

FIGURE 3.9
Types of Pure Heuristic Search.

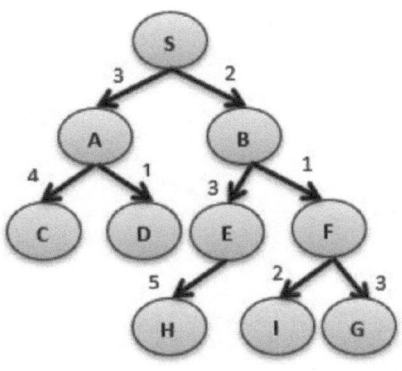

Node	H(n)
A	12
B	4
C	7
D	3
E	8
F	2
H	4
I	9
S	13
G	0

FIGURE 3.10
Example of Best-First Search.

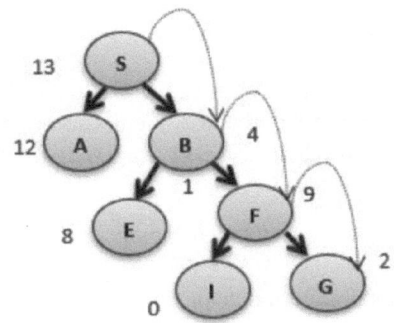

FIGURE 3.11
Solution of Best-First Algorithm.

3.2.1.5.1 Best-First Search (Greedy Search Algorithm)

In this search strategy, expansion is done towards the goal node. Here, the heuristic function $h(x)$ is used to find the distance of the destination, and $h(x)$ is lowered to reach the destination. The simple pattern is to extend the state, which has the lowest value of h.

The advantages of the best-first search are that it proceeds with fewer paths to get to the goal node, and it works well with informed search problems. The disadvantage is in the worst case, it can turn into an unguided depth-first search.

In Figure 3.10, greedy search is used to find the route from 'S' to 'G' [13]. Here values $h(x)$ of certain nodes are given in the table beside the figure. The final destination is from node 'S', it can be traversed close to either 'A' ($h = 3$) or 'B' ($h = 2$). Here 'B' is chosen because of its lower cost. In this state from 'B', it can be either moved to 'E' ($h = 3$) or 'F' ($h = 1$). Again 'F' has a lower heuristic cost, so we will choose 'F'. Eventually 'F' can go to 'G' ($h = 0$). The diagram of the full path is shown Figure 3.11.

The final path is: S -> B -> F -> G

3.2.1.5.2 A* Search Algorithm

The A* (a star) search algorithm merges the strength of a greedy search algorithm with a uniform cost search [14]. In this search, the 'summed cost' is expressed as $f(x)$. The heuristic function is defined as the aggregation of the cost in a uniform cost search, which is expressed as $j(x)$, and the greedy search cost is expressed as $h(x)$.

$$f(x) = j(x) + h(x) \tag{3.2}$$

Here, $j(x)$ is the reverse traversal cost, which is the combined cost from the root node. Furthermore, $j(x)$ is the next cost and is an approximation of the present point from the destination point. The basic methodology is to pick the node with minimal $f(x)$ value. A* strategy is optimized by arranging the node $h(x)$ cost, underrating the original cost of $h^*(x)$. The straightforward technique is to pick the node with the most reduced $f(x)$ value. A* strategy is optimized only when the cost for a node $h(x)$ is convolution and underrates the original cost of $h^*(x)$ to reach the objective. The mentioned property of A* search strategy is known as inadmissible and given by:

$$0 <= h(x) <= h*(x) \tag{3.3}$$

The advantages of the algorithm are listed next. first, A* search algorithm is the finest algorithms compared to any other searching algorithm. The second is that it happens to be a complete and optimized algorithm. The last advantage is very complex problems can be solved using this algorithm. The next discussion about the disadvantages of A* search algorithm, like that it does not produce the shortest path as it is bases on estimation and heuristics, and the next one is that it has some complexity issues.

Figure 3.12 shows the working of A* search algorithm. Find the shortest path travelled by node S to reach node G.

Beginning from the node S, this algorithm simplifies [$j(x)$ + $h(x)$] for each and every node, selecting the node with the shortest sum (tick marked in the table) and ignoring the other one. The whole of the working is shown in Table 3.1 below.

The complete path of the tress is shown in Figure 3.13.

Path: S -> A -> C -> G

The main points to remember are that this algorithm does not search for all remaining parts but returns the path that occurs first and its efficiency reckons the standard of heuristics.

3.2.2 Task Domains of AI

The domain or class of AI is classified into different task [15] categories represented in Figure 3.14 such as mundane/general tasks, formal tasks, and expert tasks.

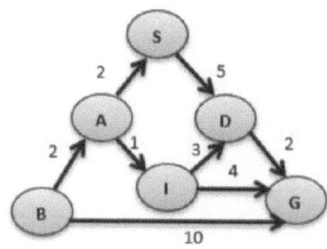

State	H(n)
S	5
A	3
B	4
C	2
D	6
G	0

FIGURE 3.12
Example of A* Algorithm.

TABLE 3.1

Path Calculation of A* Algorithm

Path	$h(x)$	$j(x)$	$f(x)$	
S -> A	3	1	4	tick
S -> G	0	10	10	ignore
S -> A -> B	4	1 + 2 = 3	7	ignore
S -> A -> C	2	1 + 1 = 2	4	tick
S -> A -> C -> D	6	1 + 1 + 3 = 5	11	ignore
S -> A -> C -> G	0	1 + 1 + 4 = 6	6	tick

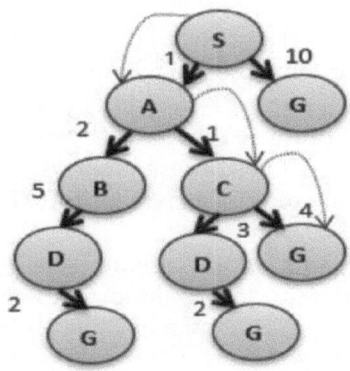

FIGURE 3.13
Solution of A* Algorithm.

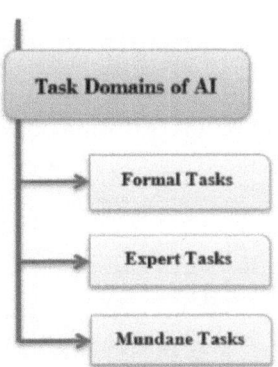

FIGURE 3.14
Task Domains of AI.

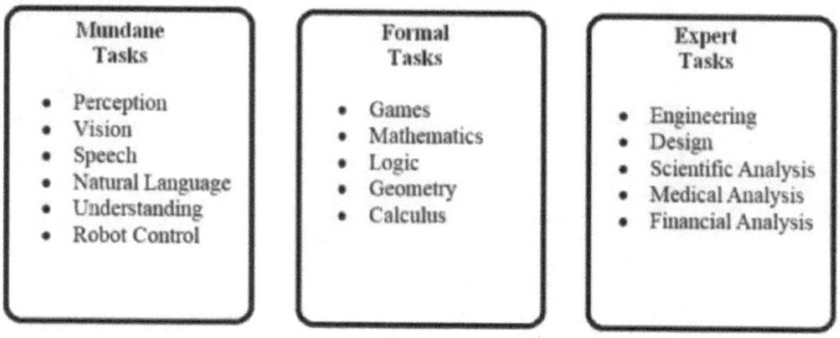

Mundane Tasks	Formal Tasks	Expert Tasks
• Perception • Vision • Speech • Natural Language • Understanding • Robot Control	• Games • Mathematics • Logic • Geometry • Calculus	• Engineering • Design • Scientific Analysis • Medical Analysis • Financial Analysis

FIGURE 3.15
Categories of Task Domains of AI.

- Mundane Tasks
  They are the common things we do every day or tasks done by humans that can be performed successfully to meet their daily needs. This includes natural language, robot control, perception, and others.

- Formal Tasks
  Such a task focuses on the application of formal logic, learning, game playing (e.g., checker and chess), and formal tasks such as theorem proving. Mathematics in the form of geometry and logics come under this task.

- Expert Tasks
  These tasks require high analytical and thinking skills as they are specialized tasks where acquired expertise is necessary. These come under the functional expert domain. Engineering, scientific analysis, and medical diagnosis come under such a task.

Figure 3.15 shows a brief overview of the application areas based on the task domain of AI.

3.3 Frontiers of AI

3.3.1 Knowledge Representation

Knowledge representation is a scope of AI [16] that is committed to demonstrating an intelligent behaviour in AI agents and the same for creating AI to solve difficult tasks. Knowledge from the real world plays a vital role in AI by signifying intelligent behaviour in AI agents. An agent will be able to act accurately on some input if, and only if, he has some experience or knowledge about that input. So it plays a vital role in the creation of an AI.

Knowledge is a familiarity, affirmation, or comprehension of a person or thing. It can be referred to as a theoretical or practical understanding of a subject; for example, information, description, or skills, which is gained through experience, learning, or discovering. It can be used to solve severe problems in the real world by performing inference processes.

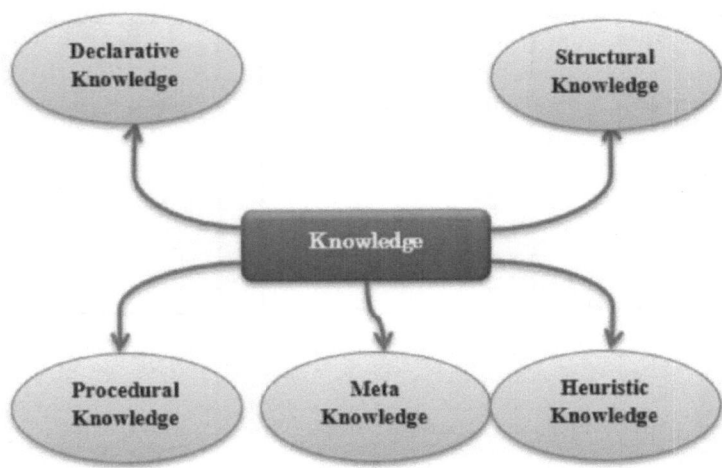

FIGURE 3.16
Different Type of Knowledge in AI.

3.3.1.1 Types of Knowledge in AI

Figure 3.16 shows different kinds of knowledge in AI. In this section, we discuss some types of knowledge including the following:

- Declarative Knowledge
  Declarative knowledge is responsible for describing things, processes, or events, and the relations between their attributes.
- Procedural Knowledge:
  Procedural knowledge is responsible for knowing how to do something. Moreover, this also includes how a specific task or skill is performed.
- Meta Knowledge:
  Meta Knowledge includes knowledge about other types of knowledge. It is used to describe things such as models or tags.
- Heuristic Knowledge:
  Heuristic Knowledge is responsible for representing the knowledge of some experts in a field or subject. These are good to work but not guaranteed rules based on past experiences with an awareness of approaches.

3.3.1.2 Relation between Knowledge and Intelligence

Knowledge from the real world takes an important part in AI, demonstrating intelligent behaviour in AI agents and the same for creating AI. When an agent has a proper experience, or knowledge or information, only then will it be able to act accurately. Figure 3.17 is a representation of intelligent decision making with the help of sensing and knowledge.

Let us suppose you have to write a code in python language, but you do not know python. In that case, you will not be able to code and react to it. In that way, the same thing is enforced by AI.

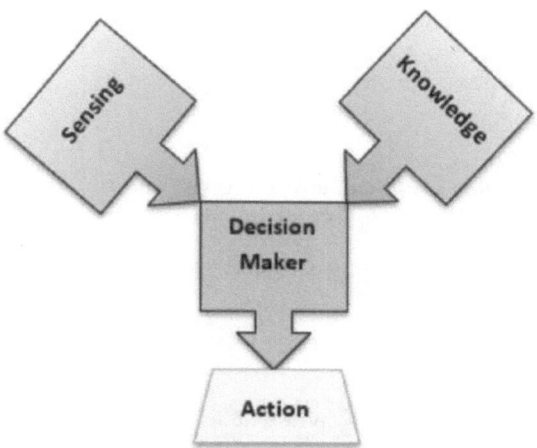

FIGURE 3.17
Intelligence Decision Making.

We can figure out from Figure 3.17 that a single decision maker is responsible for sensing a particular thing from the environment and utilizing information by taking actions. It cannot display intelligent behaviour if the knowledge part is not present.

3.3.1.3 Techniques of Knowledge Representation

There are four main types of knowledge representation using AI [17,18], which are represented in Figure 3.18 and discussed next.

- Logical Representation

The fundamental method for representing a knowledge base logically is to enroll the first-order predicate logic. In this tactic, a knowledge base (often referred to as KB) can be seen as a collection of logical formulas that comes up with a particular description of the world. Modifications in the KB result from the deletion or addition of logical formulas.

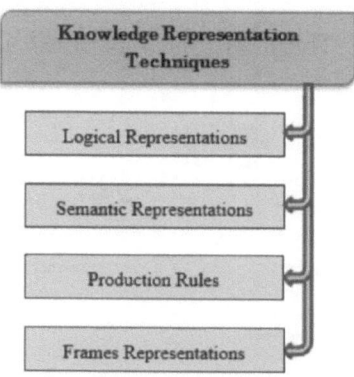

FIGURE 3.18
Knowledge Representation Techniques in AI.

Examples of logical representations are:

- Rhea and Isha are sisters: = Sisters (Rhea, Isha)
- Some Girls play cricket:

Here, the predicate is 'play (x, y)', where x = girls, and y = game. Since there are some girls only for, we will use ∃, and it will be written as:

$$\exists x\, girls(x) \rightarrow play(x, cricket)$$

- Every student respect his teacher:

Here, the predicate is 'respect (x, y)', where x = student, and y = teacher. Since there is every student, so we will use ∀, and it will be written as follow:

$$\forall x\, student(x) \rightarrow respects(x, teacher)$$

- Semantic Networks Representation

A semantic network is an approach of describing relations and properties of concepts, events, objects, actions, or situations. These are represented in the form of a network in graphic form. It consists of attributes, and the connecting edges are labelled describing the relationship with each other. These networks are top rated in AI because of their simplicity and spontaneity. They are very simple to understand and can be easily extended.

In Figure 3.19, there is a cat whose name is Tom. It is governed by robin. The cat is grey in colour. We know that all cats are mammal. We also know that all mammals are under the category of animals.

- Frame Representation

The frame is a knowledge representation system used to represent knowledge by numbers of frames connected by relationships. It works on the concept of inheritance. They are the assemblage of slots and their objects known as the slots values. They have the involvement of AI. The frame consists of slots and their recorded values to describe the distinguishable objects in the world. These attributes are different types and sizes. Attributes

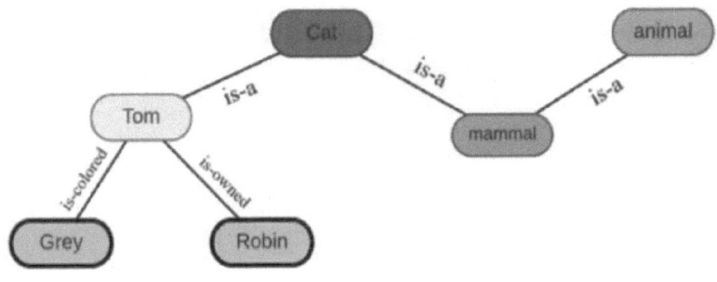

FIGURE 3.19
Example of Semantic Networks Representation.

TABLE 3.2

Frame Representation of a Cricketer

Attributes	Permeate
Sport	Cricket
Role	Wicket-keeper batsman
Batting style	Right hand bat
Bowling style	Right-arm medium
Age	38
Experience	16 years
Highest score	183*

have facets, known as names and values. These facets are characteristic of the frames that allow the restraint of the frames.

Multi frames are very useful when compared to a single frame. Multi frames contain a collection of many frames that are linked; each of them stores information related to an object or an action composed of, and joined in, the information base. The frame is also called as slot-filter knowledge representation in AI. Let us suppose we have to make a frame for a cricketer representation in Table 3.2.

- Production Rules

A production rule system consists of a knowledge base of rules and a global database, which represents the system status and a rule control structure (interpreter) for choosing the rules to execute. A production rule means a pair of, 'if scenario then reaction.' The production rule is said to be triggered if the prerequisites duplicate the ongoing situation of the word, while it is known as fired when a production's operation is accomplished. It is written as (scenario, reaction).

Example:

- IF (one works hard AND gets good grades) THEN reaction (get into a good college)
- IF (a batsman comes AND plays nicely AND does not get out) THEN reaction (he will hit a century)
- IF (eat ice cream AND chips) THEN reaction (will not be feeling hungry)
- IF (turn off the water in the shower) THEN reaction (water will stop flowing)

The major components of the production rule are:

- Global Database: The global database is a central data structure used by the production system in AI.
- Set of Production Rules: The production rules work on the global database. The system can be triggered or fired, as mentioned earlier. The application of the rule replaces the database.
- A Control System: The control system then chooses which application rule should be applied. It halts the computation once a termination condition on the database is satisfied.

3.3.2 Computational Logics

It is often necessary to develop computer programs to infer facts that are not represented but are signified by other represented facts. A brilliant robot might use logical facts implied by one other represented fact. For example, if it wants to find out how to enter the goal state in the leading place or to infer when a goal state has been reached. A database query system may have to infer appropriate information from other information in the database. Computational logic has been developed to direct such a problem. Along with it, the correlated predicate calculus expressions have proven to be a powerful means for knowledge representation for AI programs. Computational logic is thus a critical AI area.

3.3.2.1 Approaches

There are two approaches for seeking to construct a proof of the theorems.

- Semantic approach: It depends excessively on the meaning of symbols in the logical statements. When we use this approach, we aim primarily to consider all possible elucidations of the logical statements to be proved.
- Syntactic approach: Here we ignore the symbols. We use formal symbol-manipulation rules of the logical system to raise new logical statements from the older ones. This approach is more comfortable to use, especially for computers, as we can instinctively apply rules without thinking about what they mean here.

3.3.2.2 Types of Computational Logic

A logical system consists specifically of the logical statements of the system and a set of rules, both called the rules of inference of the system. Now, the computational approach to logical reasoning, often known as traditional computational logic, is divided into two parts: the complex 'predicate logic', or the more straightforward 'propositional logic.'

- **Predicate Logic**

Predicate logic involves the use of classic forms of coherent symbolism. The easiest sentences can also be represented in terms of logical formulae in which a predicate is enforced to one or more arguments. Predicate logic is, in itself, an incredibly formal kind of AI. Prepositions make declarations about the item (individuals).

The 'predicate' is the part of the proposition that makes a declaration about individuals. A predicate logic also includes a set of structured procedures for proving certain formulae can or cannot be logically obtained from others. $P(x_1, x_2, ..., x_m)$ is called a predicate of m variables or m arguments.

Example: She lives in the town.

$P(x, y)$: x lives in y.

P (Myrah, Jaunpur) is a proposition: Myrah lives in Jaunpur.

- **Prepositional Logic**

Prepositional logic is the most accessible form of logic where all statements are made by prepositions. It is limited in that they only deals with the true (T) or false (F) of complete

TABLE 3.3

Mathematical Logic Symbol

Connective	Symbol	Meaning
And	∩	Both
Or	∪	Either or both
Not	~	The opposite
Equivalent	→	If the term on left is true, then the term on right will also be true
Implies	≅	Has the same truth value

statements. A preposition is an indicative statement that is either T or F. It is a technique of representing knowledge in a logical and mathematical form.

The rules used to infer the truth or falsehood of new propositions from known propositions are referred to as 'argument forms.'

Some essential points about argument forms are the following:

- They consist of objects, functions or relations, and logical connectives.
- The logical operator, which connects two statements, is known as a connectives.
- A proposition formula that is true in any way is called a tautology.
- A proposition formula that is false in any way is known as a contradiction.

3.3.2.3 Representations of Computational Logic Connectives

The typical mathematical logic symbols and the meanings of connectives are shown in Table 3.3.

Some solutions for given problems are done with the help of deduction. Deduction means obtaining a solution by some systematic reasoning procedures. A deduction can be represented in the form of argument known as 'Modus Ponens.' Another method for proving theorems in prepositional calculus is with 'Truth Tables.'

3.4 AI Algorithms and Approaches

As we have previously discussed, AI [19] is simply a science of getting machines to make decisions and think as humans do. The development of sophisticated AI algorithms accomplished this by creating robots and machines applied in a wide range of fields like counting robotics, healthcare, marketing, agriculture, business analytics, and many more. Before we speed up, let us try to understand what ML is and how it is related to AI.

3.4.1 Machine Learning

One of the most potent and exciting technologies that comes in the real-world application is ML [20]. ML techniques drastically change the computer application and simulate human decision making using neural networks.

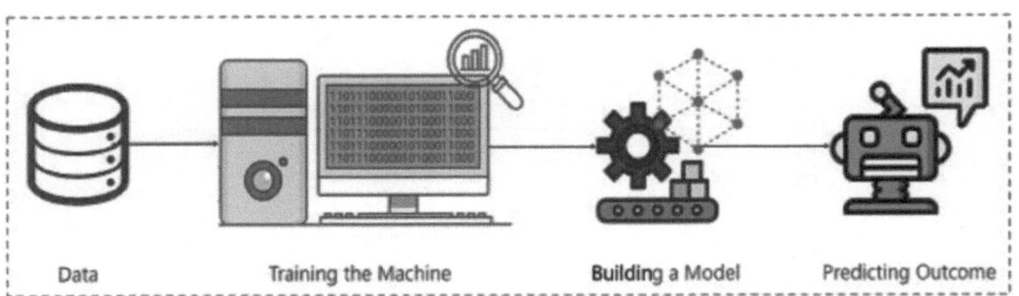

FIGURE 3.20
Machine Learning Building Model.

ML is a domain of AI, where we bring AI into the equation by learning the input data. The process of making machines learn through the provided data is nothing but ML. Devices that are trained with a massive volume of data perform the task more accurately, and it can predict the result more precisely. Figure 3.20 represents the ML steps.

In each domain, ML can possess a different nature, and based on the nature of the domain and application, and it can perform various methods. ML has lots of approaches to deal with a large amount of data. In today's life, in most areas, ML possesses its characteristics by providing information and helping in future prediction.

Application of ML includes:

- Recommender system
- Fraud detection
- Predictive analysis
- Problems of oversampling and over fitting
- Home and appliances
- Traffic alerts (maps)
- Search engine result refining

Figure 3.21 shows the development of intelligent systems that can make decisions on an autonomous basis. The algorithms learn from the past occurrence of data through statistical analysis and pattern matching. Then, it provides us with accurate results based on learned data. Data is the central spine of the machine learning algorithms. With the help of historical data, we can produce more data by training these machine learning algorithms. The whole of ML combines computer science, mathematics, and statistics.

3.4.2 AI Algorithms

In general, an algorithm uses some mathematics and logic, and takes some input to produce the output, but an AI algorithm [21] takes a combination of both inputs and outputs in order to learn and train the data and produce beneficial outputs. Algorithms in each group perform the same task of forecasting outputs on the given unknown inputs. However, here the data is the backbone when it comes to selecting the right and correct algorithm.

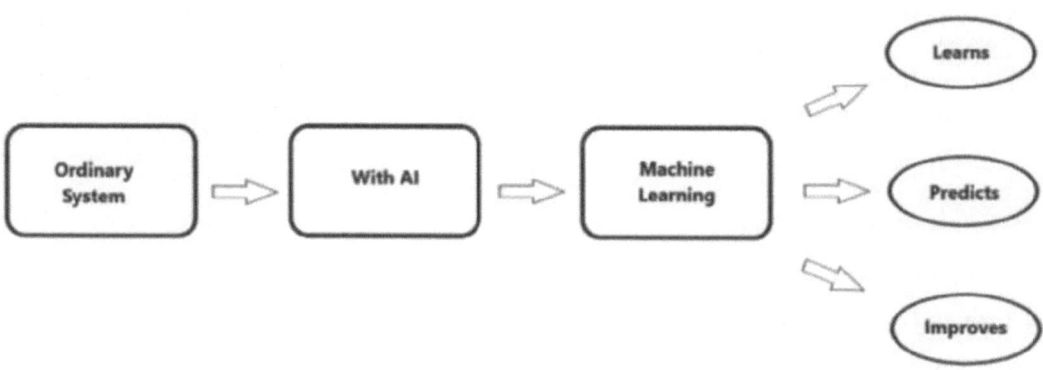

FIGURE 3.21
Machine-Enabling Intelligence through AI.

3.4.3 Types of Problem-Solving Using AI Algorithms

Figure 3.22 shows that the general type of problem solved using AI algorithms [22] are regression, classification, and clustering.

For each category of the task, there is a use of a specific algorithm. Algorithms can be used to solve different kinds of problems. Some of the essential and popular AI algorithms include:

- Logistic regression
- Random forest
- Decision tree
- Clustering algorithms
- K-means clustering
- Support vector machines (SVM)
- Regression algorithms
- Naive bayes
- Linear regression
- K nearest neighbours (KNN)

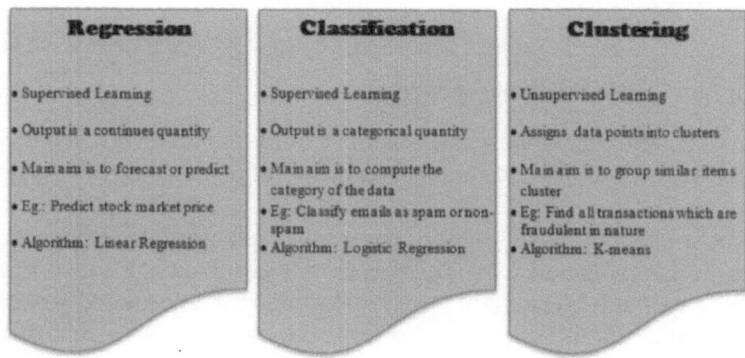

FIGURE 3.22
Types of AI Problems.

So AI includes everything that allows computers to learn decision making and how to solve problems. This field endeavours to understand intelligent entities. Computer-based intelligence has proved to be the most significant and innovative technology as it has produced many impressive products even at the beginning period of development.

3.5 Challenges Faced in Converging Blockchain and AI Techniques

3.5.1 AI and Blockchain–Technologies Integration

In informal language [23], when we talk about blockchain, it is the assigned report that accumulates all executions in the nodes in a clear and established way. The mechanics of blockchain have made everyone interested in going through blockchain app development. With its appealing and lively technology, it has hampered the professional world in a various numbers of ways.

Key features of blockchain include:

- Inflexibility
- Localized database
- Record distribution
- Data preservation
- Uncloudedness and clarity

Blockchain has always been the hunky-dory technology in the agora to invest in, but it also has few limitations. Keeping this in mind, AI appears as the correct solution to make blockchain best for its growth and evolution. It can be done in the following ways:

- Finer data management
- Improved and developed energy consumption
- More measurability
- Strengthening efficiency
- Higher security
- Brand new data gates

3.5.2 Impact of Blockchain-AI Convergence

The application of AI and blockchain [24] is almost similar in business. When talking about integrating them, various challenges come into play. Let us discuss them without wasting much time.

Blockchain technology is the assigned report that accumulates all executions in the nodes in a clear and established way. The mechanics of blockchain have made everyone interested in going through blockchain app development; it is localized, and the nodes are miscellaneous. As blockchain is mutual and free-sourced, it causes great difficulty for AI outputs to come at a single point. So the idea of merging these two technologies is still new. They are marching to a different drummer. A massive volume of money and time is required for exploring AI and blockchain to look into similar grounds.

An additional challenge associated with the merging of blockchain and AI is security. Blockchain is a mutually assigned, securely encrypted, and decentralized database that delivers AI with reliable and endless knowledge. The technology rests on cryptographic algorithms, which helps the data to be secure. Nevertheless, if one tries to use AI to change the secured data, it is very hard to decrypt the files, which results in data hacking [25].

3.5.2.1 Challenges

- The groundwork of blockchain is decentralized and quite different from that of AI. If we talk about the nature of nodes of blockchain, we find them to be heterogeneous. Hence if the blockchain is public, then it will become impossible for ML outputs to come as a single point.

- AI requires a large number of data sets and the database that blockchain form is not scalable to consume such amount of data. This implies that blockchain cannot integrate with AI in its current state.

- There are some instances where AI failed to show its foremost power such as Uber self-driving cars which have failed when they started avoiding red lights. In such cases, if the AI is decentralized, it will be challenging to manage the damage AI causes.

3.6 Conclusion

AI refers to machines that are worked to perform intelligent tasks that have generally been accomplished by people. In other words, AI is the motor or the 'mind' that will empower investigation and is dynamic from the information gathered. In the above sections, we had a detailed discussion on AI techniques and methods. Blockchain is a decentralized system of machines that records and stores information to show an ordered arrangement of occasions on a straightforward and permanent record framework. By definition, a blockchain is a disseminated, decentralized, changeless record used to store scrambled information. Blockchain and AI have developed into driving advances that power advancement across almost every industry. AI and blockchain are ending up being a significantly amazing blend, improving each industry where they're actualized. These innovations can be joined to overhaul everything from nourishment store network coordination and human services record sharing to media eminences and money-related security. The reconciliation of AI and blockchain influences numerous perspectives, including Security – AI and blockchain will offer a twofold shield against digital assaults. It's a given that every innovation has its level of complexity nature, yet both AI and blockchain are in circumstances where they can profit from one another and help each other.

References

[1] Dean, J.. 2020. "Google Research: Looking Back at 2019, and Forward to 2020 and Beyond." *Google AI* [Blog].

[2] Cioffi, R., et al. 2020. "Artificial Intelligence and Machine Learning Applications in Smart Production: Progress, Trends, and Directions." *MDPI*.

[3] Davenport, T., Guha, A., Grewal, D. et al. 2020. "How Artificial Intelligence Will Change the Future of Marketing." *Journal of the Academy of Marketing Science* 48: 24–42. doi:10.1007/s11747-019-00696-0.

[4] Russell, S., and P. Norvig. 2003. "Artificial Intelligence – A Modern Approach."

[5] Poole, D. L., and A. K. Mackworth. 2020. "Python Code for Artificial Intelligence: Foundations of Computational Agents."

[6] Bullinaria, J. 2018. *Lecture Notes for Data Structures and Algorithms*. School of Computer Science, University of Birmingham, UK.

[7] Cormen, T. H., C. E. Leiserson, R. L. Rivest, and C. Stein. 2019. *Introduction to Algorithms*. Massachusetts Institute of Technology.

[8] Sanders, P., K. Mehlhorn, M. Dietzfelbinger, and R. Dementiev. 2019. *Sequential and Parallel Algorithms and Data Structures*. Springer.

[9] Manelli, L. 2020. *Data Structures. In: Introducing Algorithms in C*. Berkeley, CA: Apress.

[10] Cheng, S., and B. Wang. 2012. "*An Overview of Publications on Artificial Intelligence Research: A Quantitative Analysis on Recent Papers.*" *2012 Fifth International Joint Conference on Computational Sciences and Optimization*. Harbin, 683–686. doi:10.1109/CSO.2012.156.

[11] Friggstad, Z., J.-R. Sack, and M. R. Salavatipour. 2019. *Algorithms and Data Structures*. Springer.

[12] Bohr, A., and K. Memarzadeh. 2020. *Artificial Intelligence in Healthcare*. Academic Press.

[13] "Artificial Intelligence Tutorial." *Javatpoint*.

[14] Gevarter, W. B. 1984. *An Overview of Artificial Intelligence and Robotics*. National Bureau of Standards.

[15] Cioffi, R., M. Travaglioni, G. Piscitelli, A. Petrillo, and F. De Felice. 2020. *Artificial Intelligence and Machine Learning Applications in Smart Production: Progress, Trends, and Directions. Sustainability – MDPI*.

[16] Araszkiewicz, M., and V. Rodríguez-Doncel. 2019. *Legal Knowledge and Information Systems*. IOS Press.

[17] Koenraad, De Smedt. 1988. *Knowledge Representation Techniques in Artificial Intelligence: An Overview*. Springer.

[18] Thomason, R. 2020. "Logic and Artificial Intelligence." *The Stanford Encyclopedia of Philosophy* (Summer Edition).

[19] Balas, V. E., R. Kumar, and R. Srivastava (Eds.). 2020. "Recent Trends and Advances in Artificial Intelligence and Internet of Things." *Intelligent Systems Reference Library*. doi:10.1007/978-3-030-32644-9.

[20] Ramasubramanian, K., et al. 2019. *Machine Learning Using R*. Springer.

[21] de Mello, R. F., et al. 2018. *Machine Learning A Practical Approach on the Statistical Learning Theory*. Springer.

[22] Quan, Z., et al. 2019. *Advanced Machine Learning Techniques for Bioinformatics*. IEEE/ACM Transactions on Computational Biology and Bioinformatics (TCBB).

[23] Zhang, G., T. Li, Y. Li, et al. 2018. "Blockchain-Based Data Sharing System for AI-Powered Network Operations." *J Commun Inf Netw* 3, 1–8. doi:10.1007/s41650-018-0024-3.

[24] Casinoa, F., T. K. Dasaklis, and C. Patsakis. 2019. "A Systematic Literature Review of Blockchain-Based Applications: Current Status, Classification and Open Issues." *Telemat Inform* 36: 55–81.

[25] Salah, K., M. H. U. Rehman, N. Nizamuddin, and A. Al-Fuqaha. 2019. "Blockchain for AI: Review and Open Research Challenges." *IEEE Access* 7: 10127–10149.

4

Blockchain for Internet of Things I

A. Reyana

Hindusthan College of Engineering and Technology, Coimbatore, Tamil Nadu, India

S. R. Ramya

PPG Institute of Technology, Coimbatore, Tamil Nadu, India

T. Krishnaprasath

Nehru Institute of Engineering and Technology, Coimbatore, Tamil Nadu, India

P. Sivaprakash

PPG Institute of Technology, Coimbatore, Tamil Nadu, India

CONTENTS

4.1 Introduction

Internet of Things (IoT) has had rapid growth in industry and research fields, but nowadays it is experiencing security and privacy vulnerabilities. This is due to the majority of devices bring applied to decentralized topology having limited resource constraints. IoT being a disruptive technology evolving embedded and cyber-physical systems provides high granularity in the network of interconnected sensors where information is collected from the real-world environment. The approach to blockchain in the IoT [1–3] will improve efficiency and deliver advanced services to major application domains. The other challenges include lack of centralized control, context-aware risks, and others that need to be addressed comprehensively.

Bitcoin, the centralized digital currency launched in the year 2008, solves privacy and anonymity issues using bitcoin as the technology behind blockchain. Blockchain today has gained popularity in applications like cloud storage, digital assets, smart city, and others, consisting of the ledger where blocks are chained. To mine blocks, peer-to-peer (P2P) nodes verify the occurrences of new transactions and broadcasts them to the entire network. Thus all nodes verify the centralized transaction validating the signature ensuring robustness. However multiple miners' usage of the same resource for a single transactions will cause transaction delay, which can be addressed by features like security, anonymity, and decentralization. Also, adopting blockchain in IoT cannot be implemented directly and it is necessary to address critical challenges including:

- Resource restriction in IoT devices leading to intensive mining
- Requiring low latency as block mining is time-consuming
- Poor scaling when the number of nodes gets increased
- Bandwidth limitations in IoT devices causing traffic overhead on blockchain protocols

IoT and blockchain thus transform concepts and create new possibilities exploring the benefits of IoT from the decentralized nature of blockchain. Currently, IoT will have 50 billion devices connected inclusive to conventional and heterogeneity equipment like smartphones, laptops, refridgerators, cars, and others. Applications like smart health, smart home, and smart cities have increased exposure to privacy and security threats harming physical security. The solutions implemented provide equivalent levels of security capable of auditing and controlling the environment. Blockchain authenticates and audits data in a decentralized manner eliminating the single point of failure and building trust among users. Today blockchain is used in many applications like logistics and supply chain, smart contracts, and many more.

4.1.1 Overview

IoT includes data processing and communication among devices in different platforms without any human intervention [4]. This computation of connecting objects happens through the internet, and today IoT is considered as an extension of the internet allowing access to smart objects as services. Blockchain uses cryptography [5,6] as a key characteristic to authorize network interactions. An example would be the following: Consider smart contracts. The scripts residing on the blockchain execute self-scripts allowing distributed workflow accurately. The objective of this chapter is to provide a description highlighting ways blockchain and IoT can be used together. Blockchain has distributed networks consisting of non-trusted members interacting with each other without any trusted intermediate parties. Recently blockchain technology has attracted the interest of stakeholders in the financial sector. In the absence of a trusted intermediate, this leads to faster reconciliation of distributed information, replicating and sharing it among the members of the network.

Here in Figure 4.1 each block is identified with cryptographic hash functions, and a link between the block is established on referring to the hash of the previous block creating blockchains. For a better understanding of a blockchain network consider a set of clients operating on the same blockchain. Here it uses private and public keys. Private key for one's own transactions and the public key is for network interactions [7–10]. The neighbouring peers ensure valid incoming transactions and discard invalid transactions. The world view update happens on adding blocks to the chain once the transaction gets completed. In the other case, the block is discarded.

4.1.2 Applications of IoT

The function of the application layer is to provide customer service. Examples are smart home, smart industry, and others. IoT [11–13] has wide applications allowing its usage to people in business and in daily life transforming the society into a smarter world. Intelligent products include smartphones, smart television, and wearable sensors. Smart health monitors body fitness controlling heart rate, and monitoring patient's treatment in homes and hospitals

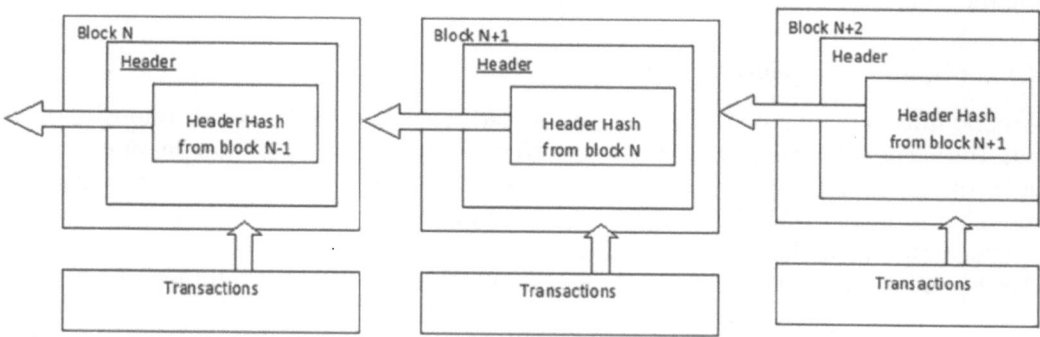

FIGURE 4.1
Blockchain Replication [5].

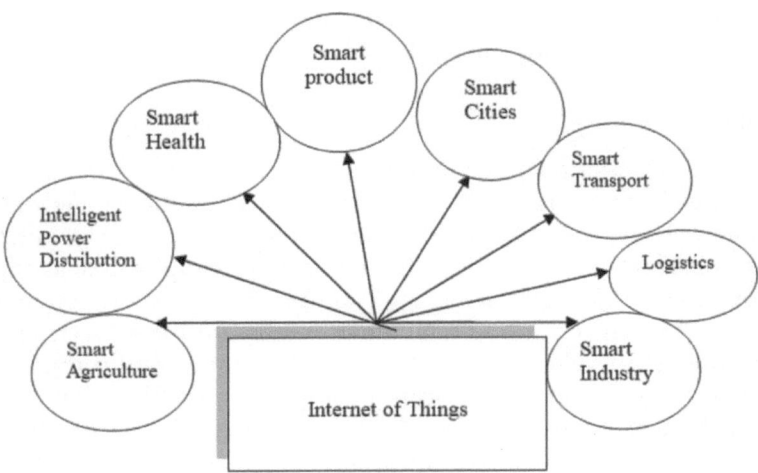

FIGURE 4.2
IoT Applications.

based on requirement. Usage of IoT (Figure 4.2) in health for diagnosis and maintaining treatment records in the long-term has become more effective nowadays. Intelligent transport notifies traffic signals, controls routes by monitoring vehicles from a remote location, and integrating intelligent platforms. Energy monitoring is done by a smart grid where power distribution is intelligently monitored in substations and residential power meters. Inventory management is done through smart e-commerce and finally smart industry to track goods, control production, inventory management, pollution control, energy-saving, and more.

4.1.2.1 Traffic Management

IoT plays a vital role in forming eco-friendly smart cities. Vehicles stuck in urban traffic bring an annoying experience in creating global warming due to more emissions. For instance, traffic lights are fixed with IoT sensors alerts on heavy traffic in urban areas. This Big Data problem can be analyzed and handled by suggesting alternative routes. Also people can be directed to parking places during their temporary stop. The parking places are monitored with IoT sensors with updates in the app installed while the person is leaving the area. The scenario could be practised in smart cities.

4.1.2.2 Waste Management

Around three million tonnes of waste is generated everyday and this is continuously increasing. The waste production should be minimized through recycling for maintaining a sustainable environment in smart cities. IoT sensors track the recycling bins giving alerts regarding the filled container [14]. This tracking process should continuously be in progress and the representatives should check to see if there's an increase or decrease in the waste levels. Waste management authority can take proper measures on illegally dumping of waste.

4.1.2.3 Smart Home and Buildings

Devices connected to IoT perform remote monitoring and manage home appliances which include heating and lighting. Smart home automation allows the user to control smart devices such as automated doors, security cameras, and electrical appliances with a smartphone with the app or any other connected devices. The temperature in air conditioner can be automatically adjusted based on the weather forecasted. In a smart home, door unlocking, security surveillance, and other security options are included. The compressor functions up to the threshold level withstanding the power fluctuation in the refrigerator. Change in lifestyle has increased the efficiency in the remote access of home appliances such as refrigerators, water heaters, and microwaves. Mobile fixed in-house automation receives voice commands and responds with the connected devices. The smart home is provisioned with different IoT devices, checks the quality of service (QoS) and formulates channel optimization. The home appliances can be managed statically or dynamically based on resident behaviour, weather condition, and other factors. The static scheduling in home appliances defines the user activities and the service charges are prepared based on the received power supply and the demand of energy in electronic appliances. In case of variations in forecasted data the user reschedules the activity to meet the energy demand. This is termed as run-time scheduling. The increase in population, automation, and modern lifestyle have caused an energy crisis due to climate change, carbon emission, fossil fuels, and other factors IoT-enabled smart homes conserve energy requirements, providing convenient and comfortable living.

4.1.2.4 Smart Grid

The power resource control is essential based on population growth. Energy consumptions are to be reduced in the building premises. The services offered in the area are to be persistently monitored and connected to the networks [15]. Once enhanced, it enables two-way communication between the sensor devices using transmission lines. Smart meters and smart thermostats suggest the best use in home appliances, minimizing utility bills.

4.1.2.5 Smart E-Health

Healthcare application performance can be enhanced by implanting human body sensors. For the patients it provides suitable treatment based on data gathered and analyzed in realtime. Advanced IoT sensors have created a revolution in the healthcare industry and have improved the quality and cost.

4.1.2.6 Smart Industry and Manufacturing

To reduce the human intervention robots are designed to automatically track and handle manufacturing tasks in a controlled manner. The industries perform automated functioning with less, or without, human intervention. Green IoT converges manufacturing systems with remote access and low downtime. The efficient data sharing among industrial firms and the factory floor improves equipment efficiency, market agility, and labour productivity.

4.1.2.7 Smart Logistics and Retail

Radio-frequency identification (RFID) and smart shelves draw attention to providing customer services. The pallet fixed on a truck send messages about the product. The interconnected physical devices in retail industry modify the supply chain digitally, transforming cost efficiency and end-to-end management of real-time remote tracking [16]. The GPS with sensors track the logistics functionality in predictive maintenance on climate changes, route optimization, and others. The option for customers to track and avail the logistics data is also provided.

4.1.2.8 Smart Agriculture

Devices are equipped with sensors for effective communication and sensing. The energy demand increases greatly making remarkable interest in several organizations. The green IoT focuses on the energy consumption satisfying a sustainable smart world. The algorithms with IoT facilitate the reduction of greenhouse effects. The devices become smarter when connected to the internet and the basic principle is classified into the below-mentioned types:

    i. Things that act on the received data
    ii. Things that gather information and send it
    iii. Things that perform both the mentioned functionalities

Thus IoT effectiveness relies on the terms of the data protected throughout the lifetime evolving blockchain into IoT. Most IoT devices depend on the centralized communication model, which is the biggest drawback. Blockchain in IoT prevents the whole network from forming a single point of failure as storage; computation among billions of devices happens in a decentralized way helping the IoT devices to scale up efficiently. Also, the integration of blockchain in IoT reduces the installation and managing cost of servers ensuring confidentiality through cryptographic algorithms in IoT networks. Blockchain protects IoT networks from man-in-the-middle attacks. Initially, blockchain application started with Bitcoin evolving into smart contracts. Improving its efficiency in coordinating applications, smart contracts made its autonomous execution in crowdfunding, mortgages, and other ways to invest. Ethereum the popular blockchain platform runs smart contracts in distributed applications, interacting with more than one blockchain. Beyond smart contracts and cryptocurrencies application of blockchain in IoT exists in various areas where smart services are involved. The applications like a wearable, intelligent transport system, logistics and supply chain, agriculture, security, and energy sector can be benefited by blockchain [14,17]. The healthcare system shown in Figure 4.3 uses the blockchain IoT (BIoT) application found in the literature review as well.

4.1.3 Blockchain Integrated IoT Architecture

Figure 4.4 explains smart home architecture where Alice, the user, is equipped with many IoT devices. These are divided into three layers namely local, overlay, and cloud storage. Transactions are pertained to a device chained by the creation of similar transactions to the ledger. The transaction, if not required, means the user can remove the device by deleting the ledger. The device communication happens only when an authorized user applies the shared key. The local blockchain performs all control access in the transactions at home. The pointer added to copies the previous block policy to a new block and adds it to the chain. Optional storage can be added to each home as shown in Figure 4.4. The overlay network consists of constituent nodes, smart devices in a home, phone, or a PC. For additional anonymity in the internet protocol (IP) layer the node in the network connects to the overlay. Delay and overhead could be reduced by grouping nodes into clusters with cluster heads in the overlay network. If the network experiences excessive delay then the node

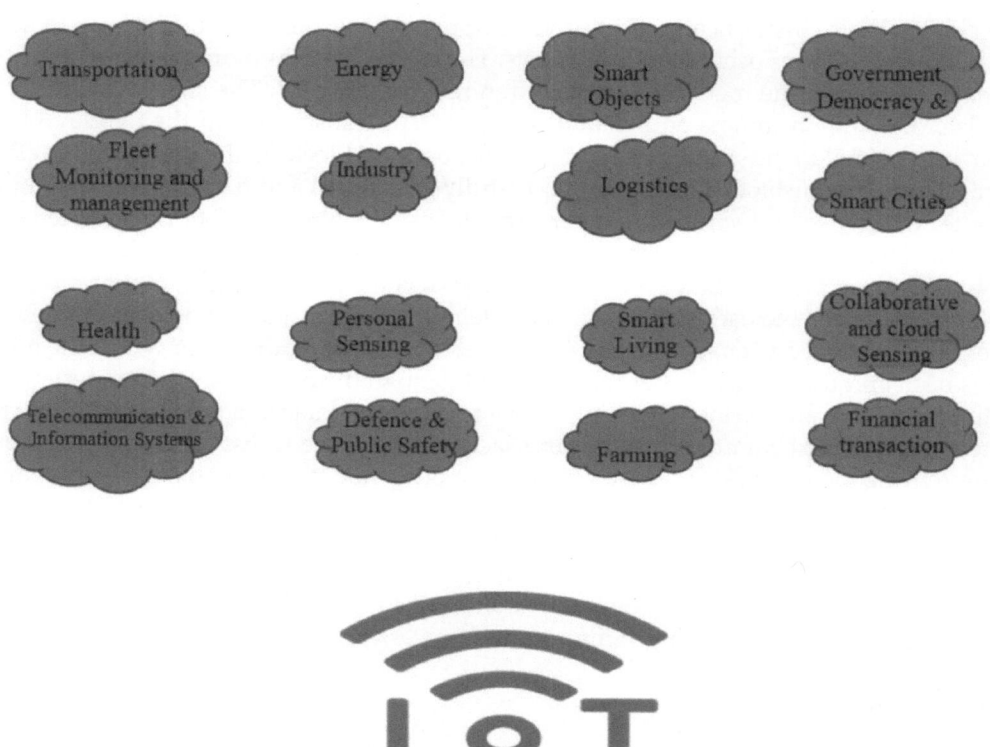

FIGURE 4.3
BIoT Applications.

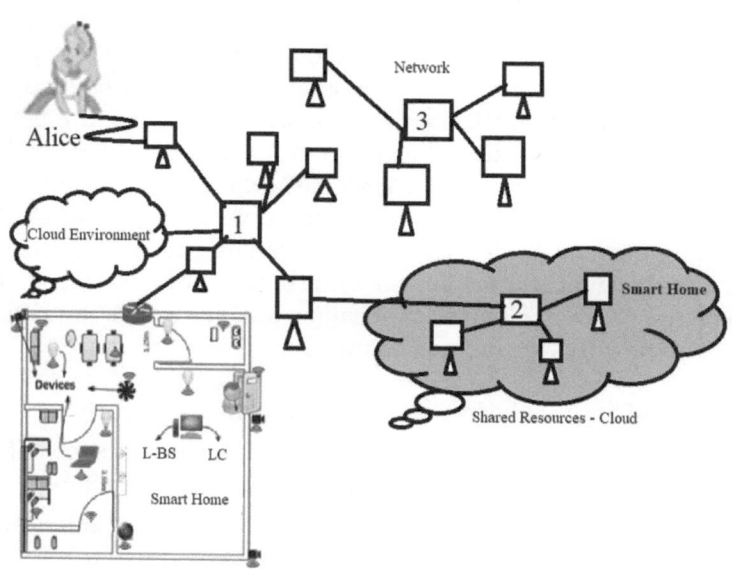

FIGURE 4.4
Blockchain-Integrated IoT Architecture [18].

can change their cluster head, and this can happen at any time in the network. In some cases, the transactions occur at higher delays. The owner that has more than one home can manage all their home together as depicted in red in Figure 4.4. The shared overlay thus consists of common miners and shared resources. All transactions in the overlay block-chain are chained to the starting transaction. The identical blocks in the user group of cloud storage have unique block number to successfully locate data and hash for the block number after which user authentication is performed. The new block number is encrypted with the Diffie Hellman algorithm using a shared key. And the person who has the shared key knows the block number. Since the only true user has the block number, it guarantees the existing ledger. It's the user preference to create many ledgers and the policies are defined here. The authorized user accesses the entire chain of data if required, for which the miner sends hash and the block number for the stored data. In the other case, the requester uses methods like safe answering when they receive minimum data from the miner. The cluster systems have blockchain that forward transactions. Also, the transactions requested and requesters are recorded by the cluster heads. The other nodes, based on their involvement in transaction communication, store the data. Monitoring a smart home in real-time and the owner's wish to access information can be done easily in ways like checking the smart thermostat's current configuration. The framework offered by IoT related to interconnec-tion of devices permit seamless communication. IoT applications focuses the IoT architec-tures middleware layer in terms of information processing. The layers are:

- a. Perception layer
- b. Network layer
- c. Middleware layer
- d. Application layer
- e. Business layer

4.1.4 Security Frameworks Using Blockchain

The leading accessibility attacks are: (1) Denial of Service Attack: Prevention of accessing the data or service from an authorized user; (2) Modification Attack: Compromising cloud storage security, seeking to delete or change the stored data. The changes could be detected by comparing the stored local blockchain hash value with the cloud data; (3) Dropping Attack: Taking control over the cluster head of a group and dropping all the blocks and received transactions. The attack could be identified when nodes do not receive any ser-vice or transaction from the network. The situation is detected when the clusters are made aware while electing the new cluster head; and (4) Mining Attack: A cluster head cooper-ates and signs a multisignal transaction for mining with the fake block. Thus security obtained by combining authorization, identification, and authentication [19].

These security frameworks are defined below:

Integrity. Expect authenticated users. No others can alter the information.

Availability. The true users can access the system whenever necessary. To uphold this service the user needs the database and communication infrastructure.

Confidentiality. It is guaranteed that unauthorized users will not gather information.

Authentication, authorization, and auditing. The identity of the user is verified to per-form a specific function in the system like storage information, owner rights, and others.

Nonrepudiation. Guarantees that after performing a specific action the user cannot deny their action. Examples would be authorized purchase or transferring of money.

Hash Functions. The mathematical function applied to generate a summary, a unique output like fingerprint data.

One way. The input computed from the hash value is tedious to find.

Compression. Hash size is represented by the small fraction of data.

Diffusion. The change of one input bit will change the hash result to 50%.

Collision. Two inputs generating same hash value leading to computational difficulty.

4.2 Blockchain IoT Paradigm

4.2.1 Blockchain-Enabled IoT Security

In IoT interactions happen between machine to machine (M2M) without human intervention. The challenge is creating trust among participating machines. However usage of blockchains acts as a catalyst in this enabling enhanced scalability, security, reliability, and privacy. Thus blockchain is deployed in the IoT ecosystem to track and coordinate devices for transaction processing. Already the existence of the search engine Shodan for internet connected devices exposes insecured devices in addition to blockchain in IoT, which enhances the system reliability eliminating the single point of failure. Data encryption in blockchain with cryptographic algorithms and hash functions provides better security services. Whereas implementation of hashing and cryptographic functions demand more processing power, which still remains a challenge in the digital economy. Trust maintenance is the initial concern in the deployment of blockchain in IoT. The information sequence gathered in blockchain is in chronological order of transactions networked by time stamping systems. It also offers real-time transactions preventing spoofing of data and tampering preventing security in industrial IoT as shown in Figure 4.5. Blockchain recording of sensor devices once deployed is called as Pindar Propounds.

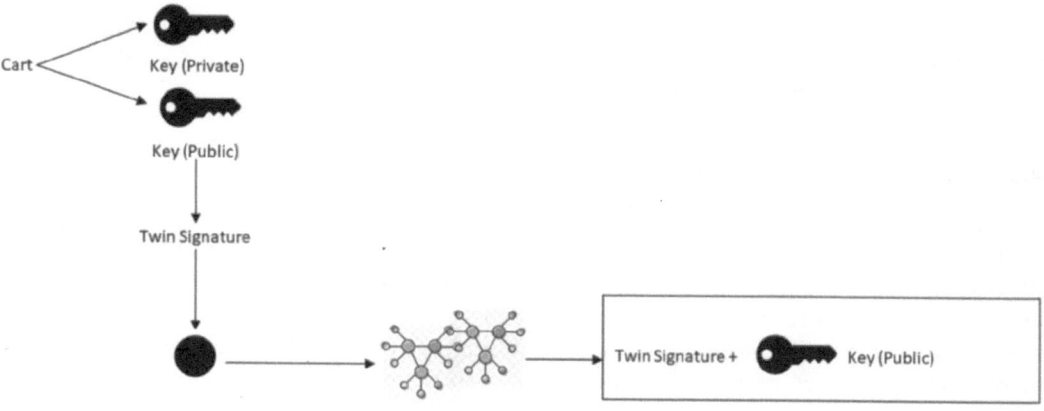

FIGURE 4.5
Signature Verfication in Blockchain [5].

4.2.2 Challenges

4.2.2.1 Storage: Capacity and Scalability

Blockchain that grows every 10 minutes at the rate of one MB raises the question of scalability and storage capacity in blockchain. The challenge is about how the copies are stored actively. A node in the network validates the blocks and transactions with significant storage size requirements. Blockchain's increase in more resources reduces system scalability. In the other case the overloaded system due to oversized chain reduces the performance introducing negative effects and increasing the synchronization timing for new users. The important feature of consensus mechanism in the distributed environment is the transaction validation. The modulation power and computational time directly affect the transaction timings between the blocks. The number of transactions increasing in the network and in turn consensus protocol has direct affect in the network scalability. Considering Bitcoin and its scalability limitations the blockchain protocol shall improve the consensus latency where Litecoin has a faster transaction capability similar to Bitcoin with improved storage efficiency. Bitcoin raises the scalability of blockchain by altering the chain-selection rule. Another development is the Inter Planetary File System (IPFS) protocol for decentralized [20] sharing and file storage in a distributed environment is an increase in efficiency of the web at the same time as removing attacks and duplicating every file in the network.

4.2.2.2 Security: Weakness and Threats

Vulnerabilities and security threats in Bitcoin protocol have been analyzed. It is observed that blockchain controls 51% of mining participants limiting the number of users. Most of the Bitcoin attacks are the spending the same coin twice. Bitcoin takes a longer time of 20–40 minutes for the transactional depth leading to a double spend attack. Similarly the user transactions are sent directly to the merchant causing racing attacks where the merchant is cheated. The protocols rely on communication are those vulnerable to the attacks violating their work from the network. Overcoming this code updation and optimization in blockchain is done using the cryptocurrency community to improve the protocols. Thus blockchain technology [18,21–24] has both soft and hard forks to improvement including updating of software protocol and functionality. However nodes keeping old rules have to be changed radically as new protocol will be compatible with the nodes of older versions. Nowadays in newer versions the nodes are computing with each other as they have to decide on the versions in the fork community. Hence fork versions are tediously dividing community leading to risk among blockchain users. Further new bugs are introduced to every day improvements in blockchain if necessary.

4.2.2.3 Privacy: Anonymity and Data Privacy

Bitcoin protocol has no built-in privacy. Its transparency is the key feature. Each transaction can be traced and audited in blockchain beginning from initial transaction. This transparency builds trust whereas the challenge still exists on anonymity as the user, as Bitcoin allows mechanisms for the user of multiple wallets in pseudonymous ways. There prevails the necessity of an effort to build stronger anonymity in Bitcoin features. Many blockchain technologies requires higher levels of privacy while dealing with sensitivity data. Many attempts tackle the anonymity problem. Hence the use of a signature in transactions making it untraceable and not easily trackable by the individual system. These services make transactions schedule smallest payments.

However these services are prone to theft like in this example: an originally proposed coin joins with Bitcoin leading to anonymity. Users' agreeing on join payments breaks the assumption on transactional inputs obtained from same wallet. Mixing servers also lack the anonymity between users based on implementation. Hence the dark wallet was developed providing complete anonymity in Bitcoin transactions. Security increase in cryptographic mechanisms randomized in mixcoin. Increased security is achieved by the advancement of coin shuffle modifying coin joins. Coin swap receives coins making payment with unconnected coins increasing the anonymity of mixed servers. Hence to deal with privacy data encryption practiced in blockchain even though the compiler translates the generic code into cryptographic primitives enabling transaction anonymity. Usage of distributed a hash table in a centralized network helps in storing data references. Privacy problems in blockchain dealt differently on providing authentication and authorization mechanism. However participants willing to preserve data privacy in private blockchains enterprise environments deploy scalable and distributed ledger using Hyperledger fabric to access a control list and services in the network. The members in the network know each other with public identities. Large amounts of data stored inside blockchain are managed relying on authorities verified by time stamps and data integrity. Thus the information from external sources is obtained, verified, and secured with blockchain.

4.3 Blockchain Technologies for IoT Environment

4.3.1 Universal Digital Ledgers

Distributed ledgers are for the most part referred to as a result of their utilization as digital forms of money, however in fact cryptocurrency and ledgers have two different technical methodologies. While considering the Coin Market Cap, a cryptocurrency is followed by most international distributed ledger organizations in not exclusively giving motivations to keep the hubs of different structures. Some distributed ledger technologies (DLTs) other than digital forms of money incorporate brilliant agreements (presented by Ethereum [25]) or file capacity. Due to rapid growth of accessibility to the internet and more gadgets incorporated with wireless technologies and sensors incorporated with them, the cost of mobile devices are going down, and cell phones are something beyond an unquestionable requirement. All these aspects are making the requirement ratio for the IoT to be high. IoT is a segment that has a wide scope of utilization, from straight-forward smart homes from clinical aspects to modern applications that makes it basic to have the best similarity of execution and security that can be accomplished. An increasingly definite clarification about DLT is accompanying the least complex structure, a distributed ledger in a database held and refreshed freely by every node in a huge network. The conveyance is quite interesting that records are not connected to different nodes in the network by a central position, rather autonomously connected and communicated by each existing node to make the independent decision of the network. The formations of DLT have revolutionized data assembling and information sharing between parties. It tends to be utilized in both basic repositories and information for exchanges. Distributed ledgers are showing that we need to care less about maintaining the database and put progressive responsibility about handling of the records transactions in a network. So far, the most concentrated sector of DLTs and blockchain is the financial area. Bitcoin began this extreme change since it is utilized as a P2P framework to pay for end product and administrations without the need of focal specialists otherwise known as banks or financial organizations.

4.3.2 Distributed Ledgers in IoT

Utilizing DTLs, blockchain, and IoT together can stay away from some major and minor issues that exist on associated gadgets as referenced in the accompanying issues that can be fathomed. Blockchain is used to track estimations of sensor information and forecast whether a fraudulent exchange takes place. IoT gadgets are intricate, distributed ledger appropriate for gadget identification, confirmation, and to secure information. A distributed ledger shields the IoT gadget's information from altering as shown in Figure 4.6. Blockchain empowers gadget independence, integrity, and trustworthiness of information without removing in efficiency. Blockchain helps to decrease the organization and process expenses of IoT since there is no middle man in the network.

4.3.3 Distributed Ledgers SOTA

DLT is a computerized record of exchanges and responsibilities for kinds of information that are duplicated and held among the entirety of the members of a shared network. Accord calculations are expected to guarantee that every hub of that network possesses a similar duplication of the ledger as different hubs. The primary kinds of DLTs or blockchains are normally two: open and private. Bitcoin is an open ledger where anybody can contribute exchanges. No third authority is expected to give authorizations. A private ledger is where all who take an interest are known and trusted and the accord procedure is overseen by a lot of members that have been pre-chosen here and there from an element called a facilitator. Right now and private DLTs are considered yet additionally a half-breed type of these two. DLT stages can be isolated as (a) permissionless, (b) permissioned, (c) with keen agreements, and (d) with exchanges as it were. The effects that these advances

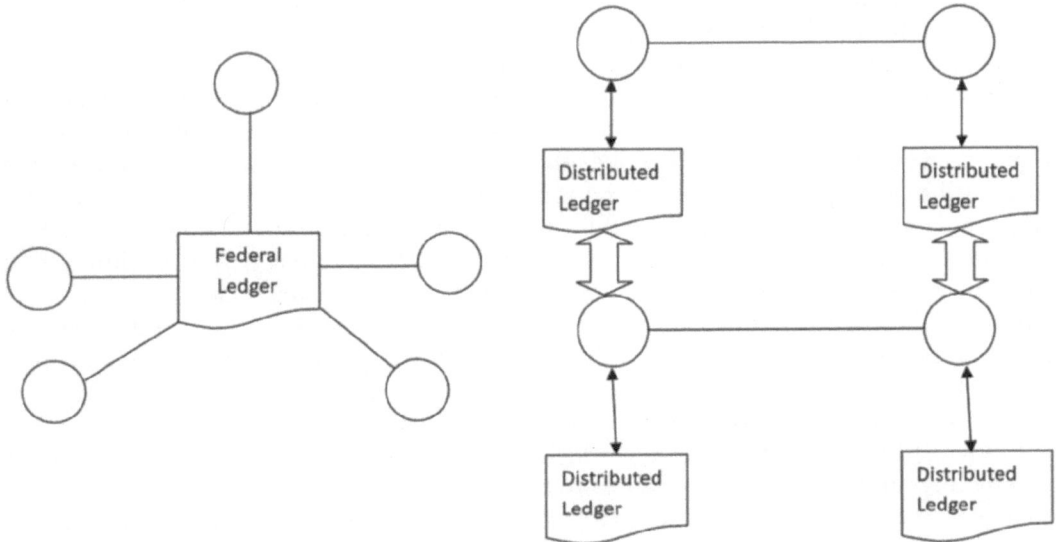

FIGURE 4.6
Ledgers [25].

bring are to upset current plans of action by evacuating the 'center man' substance, diminishing foundation costs with the need to keep up servers vanishes, and making dApps. Every one of these makes it conceivable to carefully recreate the trust that suddenly exists between two people who don't have any acquaintance with each other but who truly trade products for installment face-to-face, since it gives the advanced portrayals of indistinguishable attributes from material merchandise [24]. So if their assets are in peril they should act. It is this enthusiasm from all the members that goes about as an antifragile framework where an assault against a blockchain reinforces it, since the entirety of its members have an enthusiasm for amending the deformities that made the assault conceivable and empowering the framework to keep working. Ethereum, evolution of smooth (EOS), and IoT application (IOTA) are upsetting the DLT area rendering the potential utilizations of it practically boundless. Models like that in 'Blockchain with [IoT]' [26] show that there is a promotion level one of the adaption of advances this way, yet additionally they show the need of individuals to make and to be a piece of something new and different. One of the most fascinating parts is the accord-type calculations utilized by different DLTs to forestall misrepresentation. The Confirmation of Work (CoW) strategy is the most profoundly utilized accord calculation utilized by Bitcoin and other well-known digital currencies. It includes having PCs fathom a difficult riddle or numerical issue that places all exchanges in the following new square. It is utilized to forestall aggressors to perform DoS attacks.

4.3.4 Consensus Mechanism

Framing consensus is one of the strengths for the best possible working of a blockchain. It fundamentally comprises a component that decides the conditions to come so as to infer that an understanding has been reached in regards to the approvals of the squares to be added to the blockchain. All the minors are given a similar weight while casting a ballot and the result is published based on the majority of votes casted in the positive way of consensus mechanisms. This plan might be conceivable to execute in a controlled domain; in an open blockchain, this component would be vulnerable to Sybil attacks, since a duplication of a client with various different identifications would have the option in controlling blockchain. Clients in a decentralized network must include each block in the blockchain. This parameter is performed arbitrarily, however the issue is inclined in all attacks. PoW consensus mechanisms depend on more functions in a network, then it is more likely vulnerable to attacks. An arrangement is done such that few computations are done until answers are discovered, a procedure called mining. On account of the Bitcoin blockchain, mining comprises an irregular number (called nonce) computed by SHA-256 hash, the square header having zeroes. Subsequently, diggers need to show that played measures take care of the issue.

When the issue is tackled, it is extremely simple for different hubs to confirm that the acquired answer is right. Because of the issues recently referenced, an elective agreement technique proposed is this: a PoS consensus component has a computational force less than PoW. Therefore, diggers demonstrate consensus component intermittently. Since fewer hubs are engaged with square approval, exchanges are performed quicker than with other schemes. In addition, delegates can adjust block size and interims, and, on the off chance that they carry on unscrupulously, they can be substituted without a problem. A proof-of-activity (PoA) consensus is proposed as a principle impediment of PoS frameworks dependent on stakeage: it is amassed in any event when the hub isn't associated with the network. In this manner, PoA plans have been proposed to energize both possession and movement on the blockchain. Practical Byzantine fault tolerance (PBFT) is a consensus calculation that takes care of the Byzantine Generals (BG) Problem for offbeat situations. PBFT accepts that not exactly 33% of hubs are noxious. For each square to be added to the chain, a pioneer is chosen to be accountable for requesting the exchange. Bitcoin-NG executes a variation of the bitcoin consensus calculation planned for improving adaptability, throughput, and dormancy. The thought behind PoB is that, rather than consuming assets (e.g., vitality on an account of numerous PoW usage), cryptographic money is scorched as it is considered as costly in capacity assets.

4.3.5 Decentralization

IoT centralized architecture changed to a P2P distributed ledger ensuring robustness and scalability is offered in blockchain. This minimizes the latency of avoiding a single point of failure problem preventing an individual authority from making decisions on a distributed ledger.

4.3.6 Self-Sovereign Identity

4.3.6.1 Overview

IoT connects all individuals and objects in the physical and cyber world, by 2020 with more than fourteen billion devices are connected with actuating and sensing capabilities. When billions of devices are connected there are innumerable data resources and the challenge is the identification of entities allocating these identities so that communication happen easily. When living among group where interconnection and interactions occur, identity plays a major role in IoT systems. Digital identity is the stepping stone in the IoT environment. Without digital identity while using IoT in an untrusted environment, barely performing transactions will lead to lack of business opportunities. An identity management system follows the aspect/logic in the field of philosophy providing identity solutions under a new paradigm of blockchain circumstance for IoT. Digital identities are the keystone for online services builing mechanisms and protocols for authentication, authorization, and security exchanges in the internet era. However they rely on trusted third parties for increasing threats from internal attacks to comprise user privacy. Therefore traditional identity management systems raise many privacy concerns. Also these identity management systems suffer from long vulnerabilities and attacks leading to a single point of failure. An example is that Facebook security breaches have become honey pots for attackers. Exposing personal data leads to data breach vulnerabilities. Emergence of IoT can directly transplant native IoT environments and are of great importance in the design of an identity management system in IoT. The characteristics are:

Scalability: IoT with billions of devices demands highly scalable identity management comparatively with traditionally managed systems maintained by third party providing extremely unrealistic solutions. However, in trustless networks mutual trust relationships between different identity management systems has to be build.

Interoperability: The heterogeneous objects perform different communication capabilities leading to interoperability issues.

Mobility: Without consideration on the device location ensures connection services.

4.3.6.2 Traditional Identity Management System (IdMS)

Digital identity management systems manage users' identity information, credentials, and attributes. The figure depicts a traditional identity management system comprising three stakeholders: user, identity provider, and service provider. Those are interdependent to requests and access services relying on third party authentication protocol as shown in Figure 4.7.

Identity management systems have access to internet services and resources. In the past decades, these have isolated from centralized to federated models. On accessing internet services, all users need to register themselves in obtaining digital identities from the security domain. To deal with the problem of password management the identity management allows several service providers to reduce the number of user identities. Whereas a federated model establishes relationships between identity providers and access services.

4.3.6.3 Blockchain Identity Solutions

Blockchain had its attention on eliminating transaction intermediaries providing prominent security solutions in decentralized environments. Moreover, blockchains as distributed ledgers have permanent records in a P2P network. In addition, it simultaneously updates a network validating all members aiming to create decentralized applications (dApps). Blockchains are used in building naming systems and a secured and decentralized system. Many IT players focus on smart contracts to ensure reliability in terms of identity as shown in Figure 4.8. Identities are managed by distributed ledgers for every identity on the internet.

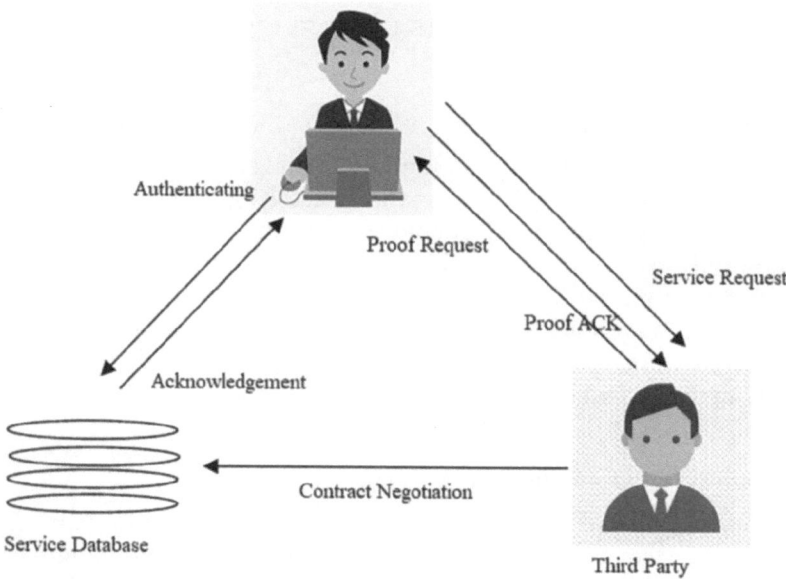

FIGURE 4.7
Identity Management System [27].

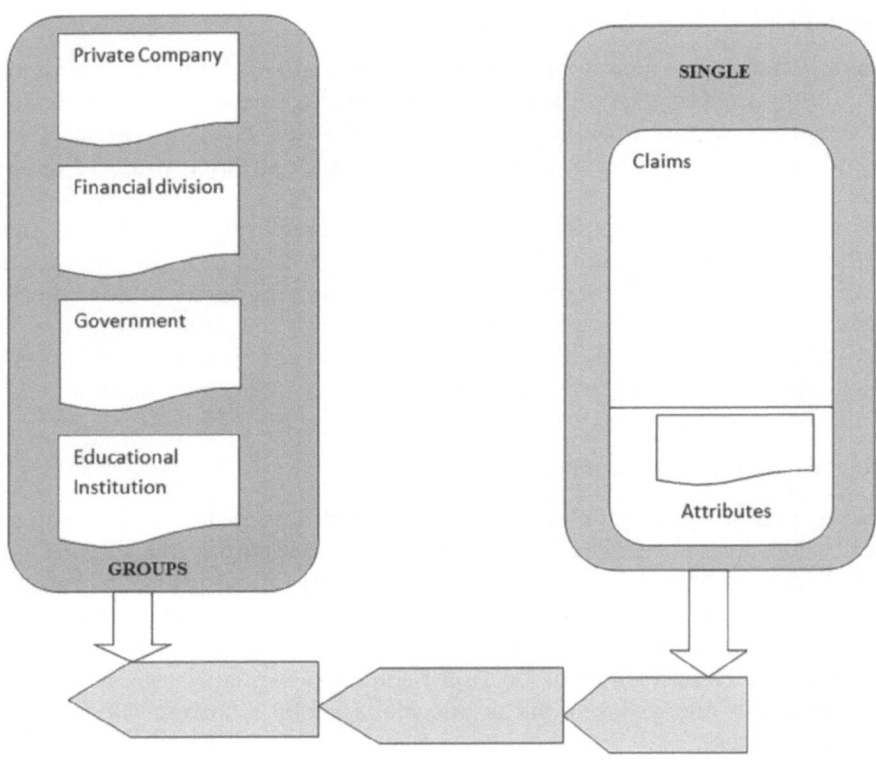

FIGURE 4.8
Overview of Blockchain-Based Identity Management Solutions [27].

4.3.6.4 Challenges in IdMS

Critical components to build effective an identity management system still remain a challenge in terms of trust, privacy, and performance. Accessing control and identity are hand-in-hand. The identity systems in IoT enable proper authorization and are designed for rapid growth of policies. A common problem existing in centralized systems is assigning of access rights, roles, and attributes that will not be suitable while scaling decentralized IoT. However blockchain couldn't address all these problems. In most cases, service and identity providers require authorized users and performance evaluations be essential to run the identity management system.

4.3.7 Smart Contract and Compliance

Blockchain supporting smart contracts are blockchain 2.0 coined by Nick Szabo. Smart contracts are autonomous programmes digitally enabled by turing language. Having a predefined set of rules and conditions capable automatic execution. The rules include contract execution, verification, and validation. And example would be a example vending machine. Blockchain 2.0 has built insecurities and features offering variability and transparency related to smart contracts. Smart contracts are now acting as a catalyst for adoption in various application domains. It enables a blockchain system with preset

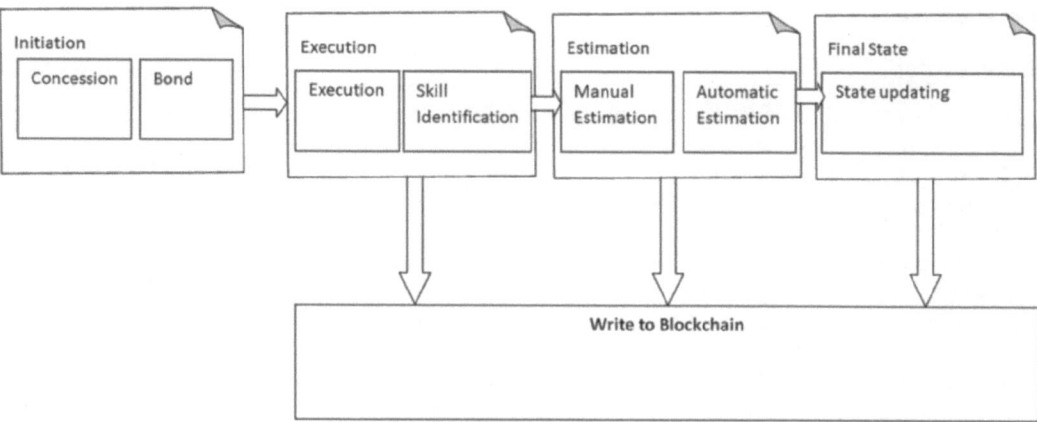

FIGURE 4.9
Smart Contract Lifecycle [28].

conditions and automatically triggers the transaction when conditions are met by materializing the contract. Smart contracts enable bidirectional transactions between two parties. The transactions are instantaneous and not subject to consensus. With the emergence of smart contracts, blockchain supports lighting applications, application affordable for a micro payment system promoting and nurturing new innovation. Also, IoT devices use smart contracts to facilitate cash flow resources in data trading. The focus on use of blockchain enables interactions of smart devices to discover and exchange messages without any centralized controlling authority. However, device communications are authenticated by Transaction Layer Security (TLS). A smart contract is an alternative approach to Bitcoin [27,28] as no blocks facilitate micropayment. In a smart contract code, blockchain stores rules of an agreement to execute the agreed terms.

A smart contract as shown in Figure 4.9 computerizes transaction protocol advances in blockchain technology on executing contractual classes automatically. The contract is recorded in blockchain guaranteeing appropriate access control. For each contract function the developer assigns access permission making the contract deterministic. Once conditions are satisfied statements are triggered to execute the smart contract in a predictable manner. Several parties involved negotiate the rights, obligations, and prohibition of contracts in discussions. The contractual agreements are drafted and verified by lawyers converted to smart contract, computerized language using logic based rule language. Smart contract has multiple rounds of negotiation in an iterative process. On necessity new contracts can be created.

4.4 Conclusion

In this chapter, it has been described that the usage of blockchain shall remove the security concerns in IoT. Further various approaches and technologies related to blockchain are explained with respect to many application domains. The open issues on integration of blockchain with IoT have been discussed for future research directions. The impact of IoT devices with blockchain is studied and analyzed, followed by technology comparisons based on IoT scenarios.

References

[1] Das, ManikLal. 2015. "Privacy and Security Challenges in Internet of Things." *Distributed Computing and Internet Technology*: 33–48.

[2] Ho, G., D. Leung, P. Mishra, A. Hosseini, D. Song, and D. Wagner. 2016. *"Smart Locks: Lessons for Securing Commodity Internet of Things Devices."* In *Proceedings of the 11th ACM on Asia Conference on Computer and Communications Security*.

[3] Amoozadeh, M., et al. 2015. "Security Vulnerabilities of Connected Vehicle Streams and Their Impact on Cooperative Driving." *IEEE Communications Magazine* 53(6): 126–132.

[4] Musaddiq, A., Y. B. Zikria, O. Hahm, H. Yu, A. K. Bashir, and S. W. Kim. 2018. "A Survey on Resource Management in IoT Operating Systems." *IEEE Access* (6): 8459–8482.

[5] Buchmann, J. 2013. *"Introduction to Cryptography."* Springer Science + Business Media.

[6] De Montjoye, Y.-A., et al. 2014. "Openpds: Protecting the Privacy of Metadata Through Safeanswers." *PLoS One* (9): 7.

[7] H. Gross, M. Holbl, D. Slamanig, and R. Spreitzer. 2015. "Privacy-Aware Authentication in the Internet of Things." *Cryptology and Network Security*: 32–39.

[8] Ukil, A., S. Bandyopadhyay, and A. Pal. 2014. *"IoT-Privacy: To Be Private or Not to Be Private."* In *Computer Communications Workshops (INFOCOM WKSHPS), 2014 IEEE Conference*, Toronto.

[9] Nakamoto, S. 2008. "Bitcoin: A Peer-to-Peer Electronic Cash System."

[10] Decker, C., J. Seidel, and R. Wattenhofer. n.d. "Bitcoin Meets Strong Consistency."

[11] Rayes, A., and S. Salam. 2016. *Internet of Things from Hype to Reality: The Road to Digitization.* Springer.

[12] Hung, M. 2017. "Leading the IoT, Gartner Insights on How to Lead in a Connected World." *Gartner Research*: 1–29.

[13] Ai, Y., M. Peng, and K. Zhang. 2018. "Edge Computing Technologies for Internet of Things: A Primer." *Digital Communications and Networks* 4(2): 77–86.

[14] Haroon, A., M. A. Shah, Y. Asim, W. Naeem, M. Kamran, and Q. Javaid. 2016. "Constraints in the IoT: The World in 2020 and Beyond." *Constraints* (7): 11.

[15] Jøsang, A., and J. Haller. 2007. *"Dirichlet Reputation Systems."* In *Availability, Reliability and Security, 2007. ARES 2007: The Second International Conference*.

[16] Alrawais, A., A. Alhothaily, C. Hu, and X. Cheng. 2017. "Fog Computing for the Internet of Things: Security and Privacy Issues." *IEEE Internet Computing* 21(2): 34–42.

[17] Buyya, R., and A. V. Dastjerdi. 2016. *Internet of Things: Principles and Paradigms.* Elsevier.

[18] Conoscenti, M., A. Vetro, and J. C. De Martin. 2016. *"Blockchain for the Internet of Things: A Systematic Literature Review."* In *Computer Systems and Applications (AICCSA), 2016 IEEE/ACS 13th International Conference*, pp. 1–6. IEEE.

[19] Reyna, A., C. Martín, J. Chen, E. Soler, and M. Díaz. 2018. "On Blockchain and Its Integration with IoT. Challenges and Opportunities." *Future Generation Computer Systems*.

[20] Skarmeta, A., F. Jose, L. Hernandez-Ramos, and M. Moreno. 2014. *"A Decentralized Approach for Security and Privacy Challenges in the Internet of Things."* In *Internet of Things (WF-IoT), 2014 IEEE World Forum on*.

[21] Yang, Y., L. Wu, G. Yin, L. Li, and H. Zhao. 2017. "A Survey on Security and Privacy Issues in Internet-of-Things." *IEEE Internet of Things Journal* (4)5: 1250–1258.

[22] Bashir, I. 2017. *Mastering Blockchain*. Packt Publishing Ltd.

[23] Mougayar, W. 2016. *The Business Blockchain: Promise, Practice, and Application of the Next Internet Technology*. John Wiley & Sons.

[24] Wood, J. 2018. "Blockchain of Things, Cool Things Happen When IoT and Distributed Ledger TechCollide."*Medium*.https://medium.com/trivial-co/Blockchain-of-things-cool-things-happenwhen-iot-distributed-ledger-tech-collide-3784dc62cc7b.

[25] T. Project. [Online]. Available: https://www.torproject.org/.

[26] Atlam, H. F., A. Alenezi, M. O. Alassafi, and G. Wills. 2018. "Blockchain with Internet of Things: Benefits, Challenges, and Future Directions." *International Journal of Intelligent Systems and Applications* (10): 40–48.

[27] He, Q., N. Guan, M. Lv, and W. Yi. 2018. *"On the Consensus Mechanisms of blockchain/DLT for Internet of Things."* In *2018 IEEE 13th International Symposium on Industrial Embedded Systems (SIES)*, pp. 1–10. IEEE.

[28] Dorri, A., S. S. Kanhere, R. Jurdak, and P. Gauravaram. 2017. *"Blockchain for IoT Security and Privacy: The Case Study of a Smart Home."* In *Pervasive Computing and Communications Workshops (PerCom Workshops), 2017 IEEE International Conference*, pp. 618–623. IEEE.

5

A Comprehensive Overview of Blockchain-Driven IoT Applications

Rajalakshmi Krishnamurthi and Dhanalekshmi Gopinathan
Jaypee Institute of Information Technology, Noida, Delhi-NCR, India

CONTENTS

5.1 Introduction

In the recent Internet of Things (IoT) era, IoT based applications play crucial position in enhancing the quality of human life [1,2]. A few such applications of IoT to mention include smart environment supervising and controlling, smart buildings, smart agriculture, smart cities, smart transportation systems, and smart healthcare systems [3]. However, the IoT system involves a heterogenous resource constraint on IoT end devices, a wide range of underlying protocols, and a tremendous volume of data generated through IoT sensor devices. Consequently, the security and privacy of IoT systems need an efficient mechanism [4]. For this, conventional centralized mechanisms of security and privacy are used through authentication, digital signatures, and encryption mechanisms that become insufficient to meet the specification of IoT systems. Alternatively, the decentralized and distributed approach of blockchain promises to be best suited for handling IOT systems.

Blockchain is a technology used to keep track of all digital transactions [5]. It is a ledger, but rather than keeping it as a centralized mechanism, it stores transactions in a decentralized, distributed mechanism that shares data across a large network of computers. This decentralization reduces data tampering. Blockchain is also referred to as distributed ledger technology (DLT). DLT are programmable by application developers and are used to maintain records of operations and pursue anything of significance such as business transactions, healthcare records, government policies, property documents. It is a network that stores information in a way that makes it difficult or impossible for the network to modify, hack, or cheat. The blockchain's primary goal is to create a creditworthy ecosystem within an untrustable distributed environment among independent participants such as an IoT ecosystem. Blockchains allow users to securely demonstrate their identities, protect digital asset ownership, and validate transactions without a high-cost intermediary.

By incorporating blockchain technology into IoT, the combination may lead to a verifiable and traceable IoT network [6,7]. This blockchain technology in IoT can record transaction data, improve system performance by providing additional security, data verification, and so on. The presence of a wide number of devices in conventional IoT applications is facing many challenges like data security, integrity, and robustness. Blockchain proposes a real-world solution for many of the limitations of traditional IoT applications. Blockchain will ensure the confidentiality of IoT data without third parties thus saving IoT devices bandwidth and processing power. In addition, blockchain can provide an IoT network with a secure and scalable platform so that sensitive information can be distributed without a centralized server. Blockchain makes transactional records tamperproof, and IoT links the physical world to the digital world by means of computers, sensors, actuator, and devices.

The three ways blockchain technology revolutionizes IoT based applications based on the way it interacts with IoT ecosystem include (i) the way it tracks IoT devices and stores the IoT sensor data. Blockchain stores the information or data in terms of blocks that are logically linked in a sequential manner within chain of blocks. To perform any change within a particular block in the chain, the change is added as new block with a timestamp and inserted at the end of the block rather than changing in the block itself; (ii) It creates trust in the data generated by IoT devices. Before adding a block into a chain, there are some steps that need to be performed. A cryptographic puzzle needs to be solved. The IoT computing devices that clear the mathematical puzzles are eligible to share the information with all the other computing devices within the distributed network. Then the blockchain network validates the appropriate solution, which is known as proof-of-work. If it is validated, the new block is attached to the existing block chain. The primary objective is to solve the combination of intricate mathematical puzzles. Next these blocks are certified by authenticating computers and guaranteed the confidentiality of each block integrated in the chain; and (iii) There is no need of intermediaries to be required, which in turn reduces time and expenditure. Blockchain enables distributed storage of all transactions in a protected, secure, valid, efficient, and transparent manner [8,9].

In general, the blockchain creation process is comprised of three stages, respectively called blockchain 1.0, blockchain 2.0, and blockchain 3.0 [10]. The key focus of the blockchain 1.0 stage targets toward peer-to-peer transactions. The popular application of blockchain 1.0 is Bitcoin. Blockchain 2.0 targets providing a trust-based decentralized framework that includes traceability and data resistance features. A decentralized blockchain framework with an intelligent contract feature was developed in December 2013, and Ethereum is the popular blockchain 2.0 application system. Blockchain 2.0 level is also known as the blockchain level of the Ethereum. Blockchain 3.0 targets performing a combination of blockchain technology with other thrust areas like the financial industry, supply chain management, and IoT. The various consensus protocols involved in blockchain 1.0, blockchain 2.0, and blockchain 3.0, are proof of work (PoW), proof of stake (PoS), and Practical Byzantine Fault Tolerance (PBFT) respectively. Hence, this chapter targets exploring and providing insight about the implications and advantages of integrating blockchain for enabling IoT systems.

The overall contributions of this chapter include:

- fundamental overview of blockchain technology such as decentralization, disturbed shared ledgers, smart contracts, cryptography techniques, and consensus mechanism
- basics IoT protocol stack layers namely application layer, transport layer, network layer, adaptation layer, data link layer, and physical layer
- various security aspects of IoT networks such as healthcare data security and privacy, healthcare IoT device access and authentication, trustable users, access policy and control of IoT devices, cryptographic keys, and hashing of data
- blockchain-based IoT applications such as a intelligence transport system, a smart healthcare system, supply chain management, an IoT ecosystem, and a smart city
- beyond cryptocurrencies and Bitcoin, the various types of block chain platforms such as public and private blockchain, hyperledges, fabrics, and R3 codra

- a decentralized framework of blockchain for healthcare IoT applications
- challenging performance factors of blockchain-based IoT applications such as scalability, adaptability, anonymity, and integrity

Section 5.2 presents the fundamental overview of block. Section 5.3 presents the IoT protocols stack, and blockchain-based IoT applications. Then Section 5.4 explains beyond cryptography and Bitcoin, the various blockchain technologies such as hyperledger, fabric, and R3 codra. Next, Section 5.5 presents a decentralized framework of blockchain for healthcare IoT applications. Then Section 5.6 addresses various challenges in blockchain-based IoT applications. Finally, Section 5.7 concludes the chapter.

5.2 Blockchain Overview

Blockchain can be described as a DTL where transactions are represented as blocks and chained together by a cryptographic hash function. It is shared by everyone in the network and once a block is added to the blockchain it is difficult to alter it. The authenticity of each transaction is assured by the digital signature on the block by the blockchain. Encryption and digital signatures make the stored data tamper proof and immutable.

According to researchers, Nakamoto developed the blockchain by assimilating the following technologies: peer-to-peer networks, cryptography, and distributed consensus [10,11]. Every participant in the blockchain network shares the ledger's copy. When a node makes a transaction, it transmits the transaction to the network. The transaction is checked by other participants in the network. It is then grouped with other transactions and forms a block. The participant that first forms a new block by agreeing to the consensus protocol set by the network broadcasts the block to the network. Peer nodes verify the block and add it to the existing block if it adheres to the rules.

Figure 5.1 illustrates the process of the operation of blockchain. The node that wishes to perform the transaction transmits the block to the network. All other network participants verify the transaction and form a new block that combines this with other transactions. This new block is transmitted to the network by the miner node with ample computational power to solve the cryptographic puzzle. Then the peer nodes, or those in the network, validate the block and add it to the existing blockchain if it meets the rules.

5.2.1 The Key Components of Blockchain Technology

The key components of blockchain technologies includes decentralized and distributed shared ledger, cryptography, consensus protocols, and smart contracts [12,13].

5.2.1.1 Decentralization

In a decentralized system every participant has equal authority. Blockchains are in essence decentralized, meaning that no single individual or community has overarching network authority. Although everybody that has copy of the distributed ledger is available in the network, no one can change it on their own. This unique blockchain feature allows users to have transparency and protection while giving control. Blockchain technology makes it

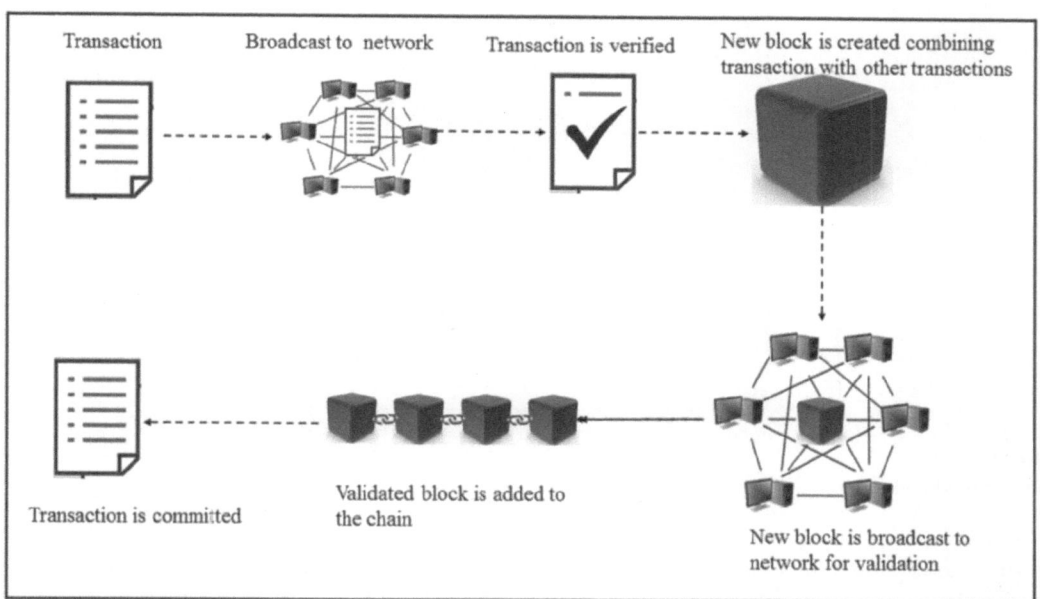

FIGURE 5.1
How Blockchain Works.

easier to introduce decentralized IoT networks such as safe and confident data sharing and record keeping. For a system like this the blockchain functions as the general ledger, holding in a decentralized IoT topology a shared record of all messages exchanged between smart devices. The advantages of a decentralized system [12] include: (i) it is less likely fail since users rely on many separate components; (ii) Blockchain is very resistant towards attacks since networks are spread across may computers and (iii) it's harder for users with a malicious rationale to take advantage of users who are using the platform for its supposed purpose.

5.2.1.2 Distributed Shared Ledger

A distributed shared ledger is a shared database as shown in Figure 5.2. It is replicated and synchronized among the participants of the decentralized network. All nodes in the network hold all the transactions and synchronize with the consensus protocol. All the participants can view transactions. All the information contained in the transactions is stored safely and correctly using cryptography and can be accessed with key and cryptographic signatures. Any update on the ledger is carried out on the consensus of all participants in the network, thus avoiding the involvement of a central authority.

5.2.1.3 Cryptography

Cryptography is one of blockchain's most significant characteristics because it allows blockchain to become immutable. Blockchain makes use of public key cryptography, digital signatures, and cryptographic hash functions to ensure confidentiality, transparency, and authenticity. Public key or asymmetric cryptography uses an algorithm for encryption and generates public and private keys while using digital signatures, too. Digital signatures use

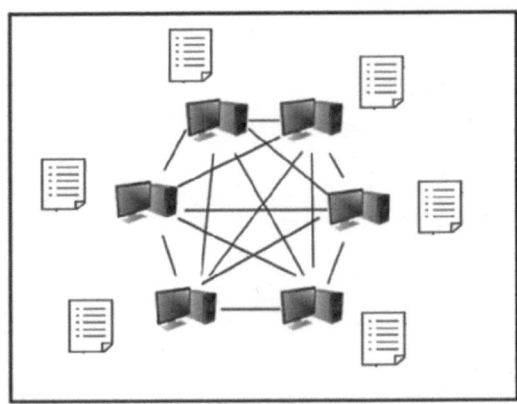

FIGURE 5.2
Distributed Ledger.

a private key with a corresponding public key to sign and validate a transaction. A node-generating transaction uses a public key to encrypt the transaction data and signs with the private key. The digital signature, transaction data, and public key are subsequently forwarded to the network. The participants in the network verify the specifics of the transaction using the public key.

A cryptographic hash function enables immutability for blockchain technology. Hash function reduces data of any length to a fixed-length alphanumeric string. Since the output is fixed length, the attacker will not be able to make out which output is created for an input string. For each transaction, known as the transaction identifier, blockchain uses the cryptographic hash function. In most blockchain protocols, the transaction identifies is a 256-bit alphanumeric string of data. Each block of blockchains is made up of thousands of transactions, and verifications of each transaction is quickly computationally difficult. Hence, processing time is minimized, ensuring the highest level of security is obtained when a little data is used for processing and verifying transactions. Blockchain uses the Merkle tree hash function [13], which takes the large number of transaction identifiers and creates a 64-character (or 256-bit) alphanumeric string known as a Merkle root.

Each block in the blockchain will have one Merkle root that verifies a particular transaction has taken place in that block or not. The Merkle tree is constructed by organizing all the transaction identifiers into pairs. If a block has an odd number of transactions, then the last transaction is replicated and paired. Suppose a block has 256 transactions, then Merkle tree will start by grouping these 256 transactions identifies into 128 hashes. A hashing function is applied to the 128 pairs resulting in 128 new cryptographic hashes. This continued until a single hash remains, which is the Merkle root. Figure 5.3 illustrates the Merkle tree and Merkle root construction. The Merkle tree represents a block containing eight transactions labelled as T1, T2, …, T8.

In the first step each transaction is individually hashed. For example, hash (T1) produces a new hash value such as H1. The new hash values are then combined with the neighbouring value to produce new hash values again. For example, hash (H1, H2) produces new hash value H12. This process is repeated until a single hash value is obtained (H12345678 in this case). The hash value thus obtained is known as the Merkle root and is stored as the block header in the block of the blockchain. The advantage of a Merkle tree structure is that it uses fewer resources leading to quick verification of data.

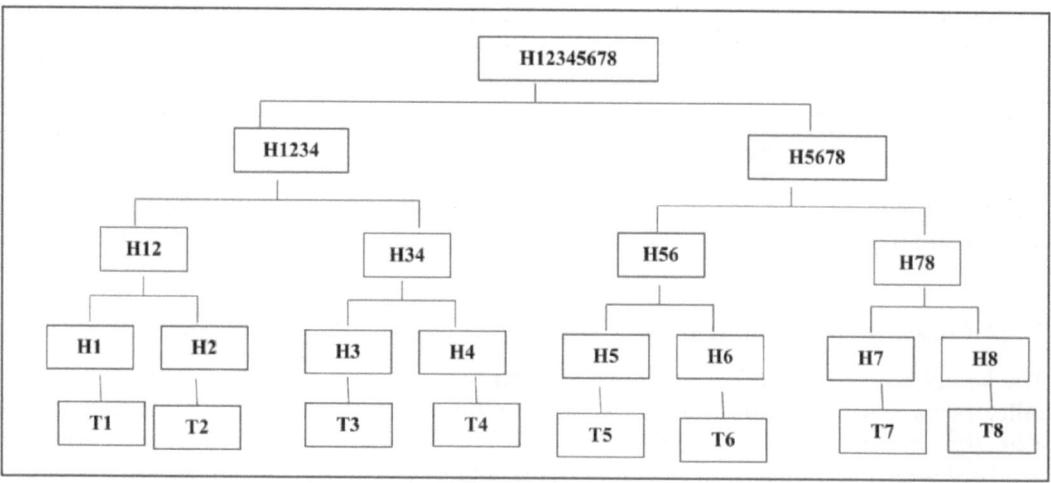

FIGURE 5.3
Merkle Tree.

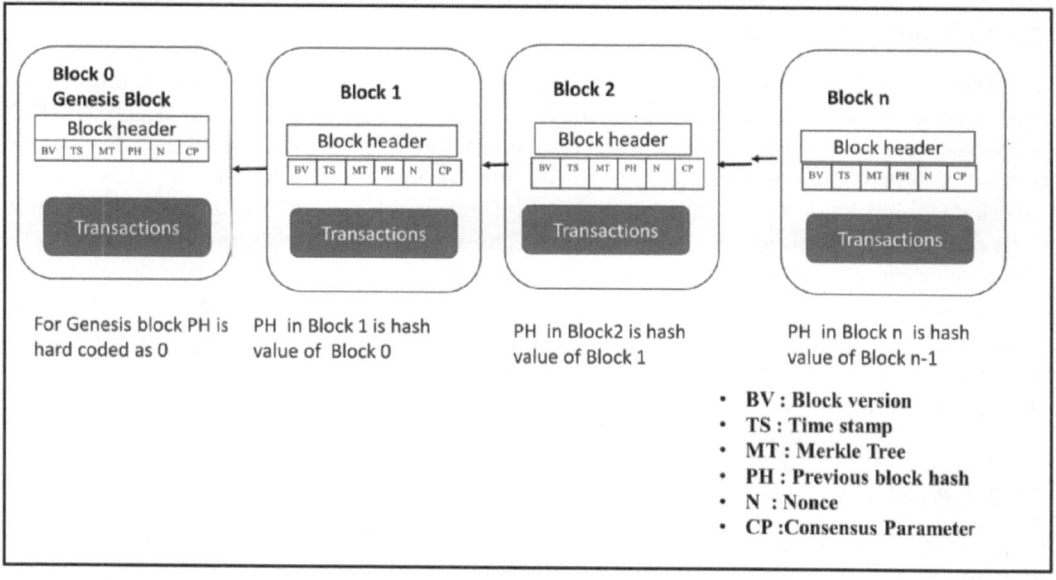

FIGURE 5.4
Organization of Block in Blockchain.

The second advantage of Cryptographic hashing is that it enables *immutability* data for blockchain technology. As the name suggests, the blockchain is arranged in a chronological order as a sequence of locks linked together as depicted in Figure 5.4. Each block consists of a block header and a transaction listing. The block header contains metadata which includes version, previous block hash, timestamp, Merkle root which is a unique identifier obtained from the hashes of all the block transactions, consensus parameters to validate the new blocks to be added, and nonce which is a numerical value to be resolved to insert the block into the blockchain. The first block in the chain of blocks is known as the block genesis. There is no preceding block to the genesis block, hence the previous hash function is hardcoded as zero.

Every block of data is linked with the hash value of the previous block, making it difficult to change the data. Suppose data has to be changed in a particular block, say block 2. It recomputes that block's hash value, but in the blocks that follow, say block 3, it holds block 2's previous hash value. Thus, a data change should regenerate header hash of all subsequent blocks in the blockchain as shown in Figure 5.4 to maintain a valid blockchain.

5.2.2 Consensus Mechanism in Blockchain

The key characteristics of a distributed decentralized blockchain network are its immutability, privacy, security, and transparency. Even though there is no central authority to validate the transactions in the blockchain, it is considered secure through the consensus protocol which is the heart of the blockchain network. A consensus mechanism means to reach agreements among the network nodes on the status of the network. It helps the validation of adding a new block to the ledger thereby entering only the authentic transactions into the blockchain. The various consensus algorithms are discussed below.

5.2.2.1 Proof of Work (PoW)

In 2008, Nakamoto proposed the proof-of-work (PoW) consensus mechanism in the blockchain especially in Bitcoin and Ethereum [14]. It solves the double-spend problem [7]. PoW is based on a mathematical puzzle that computes a value for nonce. The desirable value for the nonce should be less than the threshold value. Miners use its computing power to solve puzzles (which is also known as mining). The miner who computes a hash value lower than the threshold value first is the winner. The winner broadcasts the new block to the network cryptocurrencies as a reward. The limitation of PoW is that it wastes energy and resources on the creation of a block which consumes lot of computing power and only one miner will be rewarded based on who solved the puzzle first. PoW requires a high processing power and IoT devices that have very limited resources. PoW is thus not an acceptable choice for IoT systems.

5.2.2.2 Proof of Stake (PoS) [15,16]

PoS is an alternative consensus mechanism to PoW. The term 'stake' represents the amount of currency that the miner (or validator node) holds. The PoS algorithm picks a node as validator for the next block using a pseudorandom choice method. This method is based on various factors such as betting age, randomization, and wealth of nodes. The node that wants to be a part of the validation procedure has to pledge some currency as their network stake. The node with a bigger stake has the bigger chance. To avoid the scenario of only wealthy being selected to validate transactions some unique selection processes like randomized block selection and a coin age selection process are added to the system. In the randomized block selection process, the validators are chosen by searching for nodes with a combination of the lowest hash value and the highest stake, and since the stake size is public, other nodes usually predict the following forger.

The coin age selection system selects nodes based on length of time that the nodes were staking for their tokens. Coin age is determined by the number of staked coins multiplied by the number of days on which the coins were kept as stakes. If a node forges a block, their coin age is restored to zero, and they have to wait a certain amount of time to be able to forge another block; that prevents enormous stakes from governing the blockchain. Cryptocurrency uses this algorithm when a node is chosen to forge the next block, the validity of the block transactions will be checked, the block will be signed, and added to the blockchain. The transaction fees related with block transactions shall be charged as a reward to the node. When a node wishes to avoid being a forger, its stake will be published after a certain amount of time along with the winning rewards, giving the network time to check that no fake blocks are added by the node to the blockchain. Energy performance and health are the main benefits of the PoS algorithm.

5.2.2.3 Practical Byzantine Fault Tolerance (PBFT)

In distributed systems, Castro and Liskov developed a novel method for achieving consensus that can accommodate faulty/malicious nodes by replicating nodes/state machines [17,18]. Byzantine fault tolerance facilitates the distributed computer network to achieve a reasonable consensus correctly even with malicious device nodes deteriorating or sending out incorrect information. BFT's aim is to protect from catastrophic network failures by reducing the impact of these malicious nodes. Hyperledger, stellar, and Ripple are the three examples of blockchains that rely on the PBFT consensus algorithm. Nodes are ordered sequentially in a PBFT scheme with one node being the master, and others being referred to as backup nodes. Both nodes in the network communicate with each other with the goal of making all honest nodes come to a majority rule understanding of the state of the system. During communication between nodes, they must verify that the messages come from the specific peer node and should also verify that during transmission the message has not been altered.

5.2.2.4 Smart Contracts [5,19]

The definition of smart contracts was first proposed in 1994 by Nick Szabo, a legal scholar and cryptographer. It is one of the successful applications in blockchain technology. Ethereum blockchain technology is very popular in creating smart contracts. In the context of blockchain, it is business rule that is assigned to the transactions of the blockchain. A smart contract on blockchain contract is a piece of code that describes the terms and conditions of an agreement between parties. The commitments are enforced through the consensus algorithm when the participants install the contract. It enables the participants to (i) verify the code to ensure that it meets the agreed sections, (ii) assure that once the contract is agreed and registered in blockchain it is tamperproof, and (iii) perform in the same way for all participants. Various industries and fields like a smart city, e-commerce, and asset management use smart contracts. Smart contracts allow you to exchange value transparently, including assets, shares and properties, removing the need for a middleman and keeping the system free from dispute. In particular, smart contracts are effective in business relationships, in which they are used to decide on the conditions set by consensus between both parties. This eliminates the risk of fraud, because there was no third party involved.

5.3 IT Protocols Stack

The loT protocol stack comprises of five layers namely the application layer, transport layer, network layer, network adaptation layer, data link layer, and physical layer as shown in Figure 5.5. The IoT protocol stack supports recently popularized low power and low data rate wireless personal area network (LoWPAN) and low power lossy wireless wide area network (LPWAN). The IEEE 802.15.4 standard describes the functionalities of the medium access layer (MAC) and the physical layer. Here, the physical layer has concerns about the underlying wireless communication for the diverse frequency range and data rates of IOT devices. The MAC specifications target the channel access mechanism and time synchronization between various communication entities. However, the challenge is to map the smaller sized maximum transfer unit (MTU) of 127 bytes in IEEE 802.15.4 with huge IPv6 data packet of 1028 bytes from the network layer. Hence the network adaptation layer, 6LoWPAN, below the network supplements the IPv6 protocol for underlying low-power IoT sensor devices and enables IP communications [1–4,20].

The 6LoWPAN protocol involves routing the protocol for targeting the low power and lossy network (RPL) mechanism for communication across IoT sensor devices. In this case, each IoT device is uniquely identified and communications are possible in mesh, multicast, unicast, and point-to-point data traffic. At the transport layer, the IoT network incorporates a user datagram protocol due to the limitation on payload for resource constraints on IoT devices. Also, User Datagram Protocl (UDP) are preferred for IoT networks due to its efficiency and lack of complexity in comparison to transmission control protocol (TCP). For monitoring various statuses of nodes and discovering neighbouring nodes, the internet control message protocol (ICMP) is adapted by the protocol 6LoWPAN based IoT environment. At the application layer, message queuing telemetry transport (MQTT) and constrained application protocol (CoAP) are available to support message-based asynchronous communication. These protocols are capable of mapping onto internet-based HTTP protocols.

5.3.1 Security Concerns in IoT Applications

The significant role of IoT in the future is promising in solving real-life problems. It is envisaged to grow in terms of both hardware and software. In hardware aspects, the evolution is happening in areas such as bandwidth upgrading, cognitive network of radio-based sensor devices, and optimized utilization of a radio frequency spectrum. In software aspect, the evolution is happening in areas such as middle ware support to enhance IoT-base applications and its wide range of services [21–25].

Application Layer	MQTT, CoAP
Transport Layer	UDP
Network Layer	IPv6
Adaptation Layer	6LoWPAN
MAC Layer	IEEE 802.15.4
Physical Layer	Wireless Transmission

FIGURE 5.5
Basic IoT Protocol Stack.

According to the literature, the integration of wireless sensor networks with cyber physical system (CPS) have emerged as significant components of enhanced IoT applications. Consequently, the heterogenous nature of a sensor network and an underlying internet protocol for network connectivity in IoT leads to challenging security problems. This section discusses various security concerns in IoT networks such as data privacy, IoT device authentication, trustable users, access control of IoT devices, cryptographic keys, and hashing of data.

5.3.1.1 Data Privacy

The attack on data privacy is a low-level attack involving the violation of privacy by any illegitimate users and the denial of service using malicious activities [21]. The insecure transaction by means of software initialization and unauthentic configuration of systems are the main contributors to the data privacy attacks. The physical layer of the system network is most affected by the data privacy attack. For example, the sybil attack targets a wireless IoT environment, instantiates by malicious IoT senor nodes using fake characteristics, and targets to downgrade the essential IoT functionalities. These sybil nodes forge the device's physical address to masquerade and target to exhaust the available network resources. Consequently, the sybil nodes prohibit legitimate users for accessing the eligible system resources.

5.3.1.2 IoT Device Authentication

The primary issue with IoT systems is the identity and access management (IAM) strategy that targets to identify the authentic ownership and relationship of smart IoT devices, especially for healthcare devices [22]. For each IoT device, the identity and ownership evolve during different stages such as manufacture, retailer, and customer. It is to be noted that customer ownership needs to be adoptable in case of resale, updates, and commissions. Hence, the management of IoT devices ownership and attributes through device authentication is critical in IoT systems. Further, the characteristics and features of the IoT device need to be authenticated through device to other device, services, and users.

5.3.1.3 Trustable Users

End-to-end device communication insists on data confidentiality and integrity through an efficient authentication process as a trust model. Further, the improper data usage needs to be avoided by implementing privacy models that emphasize access policies, data encryption, and decryption mechanisms [23]. Trustable users are ensured through a three-layered security model that includes application, network communication, and services provided. According to Open Web Application Security (OWSAP) consortium, there are several vulnerabilities presented in the current form of IoT architectures, namely insecure network interfaces of IoT devices, inappropriate configurations of software, unauthentic security setups, physical damage to devices, and insecure usage of third-party firmware.

5.3.1.4 Access Control of IoT Devices

The objective of the access control mechanism is to provide support for authentication between IoT sensor devices along with underlying IP protocol [24]. Access control mechanisms need to address the dynamic switchable IoT systems for end-to-end or point-to-point authentication. Further the access control mechanism needs to incorporate sophisticated signature schemes for privacy and perform compressing sensing for data fusion of diverse IoT devices.

5.3.1.5 Cryptographic Keys and Hashing of Data

According to [25], cryptographic key management plays a vital role in authenticating IoT devices and end users. The lack of efficient security mechanisms at the IoT network layer can lead to large number of network vulnerabilities and insecure communication. The security challenge for a resource-constraint IOT network is not to compromise the efficiency of cryptographic methods for secured data communication. Further, despite the minimal overhead of security in datagram transport-level security within a Low PAN network, it must ensure an end-to-end security mechanism across IP networks. Also, there is need for efficient middleware security services that involve location-based services for IoT devices.

5.3.2 Blockchain-Driven IoT Applications

There are several advantages of using blockchain in IoT. It can reduce the single point of failure. The encryption algorithm and consensus protocols in the blockchain strengthen the IoT security as the distributed ledger allows users to audit the stored information and provides a trustworthy platform for the IoT operation. Figure 5.6 shows how IoT framework allowed by blockchain works. The IoT devices are within the same network of blockchains.

In Blockchain-enabled IoT systems, the IoT devices generate data whereas the blockchain serves as a distributed database that keeps the transaction safe and protected from malicious modification. Once the transaction is added to the blockchain network it becomes

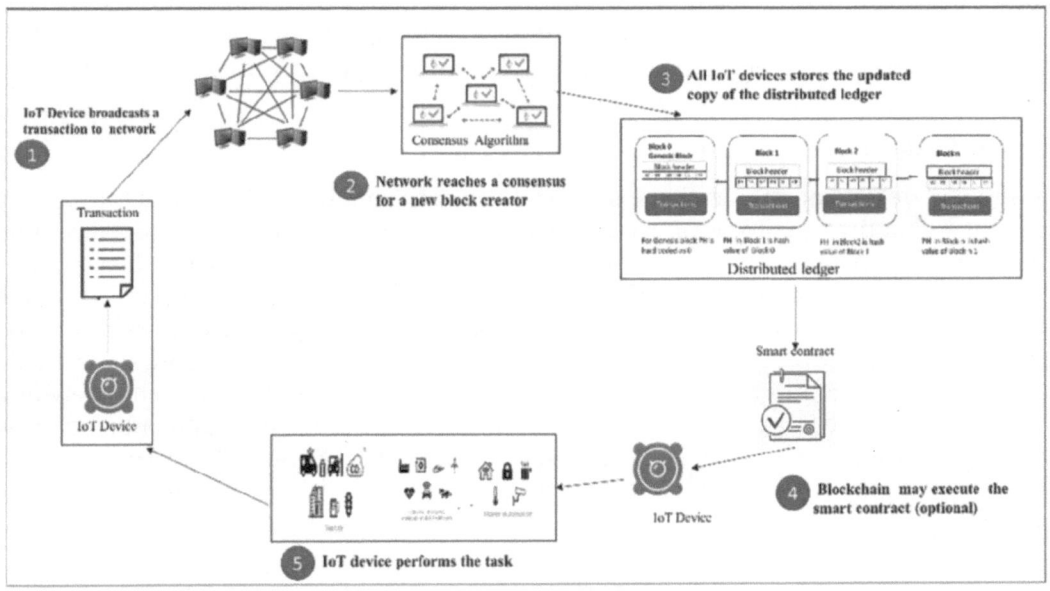

FIGURE 5.6
An Example of the Working of IoT Blockchain System.

tamper proof. The authorized user can view all the transactions and verify the authenticity of the transactions. In addition to its support as storage, it also secures data, and supports and optimizes digital trading in IoT applications. The integration of blockchain and IoT are still in the growing stage. IoT devices and application functions describe the character of the IoT-blockchain framework. Generally, many IoT applications [1,2] incorporate blockchain for digital payment, smart contract service, and data storage. Various industries like logistics companies implement blockchain for asset tracking; the hardware and software companies will use blockchain to improve the interaction between humans and IoT devices and so on. Some IoT applications with blockchain technology are explained further in this section.

5.3.2.1 Intelligence Transport System (ITS)

An intelligence transport system combines innovative various technologies using communication, electronics, computers, and sensors. The motivation behind this application is to improve road safety, reduce traffic congestion, and allow for fuel efficiency. The intelligent traffic system's key functionalities includes real-time traffic condition monitoring, identifying traffic incidents such as accidents in particular areas, and real-time monitoring of transit services. ITS discussed in [26,27] supports the driver providing information like collision warning, obstacle detection, vehicle automation, driver condition monitoring, and more. As a blockchain-based system, transaction records are the information about traffic condition transformation between vehicles. A major objective of integrating blockchain with ITS is to reduce the human interferences and allow traffic-related objects to make decisions on communication and self-execute automatically using smart contract in blockchain with respect to the traffic conditions.

5.3.2.2 Supply Chain Management Systems

The supply chain represents a complex set of interactions between a long chain of various suppliers. Tracing a series of transported goods is a dynamic process involving various parties' information. IoT has also begun to provide perceptions into how businesses can gather data while their products cross the globe. Sensors offer temperature data, position data and more, like newly discovered contracting firms. How do we restore openness and new consumer-product partnerships to enable active involvement [28–30] in product authenticity checks and moral values? The technology underlying blockchain provides an answer for us. Linked, consensus-based, and permanent registers assist in monitoring supply chain origin and transitions. The blockchain will establish a centralized ledger allowing the identification and monitoring of a good's possession across the supply chain. Shipments can be monitored at each point by using smart contracts. Whenever a product reaches a site, the merchandise can be checked, and an agreement is signed between the two suppliers trading goods. This will provide a clear and certifiable antiquity of where the product is being delivered, its condition, and whether the terms of the contract (time, day, temperature, etc.) have been met. This eliminates the need for each shareholder to monitor an asset independently via their own database – a database that does not provide accountability, coordination, or verification with all other stakeholders in the supply chain. Additionally, using a blockchain network to monitor goods will provide better transparency and commercial liability. Consumers could track where their goods came from and how they reached their doorstep. Accumulating all the data collected from IoT to a consumer-accessible blockchain will offer customers a safe, trackable, and easy way to understand where their products come from and how they get to their shop, boosting customer and producer confidence.

5.3.2.3 Smart Healthcare

Healthcare is regarded as one of key IoT applications [31,32]. Smart wearable devices offer new approaches to track patients who are not in a critical state remotely, while opening space for more critical patients in hospitals. The healthcare sector also has arrangements to track medical devices, staff, and patients in real-time. But some important problems still remain in processing, managing, and distributing patient data where blockchain can provide solutions to it. Blockchain has a wide variety of healthcare applications and uses. The ledger system enables the safe transfer of patient medical records, controls the supply chain of medicines and helps healthcare researchers access genetic codes. Maintaining medical records that contain information like medications, problems, patients past medical history, and laboratory data has a high tendency toward duplication of data and is difficult for cross validation and verification. Blockchain will enable the sharing and storing of all medical records in a decentralized manner that provides non-immutable sources and information, and is simple to verify by any approved individual. It prevents duplication of data and when the data is transacted through blockchain it is protected from tampering.

Another major advantage of using blockchain-based healthcare is that it keeps a single version of patients records shared and traceable. It also helps the medical staff and pharmacies to track patient drug intake and maintain a clear history of medication. Additionally, blockchain-based healthcare networks can also offer reliable and secure access to insurance providers, the medication supply chain, pharmaceutical firms, and medical researchers. In [33] authors discuss a blockchain-based smart contract for secure analysis and management of medical sensors using an Ethereum protocol. The sensors communicate via smart devices to record all events in the blockchain and invoke smart contracts. Smart contracts support patient tracking in real time and give patients and doctors alerts for safe medical intervention.

5.3.2.4 Smart City

Smart cities are a platform for solving urbanization's specific challenges by integrating emerging technology, innovative urban design, energy and transport management, and business planning. It is an organic convergence of smart infrastructure, smart transport, smart energy, smart healthcare, and other utilities under the premise of transparent data and decentralization. A smart city [34,35] receives and sends or signals through the internet. For example, a waste bin can send a message when it is full, sensors in the water management system can send a message if there is a leakage or a tank full condition. Sensors in the streetlight can gather the information and communicate and so on. Problems like high dormancy, bandwidth blocks, protection and confidentiality, and scalability ascend in the present smart city system architecture due to steady growth in volumes of data and number of IoT devices connected to it.

A blockchain-based solution provides a well-organized, safe, and scalable architecture by reducing the computational and storage resources to enhance the current system scenario. Blockchain in smart cities can be used for energy distribution, automate water supply, and air quality management. Blockchain in a smart city can be used for civil citizen registration to keep track of citizens. Blockchain makes this data tamperproof and sharable and also useable for holding digital identity. It can also support smart payments, user identity, transportation management, energy grid management, and more. The advantage of using blockchain in smart cities include that smart cities can interconnect to exchange data with their residents in real-time. It provides integrity over the information and efficient management of resources and many more.

5.3.2.5 IoT Ecosystem

The IoT ecosystem consists of a relation between the real world of things and the virtual world of the internet, software, and hardware platforms as well as the norms commonly used to enable such relation. The IoT ecosystem [36,37] is comprised of *sensors* to collects the required data, a *computation node* to process the data received from the sensors, a *receiver* that collects the messages sent by the other computing nodes or the related devices, an *actuator* to trigger the associated device to perform an action with respect to information gathered from the sensor and the decision taken by the computing node, and the *devices* which actually perform the task. The communication in the IoT ecosystem is purely between machine to machine (M2M). Hence, the trust establishment between the participants is still in developing stage. The adoption of blockchain technology achieves this trust by tracking the IoT devices in the IoT ecosystem and coordinates transaction processing.

5.4 Beyond Cryptocurrencies and Bitcoin

The blockchain entry into information technology with Bitcoin and cryptocurrencies changed the way online transactions were seen worldwide. The main objective for the application of blockchain is to carry out transactions or share information in a secured way. As blockchain became famous people realized that blockchain could go beyond Bitcoin. Over the years, blockchain technology has been attempting to be embraced by various industries such as banking, real estate, and politics. And because each industry operates differently, blockchain had to evolve in specific ways suitable to that industry. The different types of blockchain platforms are explained in this section.

Blockchain had been invented in a digital currency context. It is now one of the emerging technologies in financial services, supply chain banking industries, and so on. Blockchain is used by different industries in different way. For example, some industries use blockchain for digital payment but others use it for an immutable record management and for information security. Different applications need different blockchain protocols. For example, some applications work on public blockchain while others may require restricted or private blockchain networks. The blockchain platforms are categorized into three types: (i) public blockchain, (ii) private blockchain, and (iii) consortium blockchain.

5.4.1 Public Blockchain

As the name indicates a blockchain is publicly available. Anyone can join the network; for example, any participant can take part in activities like reading, writing, or participating in the network. It is a decentralized network without permission required. Data on a public blockchain is stable as data cannot be changed or altered once validated on the blockchain. Bitcoin and Ethereum are well-known examples of a blockchain to the public.

5.4.2 Private Blockchain

Private blockchains operate on access restrictions that limit certain people who are eligible to participate in the network. There are one or more organizations that manage the network, and this contributes to third-party dependence on transactions. In a private ledger, only the individuals involved in a transaction will have knowledge of it, while the others are not able to access it. Hyperledger and Ripple are examples of private blockchain.

5.4.3 Ethereum

It is an open source blockchain platform that allows anybody to build decentralized applications under blockchain technology [38,39]. It is a programmable blockchain built upon a smart contract that allows users to create a code that satisfies specific requirement. Ether is the digital currency used in Ethereum to do the digital transactions. Ether is earned through the process of mining. A miner is rewarded by a digital token of ether if it is succeeded in the validation process of the new block. The heart of the Ethereum is the decentralized virtual machine known as Ethereum virtual machine (EVM) which provides the runtime environment to execute smart contracts. All the transactions are stored locally in all nodes in the network. Every instruction executed on EVM has a cost measured in units of gas which ensures proper handling of EVM. The instruction that needs more computational resources costs more gas compared to the instruction that require low computing power. Hence, gas inspires programmers to develop quality applications by avoiding unnecessary code. Smart contracts in Ethereum are deployed by creating a special transaction. A unique identifier is assigned during this process and its code is uploaded in the blockchain. Components of a smart contract include its address, balance, executable code, and a state. Smart contracts are invoked by a transaction of sending the contract address which triggers the contract to execute the actions specified in the code. A transaction invoking the smart contract transfers a fee in the form of ether along with the input data for the function to execute from the caller to the contract. All the participants in the network then execute the contract with the current state and input data. The output is verified by the participants in the network satisfying the consensus protocol.

5.4.4 Hyperledger

Hyperledger [40] is an opensource blockchain platform project by the Linux Foundation focused on building a suite of secure blockchain implementation platforms, tools, and libraries for enterprise-grade use. It is an open source development project where people can use Hyperledger as software and platform. Simply put, Hyperledger is a program that anyone can use to build their own customized blockchain service. It is a greenhouse for blockchain opensource products. It has multiple projects that can be categorized as frameworks and tools. The different frameworks in Hyperledger are Hyperledger Iroha, Hyperledger Sawtooth, Hyperledger Fabric, Hyperledger Indy, and Hyperledger Burrow. The different tools are Hyperledger Caliper, Hyperledger Cello, Hyperledger Quilt, Hyperledger Composer, and Hyperledger Explorer. These tools can be used to make blockchain that can be private or public.

5.4.5 Hyperledger Fabric

Hyperledger Fabric [41,42] is most famous blockchain project within the Hyperledger. It uses the ledger and smart contracts to manage the transactions. It is a permissioned blockchain mainly aimed for business use. To become a member of a Hyperledger Fabric network, the participant should enroll through a membership service provider (MSP). In some cases, the competitive participants do not want to make themselves transparent. In that case, the participants can create a different channel. If two participants are creating a channel, only those participants have copies of the ledger of that channel and no one else can access it. It supports privacy for the networks using channels. The ledger in Hyperledger Fabric has two components: namely the state of the world and the transaction log. The world state is the database that at a given moment represents the state of the ledger. The transaction log contains all the transactions that have taken place to reach the current world state. When an application wants to communicate with the ledger, smart contracts are invoked that are written in a chaincode. The chaincode can be written in many programming languages such as the programming language Java, node.js, or Go.

5.4.6 R3 Corda

R3 Corda [43] is a Corda blockchain which was developed by the R3 Banking Consortium. It is a distributed ledger framework designed for financial institutions to record, manage, and synchronize financial agreements. The key features of Corda include that it does not support unnecessary global sharing of data. The transactions in Corda are validated by parties related to the transaction rather than unrelated validators. It supports various consensus mechanisms. The Corda platform supports smart contracts that connect business logic and business data to the related legal prose to ensure that the platform's financial arrangements are firmly grounded in law and are enforceable, and have a straightforward path to follow in the event of confusion, complexity, or dispute.

5.5 Case Study on Decentralized Block Chain Approach for Healthcare Systems

Blockchain promises to provide enhanced security and privacy support towards healthcare systems. In this section, we are proposing two scenarios: one is primary patient care and other is medical research. The general scenario of the healthcare system is depicted in Figure 5.7.

5.5.1 Scenario 1: Primary Patient Care

In this scenario, the following problems can be reduced by using blockchain in current healthcare systems:

- There is a time when required data can be missing when a patient needs to often visit multiple hospitals and he is not able to keep track of the history.
- It is time consuming and requires major effort to share medical records between different providers.
- When data is unavailable, the patient needs to repeat the test. This happens when a record is stored in different hospital.

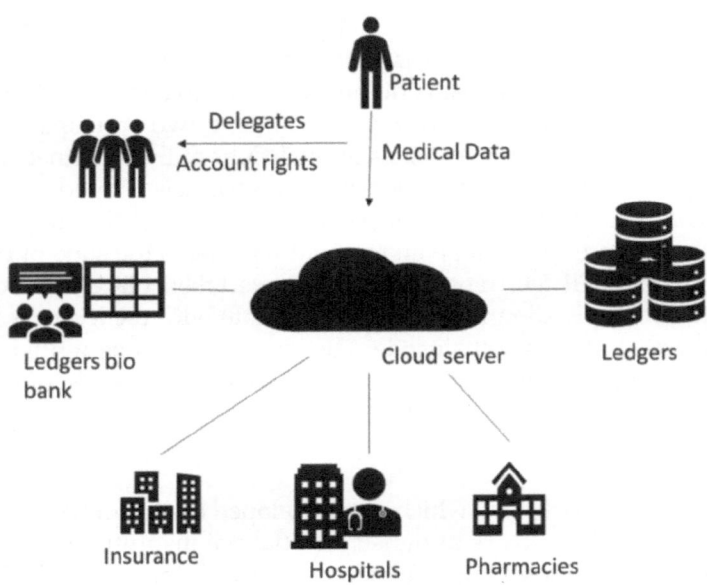

FIGURE 5.7
Graphical Representation of All Scenarios.

5.5.2 Scenario 2: Healthcare Data Aggregation for Study and Research Purpose

The essential principle of the healthcare system is to ensure that the source of the health data is through trusted medical professionals and healthcare institutions. Therefore, the authenticity, confidentiality, and integrity of data need to be achieved. The patient's privacy will be secured by using shared distributed ledger. Shared distributed ledger will provide traceability and transparency of the data aggregation process. Generally, patients do not want to participate in data sharing due to a lack of appropriate mechanism. The primary stakeholders of blockchain technology involve the networks of healthcare institutions, bio-health banks, and medical researchers. These stakeholders facilitate the entire process of gathering a patient's medical data for further study and analysis during research activities.

5.5.3 Application Scenario for Healthcare Systems

In this section, we are focusing on EMR sharing for the aforementioned Scenario 1. More specifically, we are focusing here on cancer patients that require long treatment and lifetime monitoring. We will present a prototype design and architecture of the system. Cancer treatment requires mobility of the patient; therefore, they need to secure and store data for further treatment. If a patient needs to transfer from one hospital to another, the patient must sign a consent form. Then the information is sent to the recipient. This can be complicated and inconvenient. In this case, data transfer can take time and a management person needs to introduce the patient into the system again when they receive the hard copy of the data. So, it is also difficult for patients to access data.

5.5.4 Proposed Framework of Blockchain-Driven Healthcare Systems

This framework consists of the registration service, database for storing data, nodes that manage the consensus process, and APIs for different users' roles as shown in Figure 5.8. The main function of the registration service is to register users for a different role (currently we are using only doctors and patients). In the registration process, doctors could not be malicious users; they should be verified as a medical doctor. Now, the national practitioner medical data repositories can be referred by the registration service to authenticate the identity of doctors. The registration service is to verify the authentication by the generation of a secret key pair for digital signing and an encrypting key pair for every healthcare system user. In addition, the digital signature is required between doctor and patient to transfer the medical data.

The patient's data will be stored first in the local database of the hospital, and second, in the cloud-based database that stores the patient's data and then encrypts it with the individual patient's secret key. Further, the registered healthcare person at the hospital will be able to access or upload data in the remote cloud database based on the access control policy of the individual user. The nodes follow the PBFT consensus protocol. Nodes will receive all the transactions done by users through the APIs.

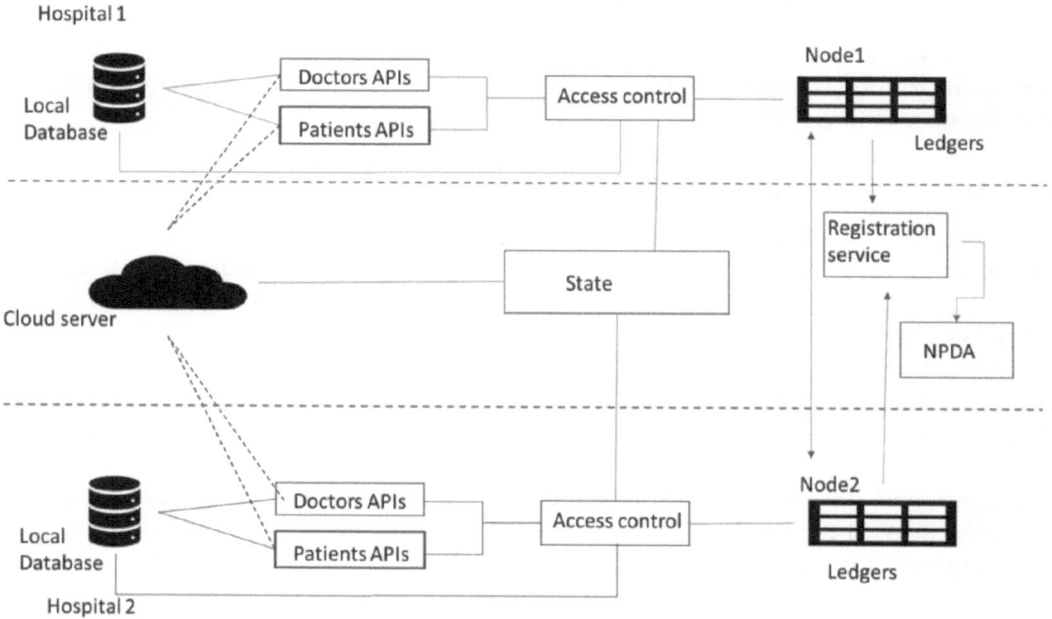

FIGURE 5.8
Architecture of a Proposed Method.

5.6 Performance Challenges of Blockchain-Driven IoT Applications

5.6.1 Integrity

In literature, several solutions have been presented for handling selfish miners attacks in resources constrained IoT networks [44]. Particularly, the selfish mining attack has been studied thoroughly due its complexity and targeting of a huge loss of integrity. Here the decision of miners to extend a block requires legitimate miners that can decrease the blockchain projected by the selfish misbehaving blockchain miners. However, the integrity of the blockchain is highly unstable due to the prevalence of selfish and misbehaving blockchain miners. This integrity hacking miner causes high computation cost and degradation in performance of the entire blockchain system. These miners perform blockchain forking that leads to a difficult consensus status, a loss of previous stored data, causes invalid transactions, and pollutes the blockchain. As a solution [9], in this case, the blockchain integrity can be maintained through proof of work (PoW), avoiding allocating a high portion of computational power to the miners, and to have a larger number of active legitimate miners.

5.6.2 Anonymity

The blockchain using Bitcoin provides pseudo-anonymity of users while performing a transaction. For this purpose, each user is identified by multiple random transaction addresses [45]. In addition, Bitcoin users can transact to themselves through changing address [46]. Here, the users can associate a changed address with their unique IP addresses. However, these methods of associating multiple random addresses with the IP address of each user lacks a high risk of anonymity due to the centralized approach of Bitcoin-user services. Also, the blockchain itself is public and hackers may analyze the network traffic of target users for a lack of anonymity. To handle this issue, one of the promising solutions is a mixing of protocols that involves two different addresses for the request of transaction and response to transactions [47].

5.6.3 Scalability

The increase in the number of blockchain transactions and applications leads to the issue of scalability [48]. Especially the adaptation of blockchain for resource constraint IoT devices is critical. The layered blockchain approach solves the scalability problem in limited ways, where the application layer and blockchain are separately performed through offloading to remote systems. However, the network-related issues for offloading secondary block operations to remote systems involves network connectivity and delays. Next PoW in blockchain consumes a large memory space within the default block size of one MB. In turn, the transaction output is invariably affected by the increase in size of each blockchain and involves complex validation of transactions. Authors in [34], proposed assigning PoW to the trust nodes of the blockchain. The miner is selected based on the space required and the low computation cost of PoW.

5.7 Conclusion and Future Scope

In the present technology era, the IoT system and its applications are involved in every aspect of human life towards enhancing the quality of human life. However, such IoT systems suffer from critical challenges, namely security and privacy due to the heterogenous nature and resource constraint of IoT end devices. This complexity includes a wide variety of protocols and standards under IoT Systems. To overcome the limitations of the centralized security and privacy mechanism, the blockchain promises to be efficient solution. Hence, this chapter addressed the fundamentals of blockchain technology such as decentralization, disturbed shared ledgers, smart contracts, cryptography techniques, and consensus mechanisms. Next, the layered architecture of IoT protocol stack was presented. Then, security requirements of IoT networks such as security and privacy of data, healthcare device authentication, trustable users, access control of IoT healthcare devices, cryptographic keys, and hashing of data were discussed. Some of the blockchain-based IoT applications were presented. Further, the blockchain technologies such as Ethereum, Hyperledger Fabrics, and R3 Corda were addressed. Also, as a case study, the decentralized framework of a blockchain for a healthcare IoT application was presented. Finally, parameters like scalability, adaptability, anonymity, and integrity as performance challenges towards blockchain-enabled IoT systems were identified. As a future scope, the blockchain application towards integration of heterogenous platforms and standards needs to be addressed.

References

[1] Atziori, Luigi, Antonio Iera, and Giacomo Morabito. 2010. "The Internet of Things: A Survey Computer Networks." *Computer Networks* 54 (28): 2787–2805.

[2] Stojkoska, Biljana, L. Risteska, and Kire V. Trivodaliev. 2017. "A Review of Internet of Things for Smart Home: Challenges and Solutions." *Journal of Cleaner Production* 140: 1454–1464.

[3] Jeschke, Sabina, Christian Brecher, Tobias Meisen, Denis Özdemir, and Tim Eschert. 2017. "Industrial Internet of Things and Cyber Manufacturing Systems." In *Industrial Internet of Things*. Cham: Springer.

[4] Mendez, Diego M., Ioannis Papapanagiotou, and Baijian Yang. 2017. "Internet of Things: Survey on Security and Privacy." *Internet Security Journal: A Global Perspective* 27(3): 162–182. arXiv preprint arXiv:1707.01879.

[5] Christidis, Konstantinos, and Michael Devetsikiotis. 2016. "Blockchains and Smart Contracts for the Internet of Things." *IEEE Access* 4: 2292–2303.

[6] Bertino, Elisa, and Nayeem Islam. 2017. "Botnets and Internet of Things Security." *Computer* 50 (2): 76–79.

[7] Huckle, Steve, Rituparna Bhattacharya, Martin White, and Natalia Beloff. 2016. "Internet of Things, Blockchain and Shared Economy Applications." *Procedia Computer Science* 98: 461–466.

[8] Fernández-Caramés, Tiago M., and Paula Fraga-Lamas. 2018. "A Review on the Use of Blockchain for the Internet of Things." *IEEE Access* 6: 32979–33001.

[9] Conoscenti, Marco, Antonio Vetro, and Juan Carlos De Martin. 2016. *"Blockchain for the Internet of Things: A Systematic Literature Review."* In *2016 IEEE/ACS 13th International Conference of Computer Systems and Applications (AICCSA)*, pp. 1–6. IEEE.

[10] Pilkington, Marc. 2016. *Blockchain Technology: Principles and Applications Research Handbook on Digital Transformations*, edited by F. Xavier Olleros and Majlinda Zhegu. Available at SSRN 2662660.

[11] Crosby, Michael, Pradan Pattanayak, Sanjeev Verma, and Vignesh Kalyanaraman. 2016. "Blockchain Technology: Beyond Bitcoin." *Applied Innovation* 2 (6–10): 71.

[12] Drescher, Daniel. 2017. *Blockchain Basics*. Vol. 276. Berkeley, CA: Apress.

[13] Al-Jaroodi, Jameela, and Nader Mohamed. 2019. "Blockchain in Industries: A Survey." *IEEE Access* 7: 36500–36515.

[14] Nakamoto, Satoshi, and A. Bitcoin. 2008. "A Peer-to-Peer Electronic Cash System." *Bitcoin.* https://bitcoin.org/bitcoin.pdf.

[15] King, Sunny, and Scott Nadal. 2012. "Ppcoin: Peer-to-Peer Crypto-Currency with Proof-of-Stake." Self-published paper. August 19, p. 1.

[16] Zheng, Zibin, Shaoan Xie, Hongning Dai, Xiangping Chen, and Huaimin Wang. 2017. *"An Overview of Blockchain Technology: Architecture, Consensus, and Future Trends."* In *2017 IEEE International Congress on Big Data (BigData Congress)*, pp. 557–564. IEEE.

[17] Li, Wenting, Sébastien Andreina, Jens-Matthias Bohli, and Ghassan Karame. 2017. "Securing Proof-of-Stake Blockchain Protocols." In *Data Privacy Management, Cryptocurrencies and Blockchain Technology*. Cham: Springer.

[18] Cong, Lin William, and Zhiguo He. 2019. "Blockchain Disruption and Smart Contracts." *The Review of Financial Studies* 32 (5): 1754–1797.

[19] Abraham, Ittai, Guy Gueta, Dahlia Malkhi, Lorenzo Alvisi, Rama Kotla, and Jean-Philippe Martin. 2017. "Revisiting Fast Practical Byzantine Fault Tolerance." arXiv preprint arXiv:1712.01367.

[20] Ray, Partha Pratim. 2018. "A Survey on Internet of Things Architectures." *Journal of King Saud University-Computer and Information Sciences* 30 (3): 291–319.

[21] Khan, Minhaj Ahmad, and Khaled Salah. 2018. "IoT Security: Review, Blockchain Solutions, and Open Challenges." *Future Generation Computer Systems* 82: 395–411.

[22] Granjal, Jorge, Edmundo Monteiro, and Jorge Sá Silva. 2014. "Network-Layer Security for the Internet of Things Using TinyOS and BLIP." *International Journal of Communication Systems* 27 (10): 1938–1963.

[23] Gomes, Tiago, Filipe Salgado, Sandro Pinto, Jorge Cabral, and Adriano Tavares. 2017. "A 6LoWPAN Accelerator for Internet of Things Endpoint Devices." *IEEE Internet of Things Journal* 5 (1): 371–377.

[24] Raza, Shahid, Simon Duquennoy, Tony Chung, Dogan Yazar, Thiemo Voigt, and Utz Roedig. 2011. *"Securing Communication in 6LoWPAN with Compressed IPsec."* In *2011 International Conference on Distributed Computing in Sensor Systems and Workshops (DCOSS)*, pp. 1–8. IEEE.

[25] Granjal, Jorge, Edmundo Monteiro, and Jorge Sa Sá Silva. 2010. *"Enabling Network-Layer Security on IPv6 Wireless Sensor Networks."* In *2010 IEEE Global Telecommunications Conference GLOBECOM 2010*, pp. 1–6. IEEE.

[26] Yuan, Yong, and Fei-Yue Wang. 2016. *"Towards Blockchain-Based Intelligent Transportation Systems."* In *2016 IEEE 19th International Conference on Intelligent Transportation Systems (ITSC)*, pp. 2663–2668. IEEE.

[27] Lei, Ao, Haitham Cruickshank, Yue Cao, Philip Asuquo, Chibueze P. Anyigor Ogah, and Zhili Sun. 2017. "Blockchain-Based Dynamic Key Management for Heterogeneous Intelligent Transportation Systems." *IEEE Internet of Things Journal* 4 (6): 1832–1843.

[28] Korpela, Kari, Jukka Hallikas, and Tomi Dahlberg. 2017. *"Digital Supply Chain Transformation Toward Blockchain Integration."* In *Proceedings of the 50th Hawaii International Conference on System Sciences*, January 4–7, Hawaii, USA. https://aisel.aisnet.org/hicss-50/.

[29] Kshetri, Nir. 2018. "1 Blockchain's Roles in Meeting Key Supply Chain Management Objectives." *International Journal of Information Management* 39: 80–89.

[30] Saberi, Sara, Mahtab Kouhizadeh, Joseph Sarkis, and Lejia Shen. 2019. "Blockchain Technology and its Relationships to Sustainable Supply Chain Management." *International Journal of Production Research* 57 (7): 2117–2135.

[31] Yue, Xiao, Huiju Wang, Dawei Jin, Mingqiang Li, and Wei Jiang. 2016. "Healthcare Data Gateways: Found Healthcare Intelligence on Blockchain with Novel Privacy Risk Control." *Journal of Medical Systems* 40 (10): 218.

[32] Pabla, Jitesh, Vaibhav Sharma, and Rajalakshmi Krishnamurthi. 2019. *"Developing a Secure Soldier Monitoring System using Internet of Things and Blockchain."* In *2019 International Conference on Signal Processing and Communication (ICSC)*, pp. 22–31. IEEE.

[33] Mahapatra, Bandana, Rajalakshmi Krishnamurthi, and Anand Nayyar. 2019. "Healthcare Models and Algorithms for Privacy and Security in Healthcare Records." *Security and Privacy of Electronic Healthcare Records: Concepts, Paradigms and Solutions*: 183.

[34] Sharma, Pradip Kumar, and Jong Hyuk Park. 2018. "Blockchain Based Hybrid Network Architecture for the Smart City." *Future Generation Computer Systems* 86: 650–655.

[35] Krishnamurthi, Rajalakshmi, Anand Nayyar, and Arun Solanki. 2019. "Innovation Opportunities Through Internet of Things (IoT) for Smart Cities." *Green and Smart Technologies for Smart Cities*, pp. 261–292. Boca Raton, FL, USA: CRC Press.

[36] Krishnamurthi, Rajalakshmi, and Mukta Goyal. 2019. "*Enabling Technologies for IoT: Issues, Challenges, and Opportunities.*" In *Handbook of Research on Cloud Computing and Big Data Applications in IoT*, pp. 243–270. IGI Global.

[37] Rahulamathavan, Yogachandran, Raphael C-W. Phan, Muttukrishnan Rajarajan, Sudip Misra, and Ahmet Kondoz. 2017. "*Privacy-Preserving Blockchain Based IoT Ecosystem Using Attribute-Based Encryption.*" In *2017 IEEE International Conference on Advanced Networks and Telecommunications Systems (ANTS)*, pp. 1–6. IEEE.

[38] Wood, Gavin. 2014. "Ethereum: A Secure Decentralised Generalised Transaction Ledger." *Ethereum Project Yellow Paper* 151 (2014): 1–32.

[39] Atzei, Nicola, Massimo Bartoletti, and Tiziana Cimoli. 2017. "A Survey of Attacks on Ethereum Smart Contracts (SOK)." In *International Conference on Principles of Security and Trust*, pp. 164–186. Berlin, Heidelberg: Springer.

[40] Cachin, Christian. 2016. "*Architecture of the Hyperledger Blockchain Fabric.*" In *Workshop on Distributed Cryptocurrencies and Consensus Ledgers* 310 (4).

[41] Androulaki, Elli, Artem Barger, Vita Bortnikov, Christian Cachin, Konstantinos Christidis, Angelo De Caro, and David Enyeart, et al. 2018. "*Hyperledger Fabric: A Distributed Operating System for Permissioned Blockchains.*" In *Proceedings of the Thirteenth EuroSys Conference*, pp. 1–15.

[42] Androulaki, Elli, Christian Cachin, Angelo De Caro, Andreas Kind, and Mike Osborne. 2017. "*Cryptography and Protocols in Hyperledger Fabric.*" In *Real-World Cryptography Conference*, pp. 12–14.

[43] Mohanty, Debajani. 2019. *R3 Corda for Architects and Developers: With Case Studies in Finance, Insurance, Healthcare, Travel, Telecom, and Agriculture.* Apress.

[44] Heilman, Ethan. 2014. "*One Weird Trick to Stop Selfish Miners: Fresh Bitcoins, a Solution for the Honest Miner.*" In *International Conference on Financial Cryptography and Data Security*, pp. 161–162. Berlin, Heidelberg: Springer.

[45] Herrera-Joancomartí, Jordi. 2014. "Research and Challenges on Bitcoin Anonymity." In *Data Privacy Management, Autonomous Spontaneous Security, and Security Assurance*, pp. 3–16. Cham: Springer.

[46] Koshy, Philip, Diana Koshy, and Patrick McDaniel. 2014. "An Analysis of Anonymity in Bitcoin Using P2P Network Traffic." In *International Conference on Financial Cryptography and Data Security*, pp. 469–485. Berlin, Heidelberg: Springer.

[47] Bissias, George, A. Pinar Ozisik, Brian N. Levine, and Marc Liberatore. 2014. "*Sybil-Resistant Mixing for Bitcoin.*" In *Proceedings of the 13th Workshop on Privacy in the Electronic Society*, pp. 149–158.

[48] Park, Sunoo, Albert Kwon, Georg Fuchsbauer, Peter Gaži, Joël Alwen, and Krzysztof Pietrzak. 2018. "Spacemint: A Cryptocurrency Based on Proofs of Space." In *International Conference on Financial Cryptography and Data Security*, pp. 480–499. Berlin, Heidelberg: Springer.

6

Artificial Intelligence for Blockchain I

Joy Gupta, Ishita Singh, and K. P. Arjun
Galgotias University, Greater Noida, Delhi-NCR, India

CONTENTS

6.1 Introduction

Smart computing can be achieved by integrating AI and blockchain. What we do right now is use a lot of computational power, and through brute force attacks we conclude the hash value, called the 'nonce.' But what if there is an AI-based approach that tackles this computing differently and efficiently. If a machine learning-based program or algorithm were fed the appropriate training data, it could glaze its skills in real time [1]. And we know that data cannot be manipulated in the blockchain. When it comes to decentralized AI, it is far more efficient than AI as it consists of the integration of blockchain and AI itself. Projects on AI are based on a centralized system, while on the other hand, blockchain is entirely based upon decentralized networks, which can be accessed by any member of that network. To be precise, blockchain is including ledger, which is distributed to every member of that network. For example, singularity net joins blockchain and AI to make more intelligent, decentralized AI. Blockchain systems can deal with a huge number of varied datasets. By creating an API of APIs on the blockchain, it could take into account the intercommunication of AI specialists [2]. Subsequently, distinct algorithms can be based on various datasets.

Monetizing data is a tremendous source of revenue for companies like WhatsApp, Facebook, and others. Currently, data is monetized by different companies without user consent. It can be used against us as well. To ensure confidentiality, cryptography plays a broad role, and along with the implementation of blockchain, which permits us as users to cryptographically monitor our information and have it utilized in the manners we see fit. It likewise lets us adapt information by and by in the event that we decide to, without trading off the individual data. As we scrutinized AI, it too needs data sets for training purposes. To fulfill this purpose, we can buy data directly from the owner or the creator straight. This makes the whole process a far fairer process than it currently is without tech giants exploiting users. These data marketplaces can also help smaller companies that are based in AI [3] as feeding or training an AI requires many expenses that are not bearable sometimes. Utilizing decentralized data marketplaces, smaller companies can access expensive and privately kept data.

As data is immutable in blockchain, and as it is used for training of AI [3], this helps researchers to make AI more capable of reaching individual decisions instantly for humans, and it becomes more typical to work with more massive datasets. Immutable datasets increase the ability of AI to process things correctly; on the other hand, AI learns something from the past and enhances its capability and caliber as well. We can then audit those decisions to make sure they still reflect reality.

'Blockchain' is the combination of the words block and chain, and it consists of blocks which later on form chains. These blocks [4] are time-stamped immutable data records managed by many computers instead of a single entity. Several cryptographic measures secure these blocks. When we mentioned that many machines manage it, we mean decentralized networks so there is no central authority. Ledger, the most crucial part of the blockchain, discloses the essential information required in the blockchain. These ledgers are immutable and shared with other users of the network. A blockchain carries infrastructure cost but no transaction cost. When purchased, whether it is of information or money, the process gets initiated by generating a new block. This block is then verified by several machinery spread over the ace net. This leads not just to create an exclusive record but an exceptional career with a unique history. Forging a solo file would mean forging the entire blockchain. That is almost impossible. This model is implemented in Bitcoin.

Blockchain [5] could be implemented in such places where the transfer of information, money, or sensitive documents takes place. Blockchain can help us in such assignments, guarantee the safety of transactions, and reduce the risk of fraudulence. Blockchain cuts off the fee-processing middle man and eliminates the necessity for the match-making platform.

6.1.1 Advantages of Blockchain

- Decentralized Network: Blockchain involves several nodes. Each node holds the exact copy of the database.
- Stability: Confirmed blocks in the blockchain are very unlikely to be reversed, implying that data manipulation is not possible in blockchain, which maintains data integrity.
- Trustless System: For any kind of transaction based on anything other than blockchain, it requires an interruption of the third party, be it a bank or different gateways. But a blockchain transaction only depends on the two parties involved.

6.1.2 Flaws in Blockchain

- Excess Use of Power: Power consumed by Bitcoin's software is approximately 2.55 gigawatts (GW), almost the same as Ireland. On the other hand, Google used 5.7 TWh worldwide in 2015. Moreover, on average, Bitcoin 'miners' consume about five times more power than they did in previous year.
- Not a Huge Distributed Computing System: When we say Bitcoin is a distributed network, we do not mean that it is a distributed computing network. At one time, all

the nodes across the world do some computation. It is supposed that these nodes gather something bigger bit by bit, but that is incorrect. All the nodes maintain the same thing. For example, in Bitcoin, all nodes verify the same transaction following the same rule and perform identical operations, record the same thing, store the entire history.

- Mining Does Not Offer Network Safety: In a blockchain, miners maintain stability and security. If we talk about Bitcoin, if there are more than 50% faulty miners, then they can alter or rewrite the previous nodes and hence the security of data disappears.

- Scalability: Weakness of Blockchain: One of the best implementations of blockchain is Bitcoin. In Bitcoin, the transaction-processing speed is kind of static as a result of there being very few users which means primarily only a few people used blockchain. On the other hand, we have Visa, which processes thousands of transactions per second. That is why when it comes to scalability, traditional ways are still ahead of blockchain.

- The Problem for Not Tech Savvy: Storing virtual currencies for a network based on blockchain could be hectic for people who are not-so-technical in real life. Usually, such technology can be useful and beneficial for someone familiar with the technology.

- Network Size: Blockchain requires a large number of users over a network.

- Anonymity Is a Threat: Too much hype for blockchain is because it is anonymous and open to all. Being open to all can be a good option, but we should not deny the fact that too much transparency poses a danger. What if a company or an organization pays someone through Bitcoin? Payment modes like Bitcoin lead to the disclosure of the amount a company has in the form of Bitcoin, which is going to be visible to everyone in that medium or channel (in this case, it is Bitcoin). Because of anonymity, people use it for illicit transactions. They have chosen their mode of transaction as Bitcoin or any other cryptocurrency.

- Overkilling Proof of Work: PoW stands for 'Proof of Work', which is a mechanism used to validate a block that is going to be added. PoW requires a lot of computing power. Every node across the same network is trying to verify the block by generating 'nonce.' The one who first finds the nonce is rewarded, and other nodes validate the 'nonce.' Now you could conclude on your own how efficient this process is. A ton of computing power goes to vainity. How practical is it? Not very. Even PoW's alternative Proof of Stake (PoS) isn't very efficient either.

6.1.3 Artificial Intelligence: Simulation of Human Intelligence

AI is sometimes termed as 'machine intelligence.' It is an outlet of computer science that emphasizes to create smart machines. What we meant is creating intelligence that makes computers capable enough to do things on their own. AI [6] helps devices to work and act like humans; that is why AI is broadly used in a self-driving car, speech recognition, problem-solving, and others. These machines are not programmed to act in different situations; they need past experiences (like data sets) for their learning purpose. Figure 6.1 shows AI has features to transform the world.

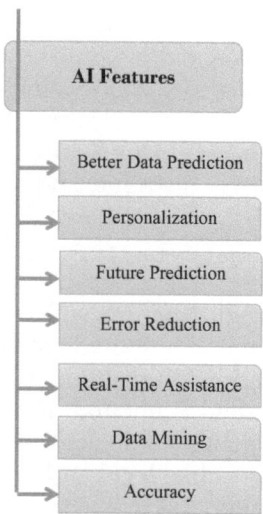

FIGURE 6.1
AI Features for Transforming the World.

6.1.4 Integration of Artificial Intelligence and Blockchain

It is estimated that blockchain costs around $600 million for validating and sharing transactions over the network. The energy consumption results in high cost, and wastage is another problem for blockchain. By wastage, we meant the mining process. Whenever a transaction takes place over a network, miners need to solve a mathematical problem that requires a lot of computing power. Such an approach is inefficient. AI [7] has the potential to handle and optimize the process and decrease the cost of mining. Blockchain holds the data from the beginning, which results in a longer and heavier chain. In such a scenario, AI can help blockchain to act and make smarter decisions about data storage and maintenance by using machine learning.

The issue with AI is its centralized nature. Since we do not know it's working, even its owner could not conclude it's decision making. It is often like a black box: no one has complete control over their created AI. And the foremost thing is that it is wholly based on probability, implying that AI could be wrong sometimes because of faulty a data feed. So, for AI to grow its full potential, it needs to have a certain level of transparency and ability to develop.

6.2 Blockchain Ledger Data Decision Using Artificial Intelligence

Blockchain mining is a waste of resource-intensive processes. To add a new block in the blockchain, miners have to validate it, which requires a considerable amount of energy. PoW protocol in blockchain resembles a lottery mechanism that is responsible for energy consumption. Proof of useful work (PoUW), on the other hand, is based on machine learning. Rather than a lottery mechanism, miners are awarded after performing honest machine learning (ML) [8] tracking work. As it is not going to be a lottery procedure, other people may contribute to the network through extra incentives. PoUW includes rewards for useful work and punishments for malicious actors; it would be an AI system along with the security of the blockchain. There are obstacles, too, in designing a PoUW system using ML. An ML task is different from hashing as it is more complex and diverse. ML tasks are heterogeneous, so it becomes difficult to verify the nature of the work done by actors on the network. In such a mistrusting environment, it is hard to distribute and coordinate an ML training process.

6.2.1 Distributed Ledger

'Distributed' [9] means decentralized in nature, which implies there is not going to be any central server; the network is entirely distributed among several nodes. Whereas 'ledger' is a mode through which data is added in real time via distributed nodes spread all over the network. However, once the data is entered, it cannot be removed or manipulated later. It is a kind of shared ledger that is not in control of any central authority. It acts as a database for different types of assets like legal and financial. Each node has an identical copy as distributed ledger always updates in real time. The advantages of distributed ledgers are following:

- Users control all of their data
- Data consistency is maintained throughout the network
- Data manipulation is next to impossible
- Insulated towards malicious attacks
- Trustless ecosystem
- Transparency along with security

6.2.2 System Overview

6.2.2.1 Environment

The environment of personal AI blockchain is comprised of proof of stake (PoS) and PoW. It is a peer-to-peer (P2P) network made up in Figure 6.2 and discussed in the following:

- Clients: These nodes are used to pay to train their models on the personal artificial intelligence (PAI) blockchain.
- Miners: They are used for training purposes. They mine a new block by calculating nonce. The training is distributed and all miners collaborate by sharing updates of their local model.
- Supervisors: They record each message transferred during a task in a log file called 'message history.' They also examine malicious behaviour during training because the environment can hold some Byzantine nodes as well.
- Evaluators: They test the final models derived from each miner and send the best one to the client. They are independent. They also divide the client's fees and pay all nodes accordingly.
- Verifiers: They verify whether each block is valid or not. They are required because computational verification is expensive and is not carried out by all nodes.
- Peers: They are not part of PAI blockchain; instead, they are using regular blockchain transactions.

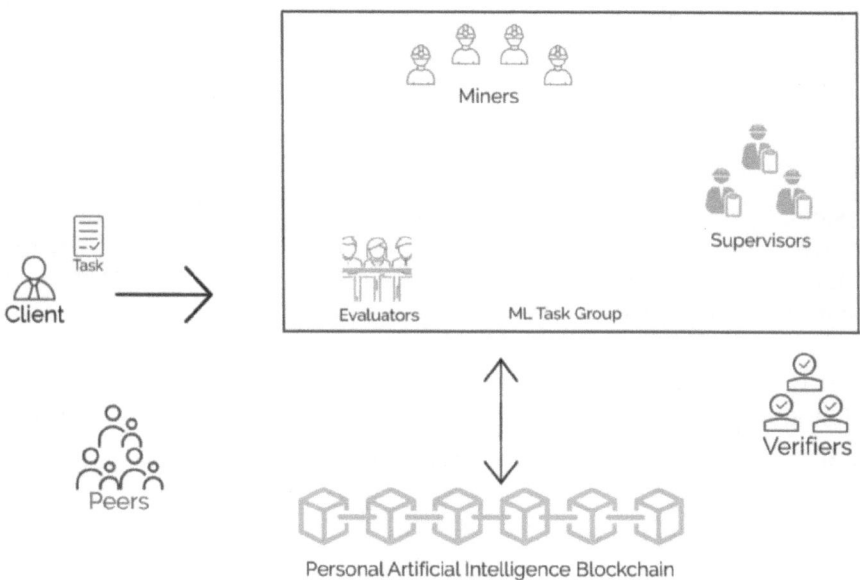

FIGURE 6.2
Personal AI Blockchain Environment Overview.

6.2.2.2 Transactions

Bitcoin transactions are digitally signed data structure that holds versioning, inputs, and outputs. A transaction is authentic only if its input refers to other unspent outputs. In PAI blockchain, the approach is different from handling training, evaluation, verification, and payments. PAI requires all extra information to be in the OP_RETURN code of one of the transaction's outputs. A transaction has to wait in mempool, a buffer of pending transactions before its inclusion in a block. Each transaction is then propagated across the whole P2P network from mempool.

6.2.2.3 Staking

Instead of rewarding nodes through PoW, we use a mechanism called 'staking.' All nodes except regular peers must first deposit some coins as collateral. The locked money is released or returned along with some extra fees if the participant finishes their work.

6.2.2.4 Tasks

A client submits a task's definition, 'T', along with some fee, 'F', as a particular transaction to the blockchain (PAY_FOR_TASK), containing:

- Description of trained model
- Optimizer
- Stopping criterion
- Validation strategy
- Accuracy loss

- Dataset information
- Performance

6.2.2.5 Protocol

In our system, blockchain ledger data decision using AI, the client broadcasts the task to the PAI network. Now, the miners and supervisors are assigned randomly by the network itself. That assignment is based on working node preferences, 'P', and the task, 'T'.

The dataset is then split into following:

- Training dataset (provided to miners to perform ML tasks)
- Validation dataset (selected from the initial dataset to validate the ML model)
- Test dataset (provided to evaluators for testing the final model)

6.3 Increasing Blockchain Efficiency Using AI

Before we talk about the increased efficiency of blockchain using AI [10], we first know if there is any need to integrate these two incredible and vast fields. If yes, then what could be the reasons for integration, followed by a conclusion.

6.3.1 Significance of AI and Blockchain Collaboration

Artificial Intelligence is an incredible field in computer science. It requires a machine learning algorithm and neural networks to improve performance. Now here comes the need for datasets. What are datasets? When you need to train your AI, you need to provide it with some sets of data. Based on the data you use to train, intelligent responses could be generated. Thomas C. Redman stated, 'if your data is bad, your machine learning tools are useless.' Trusting your AI means trusting the datasets you use to train it. You can use incredible AI algorithms [11], but if you use false or manipulative datasets for training purposes, the results may be disappointing.

Let's first understand a real-life scenario. If you own a dataset, and someone uses it for feeding purposes, technically (or practically) you're ruling the dataset because as soon as you manipulate data, the results reflect in the choices of others. For better understanding, let's use a practical example. We, the citizens, rely on big tech companies. We trust their datasets because they are for good, and they did not have any incentive to manipulate pieces of information. But we do not forget that people like us rule companies. They didn't manipulate the datasets yet!

The problem with AI is its centralized way of working. It could be understood as a black box and even the creators of it do not understand how exactly it works. As AI is based on probability, it could be right or wrong. And now, we conclude that those mistakes could be due to faulty data. So, utilization of AI at its full potential requires the concept of decentralization. Such that it provides an absolute level of transparency and ability to acquire authentic access to the data.

This issue can be solved by implementing blockchain. Blockchain is entirely based on immutable datasets. If we used these immutable datasets for training, we would create a more trustworthy AI that cannot be manipulated later.

Web 2.0 refers to websites that emphasize user-generated content, ease of use, participatory culture and interoperability which means compatible with other products, systems, and devices for end users The problem with web 2.0 is that if you're not the owner of a network, you are not permitted to read the actual interaction taking place with users through scripts because scripts are 100% private and run in their servers.

6.3.2 Blockchain Efficiency

There is a great need to change the traditional way designed for blockchain [12]. Because the demand for computing power is increasing rapidly, no matter what size the company is, this raises a challenge for company IT administrators and established data providers. Current chip architectures are moving ever closer to the bounds of what is practicable. Data centres can only become even faster and more performant by integrating circuits even more firmly into smaller spaces.

In 1965, Intel co-founder Gordon Moore proposed that every 18–24 months, the performance of processors would increase, and at the same time, the cost would reduce. This statement of his is called 'Moore's Law.' This law was true for 53 years. The problem is that the demand for computing power is growing faster than the progress in processor performance. There are cloud solutions that promise to make the most modern hardware usable as efficiently as possible for many customers. But again, there are many physical data centres present globally. Cloud data centres combine the demand of power for computing and therefore require high-performance IT infrastructures and consume massive amounts of electricity. Doubling efficiency in fixed intervals per Moore's Law is further not sufficient to meet the faster-growing demand.

One of the solutions to this problem is graphical processors units (GPUs). Compared to conventional CPUs, GPUs can perform fewer complex operations. They were initially invented to process out high-resolution images and textures. Fast GPUs are predestined for new applications that are based on AI, ML, deep learning, automation, and augmented reality/virtual reality (AR/VR) as well.

6.3.3 Benefits of Integration of AI and Blockchain

There are many benefits of such integration [12] and some of them are listed in Figure 6.3.

- Seamless Data Management: For data mining, blockchain entirely relies on algorithms. These algorithms try to find out every possible combination until they find the right one for verification purposes. Such an approach contains complexity and requires much effort. Here AI can help blockchain technology to get rid of such kind of approach. AI can make algorithms more intelligent in such a way that it makes the data management process free from faults.

- Smart Energy Consumption: If you decide to operate a blockchain, it requires a large amount of processing power. The approach we used to calculate 'nonce' in the blockchain is wholly based on the 'brute force' concept, which requires vast computing power because in brute force, every possibility for the solution gets checked whether or not it satisfies the problem's statement before validation. AI can grab the opportunity

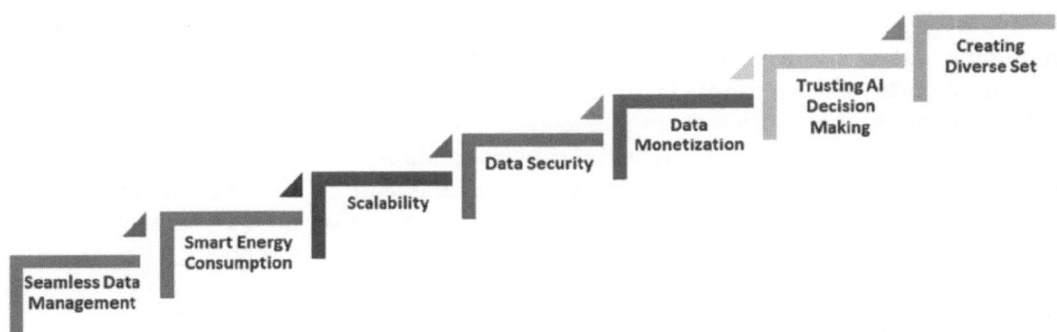

FIGURE 6.3
Benefits of AI and Blockchain.

and allow us to tackle tasks more efficiently and intelligently. We can build a machine learning model that was fed the appropriate training data. Such a model can be useful to practically polish skill in real time. Ultimately, ML reduces the number of efforts and time invested in data mining.

- Scalability: The blockchain size is increasing substantially with time. The increase in the size of a blockchain could be at the rate of 1 MB every 10 minutes. Currently, we do not have enough effective methods to optimize and eliminate data from blockchain. AI introduces a decentralized learning system that allows blockchain to be more efficient and scalable. It liberates new ways to control benefits.

- Data Security: When we combine the technologies, we have the option to provide reinforcement for profoundly delicate and important individual information of people. The progress of AI depends on the data we used to feed it; here, the data is going to be ours. Through data, AI gets to know information about surroundings, the world, and more. In short, AI continuously improves itself through input data. On the other hand, blockchain allows us to secure our data in encrypted storage on a distributed ledger, which the blockchain resembles for robust data and data security. But when it comes to application based on blockchain, blockchains need to be more secure to prevent any data leakage. For such a purpose, we have AI, which can make the app more secure through different features based on various algorithms and is trained with many data sets including features like natural language processing (NLP), blockchain P2P, linking, and image recognition.

- Data Monetization: It is one of the great sources of collecting revenue for the multinational companies such as Facebook and Google. We have discussed a little about this topic earlier such as having other companies choose how information is being sold productively for business demonstrators. That information could be weaponized against users. Blockchain permits us to ensure our information through cryptographical measures and have it utilized in the manner we see fit. This also monetizes our data, but on a tiny scale. An AI algorithm needs data to learn and develop and thus it's required to buy data sets directly from its creator through different marketplaces. This makes the whole system fairer and more transparent than it currently is. This also helps smaller companies as training an AI can be incredibly costly. For such companies that do not generate their data through marketplaces, they can access data sets.

- Trusting AI Decision: AI is based on deep learning and becomes smarter through learning; it becomes challenging for scientists to conclude how these set of algorithms came to a specific conclusion or decision. This could be because AI algorithms are capable to process an incredibly large amount of data and variables. Due to such anonymity, scientist keeps data through continuously auditing conclusions made by AI to make sure that they are still reflecting reality. Now we can conclude that AI ultimately depends upon the algorithm and data sets we used. If we feed false data, then AI cannot trust its prediction. Through blockchain technology, we have an immutable set of records of all the data, variables, and processes. With the appropriate blockchain technology, each step from feeding data to a conclusion can be observed, and we can conclude whether this data is tampered or not. It creates trust in the conclusion drawn by AI programs.

- Creating Diverse Dataset: AI is utterly centralized while on the other hand, blockchain is completely decentralized with transparent networks that anyone can access from anywhere across the globe in a public blockchain network. It provides a ledger which powers various cryptocurrencies and blockchain networks. Decentralization is trending nowadays such as singularityNET (SingularityNET is a decentralized marketplace for AI algorithms) which is is continually focusing on blockchain technology to encourage a broader distribution of algorithms and data, ensuring the creation of 'decentralized AI.' Singularity NET integrates blockchain with AI to create smarter and more efficient decentralized AI. Through blockchain, it allows us to intercommunicate between AI agents so that different algorithms can be built on various data sets.

6.3.4 Challenges of AI and Blockchain Integration

- The groundwork of blockchain is decentralized and is quite different from that of AI [13]. If we talk about the nature of nodes of blockchain, we found them to be heterogeneous. Hence if the blockchain is public, then it become impossible for machine learning outputs to come as a single point.

- AI requires a tremendous number of data sets and the database that blockchain forms is not scalable to consume such an amount of data. This implies that blockchain cannot integrate with AI in its current state.

- There are some instances where AI failed to show its foremost power like Uber's self-driving cars that failed when they started avoiding red lights. In such cases, if the AI is decentralized, it is challenging to manage the damage caused by it.

6.4 Blockchain-Based Decentralized Artificial Intelligence

Before learning further about decentralized AI [14], first, let's discuss centralized, decentralized, and distributed networks. Figure 6.4 shows how distributed, centralized, and decentralized networks look.

6.4.1 Centralized Network

It is a kind of network where all users are connected to a central server that acts as a medium for all communication and data transfer. This central server holds all the information of

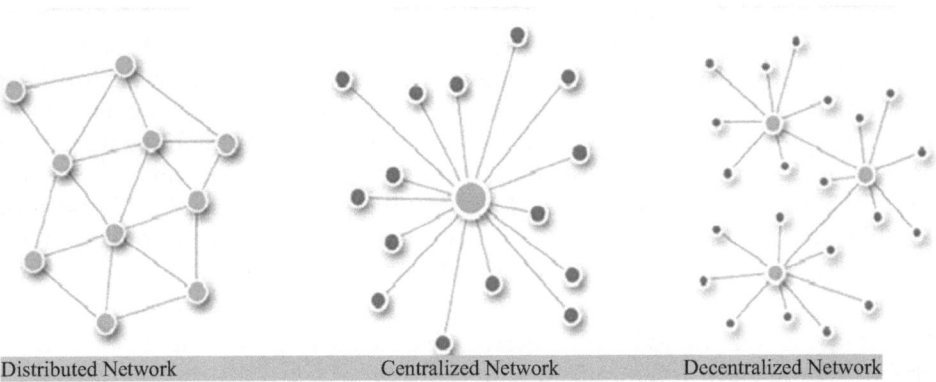

FIGURE 6.4
Centralized, Distributed, and Decentralized Networks.

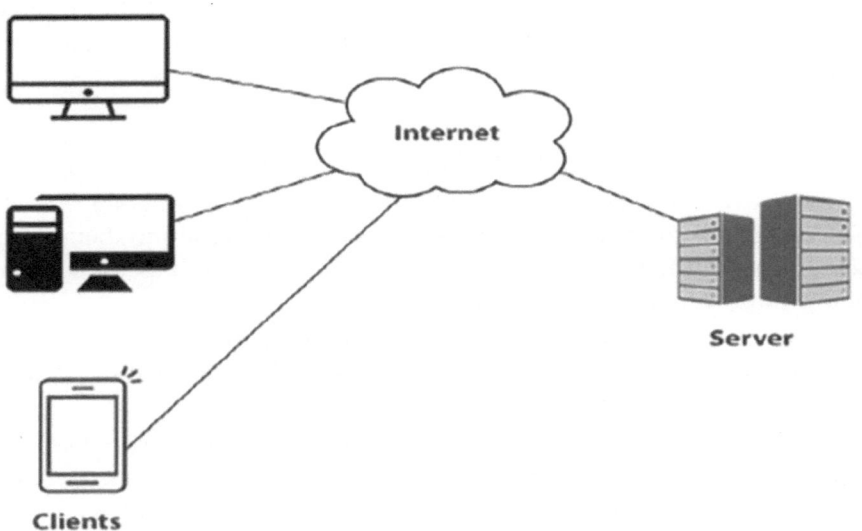

FIGURE 6.5
Architecture of Centralized Network.

the user. The centralized network is shared for instant messaging platforms. It consists of an intermediary to facilitate its operation. This means that it holds all the user's data as a bank does, and trust is paramount. The application of the centralized network is messaging. Figure 6.5 shows the architecture of a centralized network.

The advantages of centralized network are the following:

- Efficient and consistent
- Requires a slighter infrastructure, which makes it affordable

The disadvantages of centralized network are the following:

- User has to provide all their data
- It has a single point of failure which makes it unsafe, hence compromising the availability of the whole network
- Scalability limitation: it depends on a single server
- Bandwidth limitations: in the case of multiple transactions, the server could become a bottleneck

6.4.2 Decentralized Network

It distributes the workload among several machines [15]. Instead of relying on a single server, they distribute their entire server among different server stations that are situated globally. It is a 'trustless environment' where there is no point of failure. In such a network, each node is not dependent on a single server point; instead, it holds the entire copy of network configurations. The applications of decentralized networks are Bitcoins, blockchain, and Tor networks.

The advantages of a centralized network are the following:

- Ensures anonymity and privacy
- Highly scalable
- Bandwidth is never an issue as there is no single server that can create an operational bottleneck.
- High availability
- More control over resources
- Better performance
- More flexibility in system
- Less chance of failure

The disadvantages of a centralized network are the following:

- More machines are required to support the system, which raises infrastructure cost
- No regularity
- Hard to figure out failed nodes and responsive nodes
- High maintenance
- Security and privacy risk to users

6.4.3 Distributed Network

Such systems are a bit ahead of the decentralized networks. It also consists of no single server policy. Here the privilege is that the users have to decide the accessibility of information, and we can modify those privileges. It also allows the user to share ownership of data to other users. In distributed networks, processing is distributed across all nodes, but the decision could be centralized.

A real-world example is the *internet* itself. It is distributed all over the globe and consists of a tremendous number of nodes. But, while fetching or requesting data packets, users should have the necessary privileges. The applications of distributed network are multi-player online games, cluster computing, grid computing, and more.

The advantages of a centralized network are the following:

- Low inactivity
- Lesser security issues
- Manage transparency levels
- Vertical and horizontal scaling is possible
- Fault tolerant
- Fast network

The disadvantages of centralized network are the following:

- Difficult to deploy

6.4.4 Decentralized Artificial Intelligence

Decentralized artificial intelligence (DAI), also known as distributed AI, is a branch of AI. It aims to find out the solution to a problem in a distributed manner and it exploits computing resources on a large scale. Such an approach helps to solve problems that require a tremendous amount of data sets. DAI consists of autonomous learning nodes that are distributed all over the network.

DAI agents can act independently [16] which makes them loosely coupled. Furthermore, the nodes can integrate independent solutions through communication between nodes, often asynchronously. Such independence makes DAI robust and elastic; DAI systems are made to be adaptive to inherit changes. They do not require all data to be aggregated in a single location, in contrast with centralized AI. The great extension provided by DAI is that you can change the source dataset during the execution of a DAI system as it often requires hashed impressions or subsamples of enormous datasets. The characteristics of DAI are the following:

- It is a technique which includes distribution of task among agents (nodes)
- It also includes the distribution of power
- It consists of the communication of the agents

We discussed centralized, decentralized, or distributed networks enough, but we have not discussed anything about how decentralized networks are different from a distributed network.

6.4.5 AI & Blockchain: The Magnificent Unification

Blockchain is used to keep accurate records and authentication. While AI helps in decision making and recognition of patterns, blockchain and AI hold some cooperative features. Collaborating with them leads to the expansion of both technologies in such sectors that were deemed impossible in the past.

Such collaboration ensures users with:

- Enhanced security
- Decentralized data control
- Marketplace for data
- More control over data usage and models

6.4.6 Centralized versus Decentralized Artificial Intelligence

Centralized AI is becoming more challenging every day. It is a vicious cycle of 'rich getting richer', which states that only the richer and bigger companies tend to have access to large and labeled datasets. Every technology has its pros and cons, so as with AI [17]. Datasets upon which the AI model is entirely based on are so costly that small organizations could not afford to buy such datasets. The previous chart depicts the same scenario. There is a cyclic graph shown in Figure 6.6 which is based upon the traditional lifecycle of an AI solution.

The entities placed inside the circles are conceptually decentralized activities that are placed in centralized processes.

6.4.7 The Data Centralization Problem

This is where we fail to utilize AI completely. AI is a data problem through which different problems arise, like an intelligence problem. In the current era, large companies own most of the datasets, which gives birth to AI problems. Now to try to understand a situation related to what we have discussed, imagine a healthcare AI scenario that uses participant as a dataset provider. Each participant could contribute their data with the right security and privacy guaranteed. Decentralized ownership of data is a necessary step for the evolution of AI [18]. The participation of every person and assurance of security and privacy is

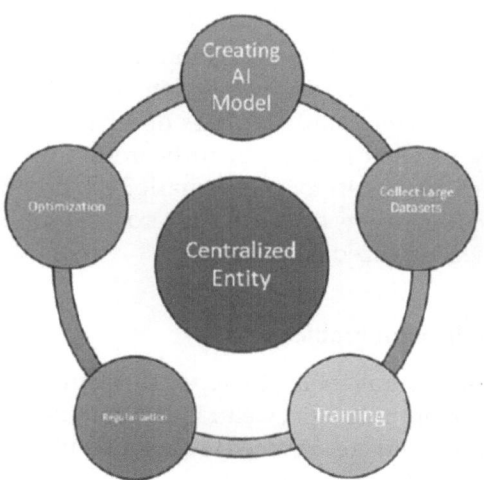

FIGURE 6.6
Lifecycle of an AI.

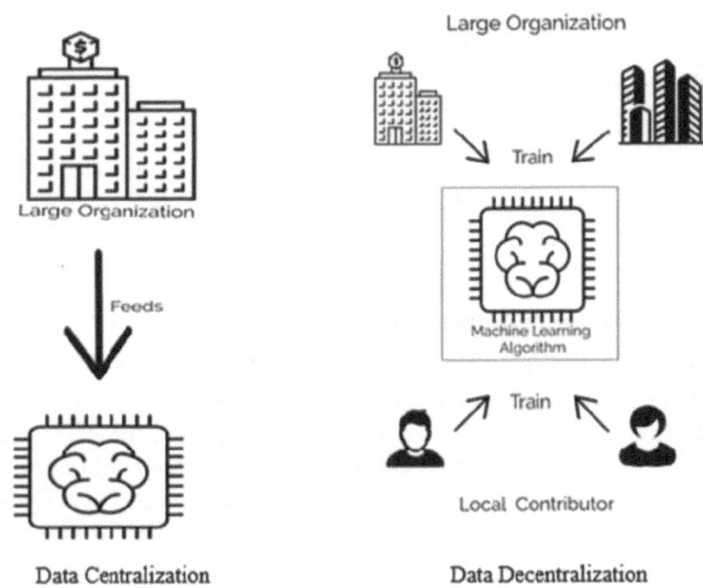

FIGURE 6.7
Data Centralization and Data Decentralization Real Time Scenarios.

required to decentralize datasets. As far as we discussed, blockchain is the technology suitable for such purpose. Blockchain is an exciting technology and is meant for data vulnerability and a distributed network. If we want the contribution of each participant, it implies it needs to be distributed not to be decentralized. All these requirements could be found in the blockchain itself. Figure 6.7 shows the two scenarios that are data centralization and data decentralization.

6.4.8 The Model Centralization Problem

Other than data centralization, another problem that exists is model centralization. Let's understand what is meant by 'model centralization.' Suppose different companies are trying to build an AI solution [19] for a specific problem and a few of them generate a model that has its uniqueness. Now the question arises of what would be better if we adapted each model individually, or what if data scientists around the world could propose and objectively evaluate different models for this scenario? Wouldn't that be great? The decentralization of models and algorithms drastically improve AI solutions over time. Figure 6.8 shows centralized and decentralized AI models.

6.4.9 Some Other Centralization Problems

Other than model and data centralization, there are still some problems left like centralized training and regularization optimization. Centralized training implies that training should be done by the same group who created the AI model themselves. Let us train the model by making it available on a decentralized network (where one's decision is independent of others) so that decentralized training could be made possible. Training of AI is an essential feature of the AI solution, which is affected by centralization.

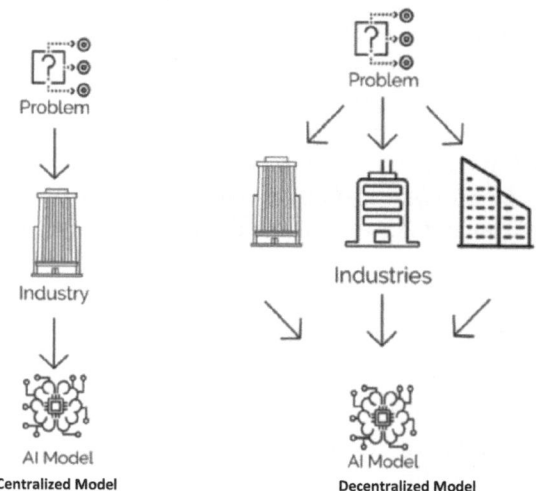

FIGURE 6.8
Centralized and Decentralized AI Models.

Another problem is of centralization is regularization optimization. Generally, AI solutions are meant to solve problems. The problem may belong to an organization, or it could be a social problem. For fulfilment of such purpose, we need an AI solution to perform correctly. But how do we know it is functioning successfully or not? Is it improving or deteriorating by the passage of time? We use the same data scientist for the regularization of AI models, which were used to create it. What if we could use a decentralized network of AI experts to find pros and cons, vulnerability, and to try to improve it continually. AI regularization and optimization are congenitally decentralized methods that are forced to adapt with decentralization.

6.4.10 Trust Problem with Centralized Big Data

In the current era, if the race is about collecting, storing, and analyzing as much data as possible, then companies like Google, Netflix, and Apple are going to be at the top in the list. The case is that the richer become more vibrant and powerful, and the barriers to innovation become even higher [20]. It also puts the trust of society in 'black boxes' due to the close nature of big AI companies.

AI solutions work in three essential layers:

1. Data repository
2. Algorithm
3. AI interface

Trusting an AI decision requires 100% confidence in:

- Integrity and security of the data
- Responsive machine learning algorithm
- AI's interface

Most of the AI models are currently centralized in nature and being centralized in nature forces users to blindly trust each layer without knowing what is going on behind the curtains.

In previous times an AI was used named COMPAS. It was used in many courts of law in the United States. Later, it was concluded that COMPAS recommended longer prison sentences for Black individuals and not white individuals with all other data points being equal. As a result, AI makes racially-biased decisions, which no one is able to explain. Even its parent company could not explain. Such biased AI models can cause severe destruction. If COMPAS was a decentralized AI, any of the data scientists would be able to figure out the exact issue.

6.4.11 Collaboration between Blockchain and AI

Blockchain holds a tremendous amount of data. Within a few years, blockchain holds enough data to beat big data with a large amount of security [21]. Blockchain helps shift the power from those who own big datasets to those who build smart and useful solutions and algorithms. There are three live projects running based on decentralized AI and blockchain; they are the following:

- Ocean Protocol: It aims to create a 'decentralized data exchange protocol and network.' It incentivizes the publishing of datasets for training purposes of AI models. In laymen language, if you upload valuable data over an Ocean network and if someone else uses that data to train an AI, you are going to be compensated.

Google is among those companies which hold an enormous amount of data. Google has its nest owners across the world. By the word 'nest owner,' I mean users of Google products. Now, what happens is just like me, data from another several nest owners gets uploaded to Google. From that data, Google could create a robust dataset that could be helpful for building AI. The data used, which is mine and yours, has value but Google gets it for free.

What if you get to compensate for your data? With Ocean Protocol, you can license your data and get some Ocean tokens for it. All of the data you are giving away for free now has

- Data integrity (origin of data is known)
- Clear ownership
- Cryptocurrencies and blockchain measures to buy or lease it
- SingularityNet: You all must have heard about 'Sophia,' the AI-integrated robot owned by SingularityNet. Suppose you developed an AI algorithm that could help marketers, the government, and society as well. That is where SingularityNet focused at the AI level. You can make your AI model available for others. For example, if you have built a model to study energy consumption in India, it can be integrated with other complementary models, creating even more powerful and precise AI.

Since the ownership of the model is clear, it implies its intellectual property is protected. Whenever the model gets used, you are going to be compensates in SingularityNet's adjusted gross income (AGI) tokens.

- SEED: It focuses on interface level and ensures us to trust the bot in our lives. One of the best examples of a bot is Amazon Alexa as it is being used throughout the world. Even trusting a reputed company like Amazon does not give you the certainty that its bot has not been hijacked. The solution for this is integrating it with the SEED network.

SEED network is an open source and decentralized network where we can manage, view, and verify all bot interactions. Whenever you interact with a bot, you use your data in the form of interaction to feed it. Should you be compensated for feeding? SEED says you should, and it secures your asset rights in the blockchain.

6.4.12 Blockchain-Based Platform for AI

So far, we discussed and may conclude that AI exposes user data privacy during training, and it requires holding high cost for training, which is becoming a hurdle to the development and evolution of AI [22]. The issues could be data privacy, ownership, and exchange, and model privacy, which is challenging solve with a centralized paradigm of machine learning or federate learning. As a result, a blockchain-based training paradigm could be used, which aims to train a model with distributed data and to reserve the ownership of data and the interest of the trained model. In this approach we use blockchain as a base architecture in which we abstract different actors (i.e., model provider, data provider) taking different actions to archive its own target such as realizing and distributing encrypted model training by federating learning with different actors, setting a smart contract as model training infrastructure, and setting up a notification server. Pricing of training data is set according to its contribution and therefore it is not about the exchange of data ownership.

With the expanding usefulness of deep learning, the importance of data has grown even more. However, in the current centralization paradigm, data is collected from the end user and uploaded on the remote server for data analysis and modelling. During the process, there comes a notable difference between data and model because of the following issues:

1. **Centralized Cost:** Training a model with a large amount of data sustains a very high cost. Model developers must be able to afford the expenses of training and storage for the model training. Although big companies can afford this training, such high-cost training comes across small companies as a barrier.

2. **Security & Privacy:** After all the data is collected from users uploaded on the servers, maintenance and security costs are incurred. Moreover, such centralized data storage has server privacy issues. Users have to provide their valuable data to a third party and even to malicious parties in case of if the server got hacked.

3. **Ownership:** As soon as the user provides its data to the server, it loses ownership over it and cannot further control the transmission of data. Therefore, it is necessary to set up a reasonable user incentive in such a paradigm.

4. **Single Point of Failure:** A centralized modelling schema often faces a problem of eventually being litigated out of existence or a failure of the server.

To solve such challenges, federate learning plays a significant role. It reserves the privacy of the data from the data owner. In this way, we do not require collecting a large amount of data and storing it on a centralized network. Instead, data remains with its creators. Meanwhile, researchers move to model towards them.

6.4.13 Blockchain Contribution in Decentralization

- Blockchain is the base of infrastructure: Blockchain can be useful as a base infrastructure where everyone can participate. We can make AI models highly decentralized, which is helpful to train an algorithm with different king of datasets, assuring ownership of digital assets.

- Implementing smart contract: A smart contract is going to act as a modelling infrastructure to publish a modelling task, a training and aggregated command, and a rewarding strategy.
- Model protection: Instead of moving data toward the model, we are deploying a model over the blockchain model and moving it toward data. Besides, we should provide security to models. This can be done by applying homomorphic encryption and multi-party computation.
- Lowering training cost: We use federate learning to train a model, which reduces the training cost to a great extent.

6.4.14 Design Overview

In this section, we present DAI [23] as a new paradigm of ML. The concept of DAI is based on three factors as it is shown in Figure 6.9.

- **Data Provider:** Data providers can be end users, companies, or any organizations willing to utilize their data for exchanging services or other incentives as shown in Figure 6.10. They collect data from sensors or other data resources. After collecting data, it is evaluated and provided the quality measure, as well as its schema. Data providers grant their data for modelling purposes, which does not mean the data has been given out to other parties.
- **Model Provider:** It is one's own model. The model provider develops and distributes a machine learning model to utilize and train it with the help of data providers as shown in Figure 6.11. It can be a raw model or pre-trained model; it doesn't matter. Alongside initializing a model for training, the model provider evaluates and updates the model in accordance to test data, schema, and reward plan for training data, in

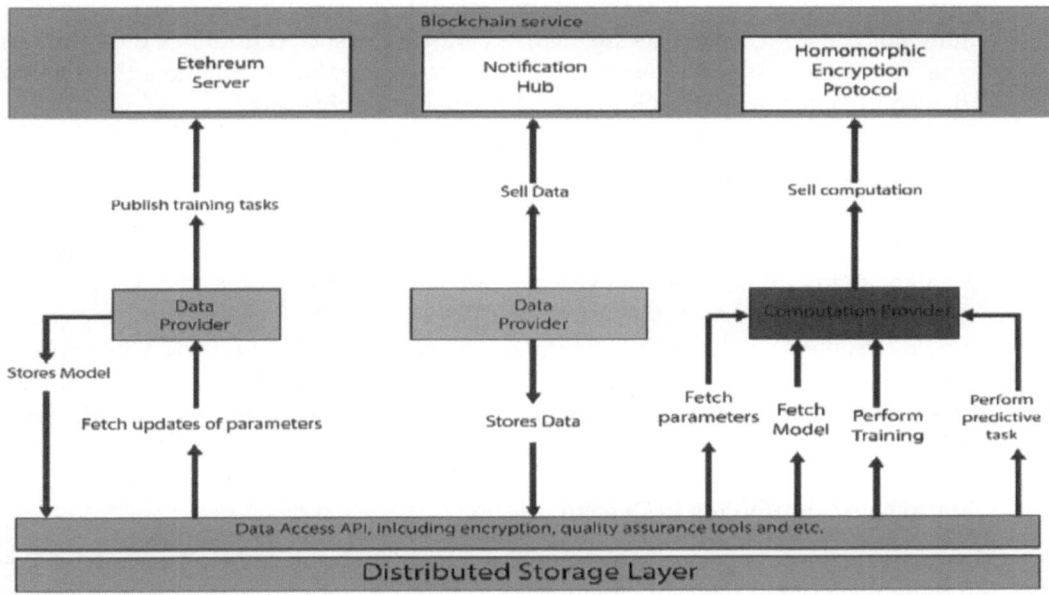

FIGURE 6.9
Block Diagram of Decentralized Artificial Intelligence.

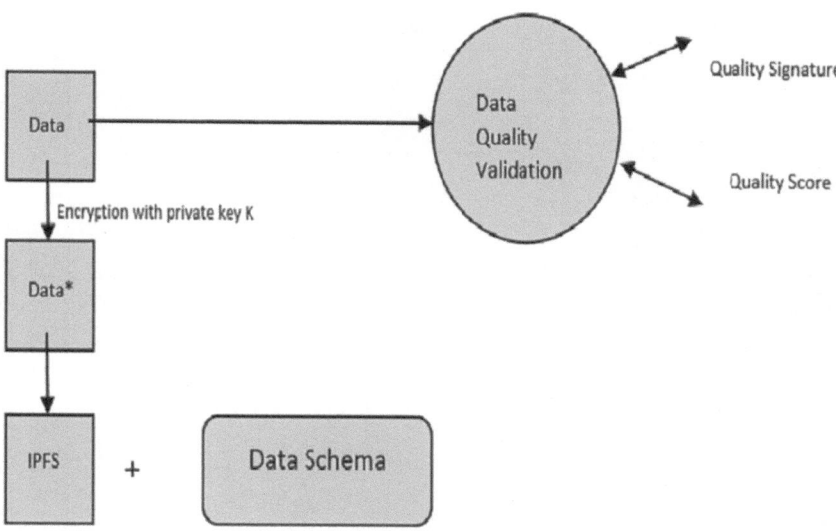

FIGURE 6.10
Data Collection through Data Providers.

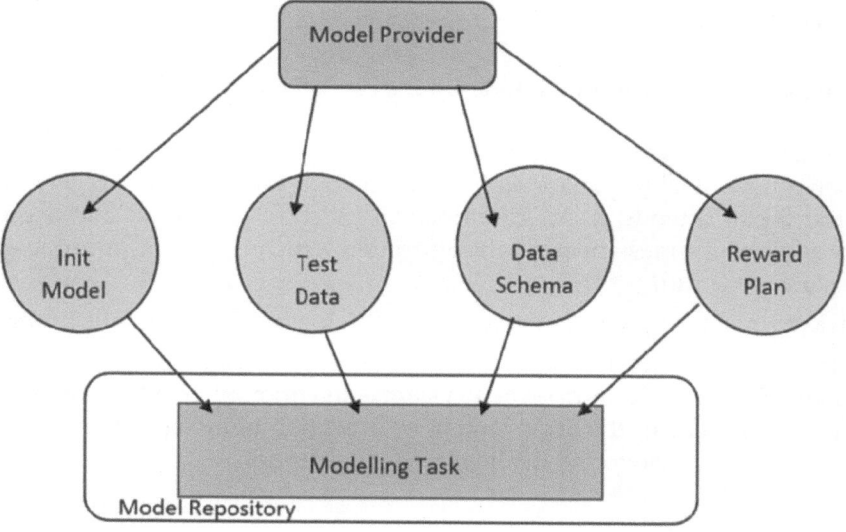

FIGURE 6.11
Model Selection from Model Provider.

addition to the model itself. While in DAI, they only need to provide the model through smart contracts to computation providers.

- **Computation Providers:** It is the node in the network that runs the smart contract for model training. It ensures that both data and models are protected. A model training task is distributed among multiple computation providers as a federated learning task. A computation node in a network is independent meaning even a data provider could become a computation provider if it has a high-performance GPU cluster. The working model of computation provider is represented in Figure 6.12.

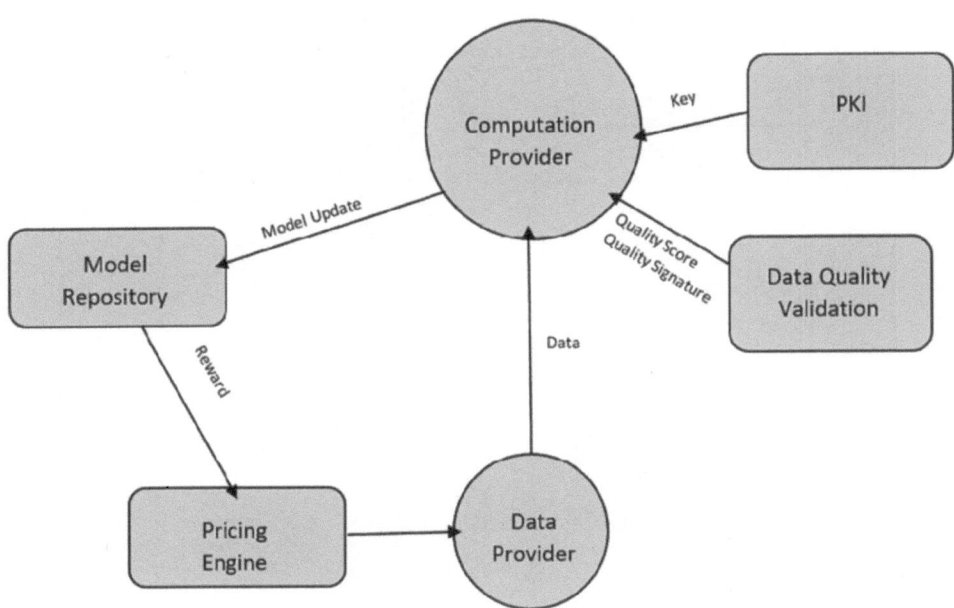

FIGURE 6.12
Block Diagram of Computation Power.

There are some base supporting nodes in the blockchain network:

1. **Notification:** It takes the responsibility of communication and event triggers. When a new block is generated in blockchain, a notification hub generates a new Goroutine to listen to it. A Goroutine is a function or method that executes independently and simultaneously in connection with any other Goroutines present in the same program.
2. **Blockchain:** It records the address of the participating nodes and their data or model.
3. **Smart Contract:** It is the infrastructure followed under distributed model training. A smart contract is comprised of a system convention planned to digitally facilitate, verify, or implement the arrangement or execution of an agreement. It permits the presentation of trustworthy dealings without outsiders.

6.4.14.1 Training Process Design

Now we will discuss how the workflow of model training on a blockchain-based platform, DAI, would take place where all kinds of actors participate and collaborate with others to archive and train a secure model with encrypted data in a blockchain context.

1. **Blockchain Infrastructure:** Blockchain Infrastructure refers to the fact that in this system, the communication bridge and context is going to be based on blockchain. There is some measure that needs to be taken to ensure that all actors can freely join and exit the platform at any time and equally participate in this data transaction network. Moreover, this is a public network with token trades; hence PoW is used to ensure the highest level of security.

2. **Preparation:** As soon as an actor joins the network, an Ethereum account should be generated to execute actions. Then, the data provider chooses to sign an offline digital signature to protect the private key instead of signing with remote procedure call (RPC). Then it generates a data schema file to upload Inter Planetary File System (IPFS) protocol, which returns a hash string. After getting the hash; the data provider generates an offline transaction. It gets passed to the Ethereum server and waits for mining. During the process, the notification hub receives this action and records it.

3. **Model Provisioning:** We need to secure the model, model provider, and data provider as well; for such purpose, we need to encrypt the initial model whether it is trained or untrained. The encrypted model then gets uploaded to the IPFS and sends the hash to the Ethereum. Alongside a smart contract generated for the modelling task. The modelling task contains four measures:

   a. Test data for evaluation

   b. Accuracy of the model with test data

   c. Data schema for training data

   d. Rewarding strategy

4. **Model Training:** After the publication of the modelling task, a smart contract is generated to match data providers. Now, if we have N data providers, the task is going to be divided into N sub-tasks, with an identical initial model for different training data. In short federate learning is implemented here. While the task is carried out on computation providers, they too ensure the security of model training. Encrypted data and the model are then passed to computation providers, and they train the model locally.

5. **Aggregate Model:** When local training is done by the computation providers, then the updated model gets transferred to IPSF and sends the hash to the Ethereum. The updated model then is downloaded by the model provider from IPFS by hash. Model providers aggregate the model and get a steady global model.

6. **Assigning Reward:** This is the central part of the whole infrastructure. The model provider now evaluates the updated models based on performance or accuracy with the test data and assigns a reward to the data provider and computation provider according to the contribution factor.

6.5 Sensitive Blockchain Backup Using AI

Storing data on the blockchain is not as easy as it sounds. It is comprised of several challenges. Blockchain is supposed to transform regularly, from healthcare to education. However, blockchain's enthusiasts forget to mention crucial details like it is way too expensive to store personal data on the blockchain. The majority of startups based on blockchain still adhere to Ethereum and its ERC20 token standards, which states that you need to pay for gas; whenever you transact on their platform you need to pay for gas also.

Blockchain will not be able to interrupt any real-world industry unless the problem of data storage is fixed [24]. Blockchain's distributed networks were not created to manage supermarket supply lines or agricultural loans. A distributed ledger system stores its files across many machines to create redundancy in case of failure.

Whereas on the other hand, AI is vastly used all around the globe to reduce the repetition of tasks, AI has become mainstream.

6.5.1 Types of AI

- **Artificial Narrow Intelligence**

The real-life implementation of artificial narrow intelligence (ANI) is Google's RankBrain algorithm and Apple's Siri. ANI means that intelligence is concentrated to narrow parameters and contexts. In simple terms, Google's ANI cannot do much more than rank pages. Even when it is said that Siri does not understand a request, the fault is not of Siri; in fact, its AI is narrow.

- **Artificial General Intelligence**

It is better than ANI and also known as human-level AI. The fascinating fact about AI is the idea of machines imitating the powerful human brain. Few experts also claim that artificial general intelligence (AGI) can go far beyond human capacities by its ability to process and analyse a vast amount of data at an incredible speed.

- **Artificial Super Intelligence**

Machines such as artificial super intelligence (ASI) should be operable beyond human intelligence in all aspects. It would require decades to come up with solid research on ASI machines.

6.5.2 The Working of Artificial Intelligence

Suppose you have assigned a task to an AI machine. It analyzes through the neural network. Neural networks are more like the human brain and are comprised of a vast number of nodes, which helps in recognizing patterns that are numerical and contained in a vector. It processes every data and enhances its certainty. Through each data input, it predicts something that gives to the next layer as input and finally we get the output from output layer. Figure 6.13 shows us a brief idea of a neural network.

6.5.3 AI and Virtual Machine Backups

Virtualization is a kind of revolution that every country has faced and then evolved. It made so many options available for the backup industry. Now IT companies are

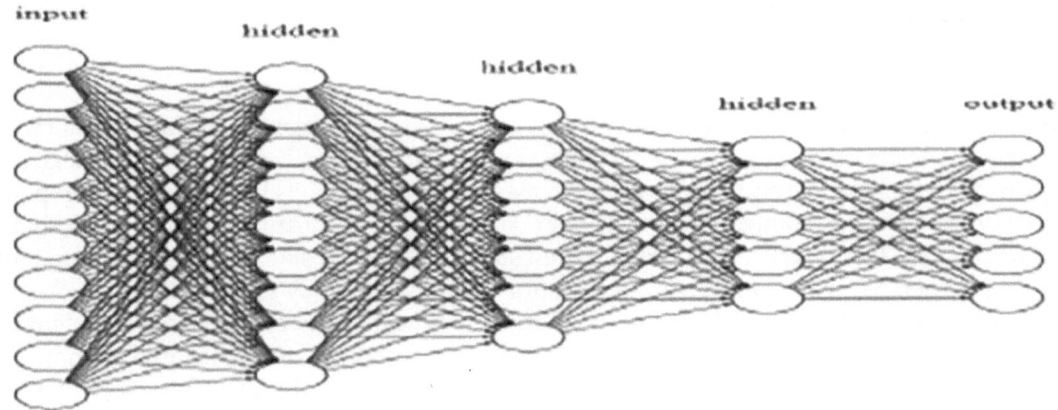

FIGURE 6.13
Basic Model of Neural Network.

comprised of a software-defined datacentre with server virtualization. They also make sure to modernize their strategy toward data protection, including more backup and recovery options. We can undertake that ML, AI, and predictive analytics capture and make backup administrators quicker and more creative at their jobs. Systems are now intelligent enough to detect which versions of file and application recovery points to roll back after an attack. AI will force predictive learning algorithms and frequently perform favorable recoveries, removing outages even before an end user can detect them.

6.6 Combining Blockchain and AI for Monetization of Data

The users on blockchain have full control of their data in terms of granting or revoking access [25]. Presently, data monetization is done by centralized platforms. Using blockchain, people, too, can monetize their data. This implies that people are more willing to provide and share their data, and thus data produced can be used for the development of AI systems. AI can flourish well in a democratic environment where data and model providers can interact with each other without relying on intermediaries. Next, we are going to discuss how we can optimize data using blockchain and AI individually, and then combined.

6.6.1 Data Monetization

Data monetization helps to raise revenue. Successful companies like Google, Amazon, and Facebook, have adopted data monetization and made it an essential part of their strategy. Monetization of data can be done in two ways as shown in Figure 6.14.

The benefits of data monetization are the following:

- Optimize the use of data
- Increase operational productivity and efficiency
- Reduce operating costs
- Boosts profitability
- Identifies and migrates risk
- Enhance insights into how best to improve products
- Improves understanding of customers
- Strengthens customer trust

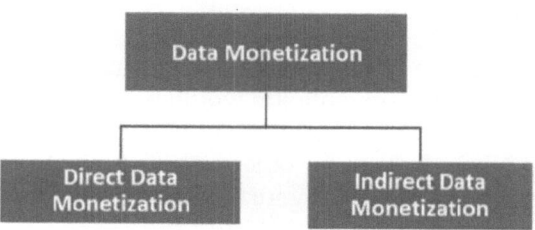

FIGURE 6.14
Types of Data Monetization.

6.6.1.1 Direct Data Monetization

As its name suggests, it involves selling access to your data directly to third parties. Selling data can be done in raw form or in a form that has already been processed and transformed into analysis and insights. For example, it may be a contact list of potential market influencers or finding an impact on buyer's industries and businesses.

6.6.1.2 Indirect Data Monetization

It is an exciting kind of data monetization. Firstly, the data is optimized. This involves analyzing data, which reveals insights that can be helpful for an organization's business performance improving. This optimized data can be used to find out public interest, how to reach customers, and understanding behavior so that one can enhance their sales. Data can also be useful to avoid risks, save costs, and streamline operations.

6.6.2 Data Monetization with Blockchain Technology

Consumer data can be beneficial for collecting substantial revenue. At the same time, companies should be responsible for managing data. Blockchain technology can help to restore public trust in a companies' managing user data. Blockchain is a platform that enables security and the immutability of distributed ledger all at the same time. It treats data responsibly, making transparent transactions and validations, along with enabling digital marketers to measure the success of marketing efforts. The implementation of data monetization could increase the revenue of the business, but the strategy needs to be effective.

6.6.2.1 Recognize Different Types of Data and Their Use Cases

There are different sources an organization can collect data from; it can be social media, IoT devices, cloud, or web. Each organization is steadfast in accumulating specific types of data. For example, data of patients, doctors, treatments, and cures, can be useful for healthcare organizations. So before collecting data, business leaders should ask these questions:

- What type of data are they seeking?
- What kind of perceptions can be planned from the collected data?

Answering these questions, a business leader can conclude innovative use cases for collected data. A business leader should find out all the use cases of the collected data to maximize their data revenue. Also, organizations should make sure that they are sharing anonymized data to evade privacy violations.

6.6.2.2 Identify Prospective Buyers

After going through several use cases, organizations should section their collected data based on various use cases. Such segmentation would be helpful for buyers. These buyers could be startups or well-established organizations. For example a retailer collects extensive customer data that can be helpful for other organizations to predict the interest and reviews of its consumers. In this way, an organization may come to know about the quality and feedback of its products.

An explanation of business can be done by such resources. Such organizations may require simplified data so that they can focus more on mergers and acquisitions in their target market.

Once the business leader acknowledges their forthcoming buyers, they have to decide whether they would avail data through a blockchain-based data monetization company or independently.

6.6.2.3 *Select a Suitable Blockchain-Based Data Monetization Company*

Blockchain is a potential technology. It consists of democratizing data monetization and sharing. Considering such potential, several entrepreneurs established blockchain-based companies specialized in data monetization. These companies are used to sell data composed of various technologies like IoT sensor, AR/VR, and bigdata.

6.6.3 Data Monetization with Artificial Intelligence

Data on its own does not hold any value at all, no matter whether it is big data, unstructured data, premium data, market data, or consumer data. Only after monetization, data can become useful. Monetizing data has become the top priority of almost every organization. The user's organization has lots of data and knows how beneficial it is, but they do not know the exact measure of how to monetize data effectively. The implementations of data monetization with AI are the follows:

6.6.3.1 *Creating New Digital Services*

It is a very efficient way to monetize data with many benefits. AI applications could monitor many sources – structured or unstructured – around the globe to provide accuracy, real-time monitoring, and an alert system.

6.6.3.2 *Addressing Customer Churn*

Companies lose their consumers to competitors because of better services and products. This problem could be solved by either providing consumers with a first-class customer experience or by solving any issues that occur before it affects consumers. For such a purpose, use of unstructured data could be beneficial, using its insight to know consumers better and to gain a deeper understanding of the marketplaces they operate in.

6.6.3.3 *Unlocking Value and Insights*

Financial services are curious about premium data as it is essential in learning and training AI models. This data is used to compete, improve in trading, and to enhance predictability and operationality of AI. AI can be helpful in better understanding unstructured data compared to humans. We require first to convert it into structured form whereas AI can train itself on the raw or unstructured data and could find out the pattern from it for deeper insights.

6.7 Trusting AI Decision Making for Blockchain Environment

AI comes into existence to do such tasks for which it is not programmed at all. It only requires data for its processing. The rapid increase in computational supremacy and the escalation of big data are now boosting AI to encourage its broad acceptance and applicability in various fields [26]. But the absence of clarification and explanation of a decision made by the AI algorithm is a significant drawback. The fact is that regardless of how great

AI is, you and I, like other people, will not prefer to use it if we do not trust it. Deep learning does not propose reasoning and control over its core processes or outputs. Current AI implementation as a black box may poison learning or the interface processes by adversarial attacks, and also, they are often subjected as bias. AI is such incredible technology, but still, it is not broadly adopted because no one can explain the decisions made by the computer, even the creator of the machine learning algorithm itself. So, what could be the solution to such a problem?

What if we try to capture each step made during the decision-making process, then AI can gain public trust much sooner. Blockchain can help to gain trust by making AI more coherent and transparent.

Explainable AI or explainable AI (XAI) is a new trend that provides explanations of their AI decisions. We are going to propose a framework that helps us to make more reliable and explainable AI by implementing the concept of blockchain, smart contracts, trusted visions, and decentralized storage.

It requires 100% certainty when it is about trusting AI reliability in such fields:

- Integrity and safety
- Machine learning algorithms
- AI's interface

The following are the blockchain-enhancing XAI features:

- Transparency: All transactions get stored in distributed ledger among each node of the network and are openly auditable, append only, and clear. Throughout the state, logs of the transaction and function calls are kept in a tamper-proof, secure, decentralized style that is available throughout the network.

- Immutability: The blockchain ledgers are immutable, and alteration in previous data is not possible. The ledger is encompassed with timestamped blocks. These blocks are secured with cryptographic techniques like SHA-256 and SHA-512. Every block holds a group of information and positions of the previous blocks in the form of hash.

- Smart Contracts: It is a set of codes that allows a participant to interact with others, and it also governs the accomplishments of prescribed and business ideas in a computerized, reliable, and decentralized manner. Each implementation outcome is verified and validated by all mining nodes and settled by the widely held nodes.

- Traceability and Non-Repudiation: Each participating node must sign each transaction cryptographically. Each autographed item then gets verified by miners, and if they validate it, the mining node is then added to the blockchain.

6.7.1 State-of-the-Art

The European Union's GDPR, or new General Data Protection Regulation, has proposed several suggestions for AI-based decision schemes [27]. In many cases, a person has the right to get clarification of the conclusion made by an algorithm itself. Suppose there is an automatic refusal of an online credit card applications, the person has the right to know why their application got rejected. GDPR may help designing algorithms and evaluations frameworks that avoid discrimination and are explainable. A taxonomy of the XAI methods are showed in Figure 6.15.

Some points and terminologies discussed are as follows:

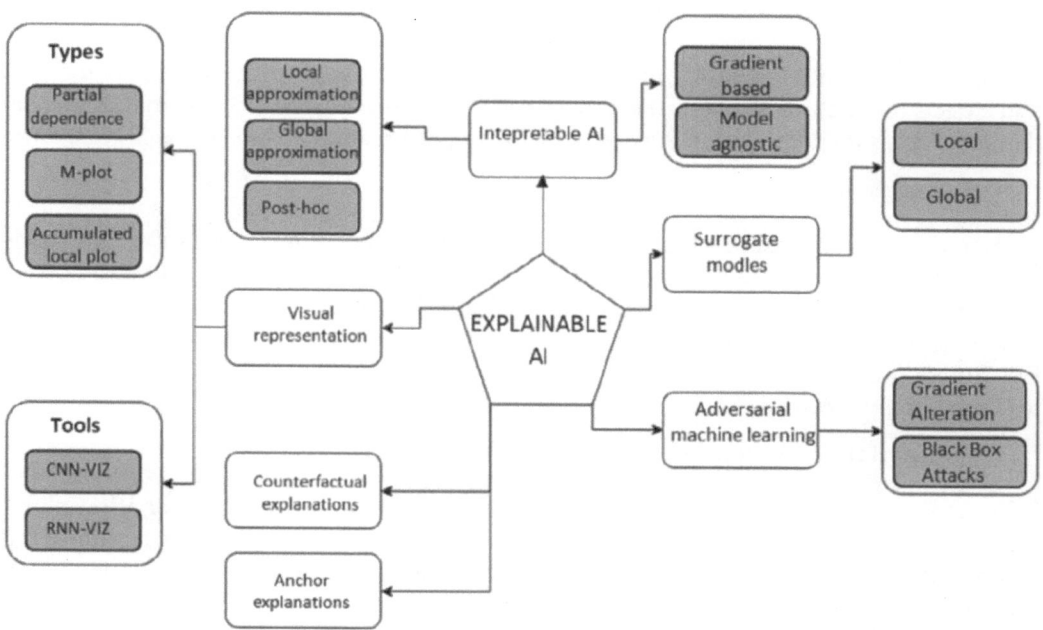

FIGURE 6.15
Taxonomy of the XAI Methods.

- **Interpretable AI**

Numerous machine learning models are ultimately interpretable as they mainly use features for their training. Decisions made by sparse linear models are shallow decision trees are understandable as these can be easily transformed into if-else conditions. There are still some machine learning models that cannot be explained like random forest-boosted trees and deep neural networks. These models are either post-hoc or intrinsic, in other words, whether they are skilled in classification and regression models, or they are skilled in resolving a task in an explainable manner, respectively. Some of the models are global approximations of learned models, while some other models are explainable for a single prediction.

Model-sceptic methods consider the machine learning models as black box functions. XAI also studies the query of inventing feasible criteria for estimating the standard of clarifications [28]. Although the typical customer of clarifications is the human end user, sometimes it is even tougher for the human to determine inadequate explanations; for such scenarios, fidelity metrics are required. This is specifically significant when a similar estimate is given unreliable explanations, which is also acknowledged as the Rashomon effect.

- **Visual Interpretation**

It is related to depictions and shown the essential features concerning the final decision made by the model. The Partial Dependence Plot (PDP) works by relegating the prediction of the machine learning model to show the connection among an estimated conclusion and a subclass of the features. M-plot is another and more efficient option in which it considers conditional probabilities in the dataset, thus evading unusual combinations of features.

- **Feature Contribution**

Means that are not inherently visual are more significant in the situation where machines make decisions instead humans. It is considered that when features are vital in a league, prediction value is the total recompense, and feature representation is done by its *Shapley value.*

- **Surrogate Models**

A model-sceptic considers ML as a box and tries to pretend it is using a substitute model. The purpose is to discover a function 'G' between a bunch of interpretable functions 'G' that most excellently matches the unique function 'f'. Local interpretable nodel-agnostic explanation (LIME) is a way used to depict substitute models locally by diminishing the loss function. It took a case say 'x' and looks for a function 'g', that is actually comparable to 'f' in corresponding to 'x'.

- **Counterfactual Examples**

In our context of making AI decisions trustworthy, a counterfactual explanation plays a significant role. It is a declaration of how the globe would be supposing it is corresponding to a required result. It is a casual disagreement of the form 'if A has not happened, then B would have to happen.' To be precise, we try to find the least modification to the feature standards, which results in flipping the entire output prediction.

- **Adversarial Machine Learning**

Counterfactual descriptions could be used as combative models to fool the system. It is observed that a slight change in the pixel value of an image could lead a neural network to generate wrong estimations. So, we found that changing even one pixel could be enough to fool the entire system; sometimes these changes are not visible to human eyes at all.

6.7.2 Blockchain-Based Framework for Trustworthy AI

XAI can be considered as promising, explaining, and trustworthy. Providing a description or fundamental explanation means that a simpler model can be developed. In an interview, one of the deep learning innovators suggested that insisting on making an explainable AI system would be a disaster. Even if humans are asked to explain their decisions, most of us would end up making a story; neural nets are similar. They learn through the training datasets, which contain billions of information and data. Auditing such a system to verify the prediction outcome, along with an explanation, would make AI more trustworthy.

In our methodology, we permit framework to generate an explanation of their predictions either directly or by taking help from model-skeptic explanation systems. A good or reasonable explanation would be considered and then contributed to the reliability of decision making. The comprehension of clarification frameworks, including the choice ends, should be checked in a permanent, carefully designed, and decentralized way, and in a way that can be followed and followed high unwavering quality and flexibility (versatility is durability against malignant assaults). In case a node tries to tamper the data in a previous block, its hash code is going to be changed entirely, and thus the chain is broken. Ultimately entire chain of hashes is broken as it does not lead back to the genesis block. Reliability comes from the fact that each node in the network has a full copy of the ledger, and hence even if a node fails, the blockchain remains unaffected.

6.8 Conclusion

The alliance of AI and blockchain is still in an undiscovered stage. In spite of the fact that the assembly of the two innovations has started, a blend could possibly use information in ways never before thought possible. AI algorithms are dependent on datasets, and blockchain secures this data and permits researchers to check each step. AI holds incredible potential. But it should be modified and developed with utmost precautions – blockchain can significantly assist in this. We found that these two incredible technologies can benefit each other. Whether it's a hectic and disastrous amount of energy consumption or working and trusting an AI decision, such integration could birth an impenetrable technology that could be accessible and beneficial for anyone. Merging them also helps us to a secure accumulation of data. We have seen that blockchain-based DAI is a collection of various concepts and components that help to create a better infrastructure for a new AI ecosystem, which aims to train an ML model with distributed data along with reserving ownership of a data model and data provider.

References

[1] Xu, M., X. Chen, and G. Kou. 2019. "A Systematic Review of Blockchain." *Financ Innov* 5 (27). doi:10.1186/s40854-019-0147-z.

[2] Giri, C., S. Jain, X. Zeng, and P. Bruniaux. 2019. "A Detailed Review of Artificial Intelligence Applied in the Fashion and Apparel Industry." *IEEE Access*. 1–1. doi:10.1109/access.2019.2928979.

[3] De Lemos, R., and M. Grze. 2019. "*Self-Adaptive Artificial Intelligence*." *2019 IEEE/ACM 14th International Symposium on Software Engineering for Adaptive and Self-Managing Systems (SEAMS).* doi:10.1109/seams.2019.00028.

[4] Ishaani, P. 2019. "Introduction to Blockchain Technology." In *Cyber Security in Parallel and Distributed Computing*. Scrivener Publishing LLC.

[5] Casino, F., T. K. Dasaklis, and C. Patsakis. 2018. "A Systematic Literature Review of Blockchain-Based Applications: Current Status, Classification and Open Issues." *Telematics and Informatics.* doi:10.1016/j.tele.2018.11.006.

[6] Baldominos, A., and Y. Saez. 2019. "Coin.AI: A Proof-of-Useful-Work Scheme for Blockchain-Based Distributed Deep Learning." *Entropy* 21 (8): 723. doi:10.3390/e21080723.

[7] Sharma, A., M. K. Sharma, and R. K. Dwivedi. 2017. "Literature Review and Challenges of Data Mining Techniques for Social Network Analysis." *Advances in Computational Sciences and Technology.*

[8] Tanwar, S., Q. Bhatia, P. Patel, A. Kumari, P. K. Singh, and W.-C. Hong. 2019. "Machine Learning Adoption in Blockchain-Based Smart Applications: The Challenges, and a Way Forward." *IEEE Access.* 1–1. doi:10.1109/access.2019.2961372.

[9] Cao, B., Y. Li, L. Zhang, L. Zhang, S. Mumtaz, Z. Zhou, and M. Peng. 2019. "When Internet of Things Meets Blockchain: Challenges in Distributed Consensus." *IEEE Network*. 1–7. doi:10.1109/mnet.2019.1900002.

[10] Kamel Boulos, M. N., J. T. Wilson, and K. A. Clauson. 2018. "Geospatial Blockchain: Promises, Challenges, and Scenarios in Health and Healthcare." *Int J Health Geogr* 17 (25). doi:10.1186/s12942-018-0144-x.

[11] Rani, N., and H. Kaur. 2018. "An Empirical Study on Data Mining Techniques and Applications." *International Journal of Research and Analytical Reviews.*

[12] Holzinger, A. 2018. *"From Machine Learning to Explainable AI." 2018 World Symposium on Digital Intelligence for Systems and Machines (DISA),* Kosice, pp. 55–66. doi:10.1109/DISA.2018.8490530.

[13] Xing, B., and T. Marwala. 2018. "The Synergy of Blockchain and Artificial Intelligence." *SSRN Electronic Journal.* doi:10.2139/ssrn.3225357.

[14] Nebula AI Team. 2018. "Nebula AI (NBAI)—Decentralized AI Blockchain Whitepaper."

[15] Gheorghe, A., C. Crecana, C. Negru, F. Pop, and C. Dobre. 2019. *"Decentralized Storage System for Edge Computing." 2019 18th International Symposium on Parallel and Distributed Computing (ISPDC),* Amsterdam, Netherlands, pp. 41–49. doi:10.1109/ISPDC.2019.00009.

[16] Harris, J. D., and B. Waggoner. 2019. *"Decentralized and Collaborative AI on Blockchain." 2019 IEEE International Conference on Blockchain (Blockchain).* doi:10.1109/blockchain.2019.00057.

[17] Salah, K., M. H. Rehman, N. Nizamuddin, and A. Al-Fuqaha. 2019. "Blockchain for AI: Review and Open Research Challenges." *IEEE Access.* 1–1. doi:10.1109/access.2018.2890507.

[18] Mamoshina, P., L. Ojomoko, Y. Yanovich, A. Ostrovski, A. Botezatu, P. Prikhodko, … A. Zhavoronkov. 2017. "Converging Blockchain and Next-Generation Artificial Intelligence Technologies to Decentralize and Accelerate Biomedical Research and Healthcare." *Oncotarget* 9 (5). doi:10.18632/oncotarget.22345.

[19] Harris, J. D., and B. Waggoner. 2019. *"Decentralized and Collaborative AI on Blockchain." 2019 IEEE International Conference on Blockchain (Blockchain),* Atlanta, GA, USA, pp. 368–375. doi:10.1109/Blockchain.2019.00057.

[20] Shala, B., U. Trick, A. Lehmann, B. Ghita, and S. Shiaeles. 2019. "Novel Trust Consensus Protocol and Blockchain-Based Trust Evaluation System for M2M Application Services." *Internet of Things* 7: 100058. doi:10.1016/j.iot.2019.100058.

[21] Makridakis, S., A. Polemitis, G. Giaglis, and S. Louca. 2018. "Blockchain: The Next Breakthrough in the Rapid Progress of AI." *Artificial Intelligence—Emerging Trends and Applications.* doi: 10.5772/intechopen.75668.

[22] Nassar, M., K. Salah, M. H. Ur Rehman, and D. Svetinovic. 2019. "Blockchain for Explainable and Trustworthy Artificial Intelligence." *Wiley Interdisciplinary Reviews: Data Mining and Knowledge Discovery.* doi:10.1002/widm.1340.

[23] Koch, F. L., and C. B. Westphall. 2001. "Decentralized Network Management Using Distributed Artificial Intelligence." *Journal of Network and Systems Management* 9: 375–388. doi:10.1023/A:1012976206591.

[24] Hussien, H. M., S. M. Yasin, S. N. I. Udzir, et. al. 2019. "A Systematic Review for Enabling of Develop a Blockchain Technology in Healthcare Application: Taxonomy, Substantially Analysis, Motivations, Challenges, Recommendations and Future Direction." *J Med Syst* 43 (320). doi:10.1007/s10916-019-1445-8.

[25] Ouchchy, L., A. Coin, and V. Dubljević. 2020. "AI in the Headlines: The Portrayal of the Ethical Issues of Artificial Intelligence in the Media." *AI & Soc.* doi:10.1007/s00146-020-00965-5.

[26] Barredo Arrieta, A., N. Díaz-Rodríguez, J. Del Ser, A. Bennetot, S. Tabik, A. Barbado, … and F. Herrera. 2019. "Explainable Artificial Intelligence (XAI): Concepts, Taxonomies, Opportunities and Challenges Toward Responsible AI." *Information Fusion.* doi:10.1016/j.inffus.2019.12.012.

[27] Rai, A. 2020. "Explainable AI: From Black Box to Glass Box." *J of the Acad Mark Sci* 48: 137–141. doi:10.1007/s11747-019-00710-5.

[28] Adadi, A., and M. Berrada. 2018. "Peeking Inside the Black-Box: A Survey on Explainable Artificial Intelligence (XAI)." *IEEE Access.* 1–1. doi:10.1109/access.2018.2870052.

7

Artificial Intelligence for Blockchain II

Ch. V. N. U. Bharathi Murthy and M. Lawanya Shri
Vellore Institute of Technology, Vellore, Tamil Nadu, India

CONTENTS

7.1 Introduction

As we all know, blockchain is a promising technology for building secure data and it can share immutable data between trusted parties. With the introduction of blockchain cryptocurrency, it can automate payments. It facilitates access to its ledger based on the authorization, can records all transaction flow in the data logs, and the ledger is updated and maintained by all parties within the chain. With the help of smart contracts, which is blockchain 2.0, payment automation can be done without the intervention of middlemen. Instead, artificial intelligence is a way machines show intelligence through training [1]. The description given by Peter Norvig [1], who is a top figure in the AI field, is defined as

> the study of agents that receive percepts from the environment and perform actions. Each such agent implements a function that maps percept sequences to actions, and we cover different ways to represent these functions, such as reactive agents, real-time planners, and decision-theoretic systems. We explain the role of learning as extending the reach of the designer into unknown environments, and we show how that role constraints agent design, favoring explicit knowledge representation and reasoning. We treat robotics and vision not as independently defined problems, but as occurring in the service of achieving goals. We stress the importance of the task environment in determining the appropriate agent design.

The integration of blockchain and AI solves many flaws by all technical means [2]. AI relies on algorithms that learn, infer, and make decisions. These algorithms work efficiently when the data collected is taken from a reliable and trusted source. Blockchain offers a trusted and secure data source in which all the data is stored, and this data will be validated and accepted by all the mining nodes before the transaction. The data stored in the blockchain is also tamper-free and secure. So, making the blockchain data available to AI algorithms will produce efficient results in various fields like medical, banking, trading, and more [3].

There are many advantages of combing AI and blockchainand they are the following:

1. High Efficiency: The business processes involve multiusers that are accepting multiple stakeholder orders that cannot be efficient due to multiparty authorization. By combining AI and blockchain we can solve this by enabling decentralized autonomous agents (DAOs) for automation and transactions among multiple stakeholders.

2. Improved Trust in Robotic Decisions: It is difficult for customers to gain trust and understand the decision made by AI agents. As blockchain is a trusted technology for its decentralized, transparent, and distributed ledger, it makes it trustworthy to the customers. There is also no need for a third party.

3. Enhanced Data Security:Blockchain technology is well known for its trusted data. The mining nodesverify and validate data before recording them into the block. The present data is secure and trusted. When artificial intelligence algorithms use the block data, they make it more secure and trusted.

7.2 Performing Data-Centric Analysis and Information Flow by Integrating Blockchain and AI

We can make data secure and trustable by integrating AI and blockchain technology, and we can make use of data in many major fields like medicine, finance, and trade. We can store heterogeneous data in a prominent way and we can send data to an authorized party in a secure way. Let us consider an example for performing a datacentric analysis and information flow for cardiovascular medicine by integrating AI and blockchain.

7.2.1 AI and Blockchain Use in Cardiovascular Medicine

AI is an advanced statistical technique that makes predictions possible by classifying complex data. It predicts through voice, face, and image recognition. In current cardiovascular medicine, AI accesses cardiac function through image recognition by inferring cardiac rhythm from the electrocardiogram (ECG) and predicts decisions. Currently, AI is trying to improve novel cardiovascular solutions by developing outcome predictions. There are some problems withnovel solutions a lack of trust, and large and heterogeneous data sets with a recorded chain of ownership.

Secure and traceable data exchange can be facilitated by blockchain [4]. Blockchain performs transactions from an individual to another individual without the need for middlemen and also it enables decentralization. Here, we can trace the data owner by blockchain, which is a data-centric model among all participants. All the nodes in the network can verify and validate all the data present in the block. The nodes validate data with the help of consensus and they create a cryptographic hash for all transactions. The transactions are stored in a public ledger, but the identity of the participants and content of the data is not revealed. This is one of the reasons that the owner of the data will have the authority to provide access to selected providers or companies. They may not necessarily use the same platform. The blockchain smart contracts can also be used to check the progress rate and enforce immutable legal agreements among different parties.

Blockchain introduces a technological advancement in the form of incentives. These incentives develop a term called 'health cryptocurrency' where rewards would be given to the original data owner for sharing and also to the providers who helped form a diagnosis. Administrators that have increased patient satisfaction should also be committed to benefits. In the study, rewards are granted to data owners who aid in the creation of the database.

7.2.2 Integration of Blockchain with AI in Cardiovascular Medicine

Integrating blockchain with AI would drastically improve the quality of data predictions. Blockchainis emerging to create a platform for AI training, clinical trials, and regulatory purposes to improve health data structures [5]. Integrating data is challenging from various cardiovascular sources such as ECG sensors, monitoring devices based on hospital electronic medical records, genomics, and non-clinical data, but it can be resolved by blockchain.

Blockchain can trace the separate block types and create detailed information. So, the difference in The Treatment Of Preserved Cardiac Function Heart Failure with an Aldosterone

Antagonist (TOPCAT) test results [6] can be easily monitored when blockchain collects the data. In this way, partitioning the data can improve the critical solutions for AI applications, specifically for drug metabolism or disease propensity. The health cryptocurrency model may also provide incentives to the contributions of rare data. Blockchain-based AI can improve novel specific outcomes. There are several applications being developed by integrating blockchain with AI. A company called AHA took a partnership with the Open Health Network for developing blockchain-based AI products, such as Patient Sphere which personalizes patient-treatment plans.

7.2.3 Challenges

The integration of AI and blockchain will lead us to some challenges. Mainly, the application of blockchain that belongs to healthcare has to be secure. Ethical issues also might be raised with this integration. Even though the details of the stakeholder are private, it might not be desirable to access the medical data publicly in the blockchain ledger. Apart from these, consensus and professional guidelines for society and regulators need to consider issues such as data security, integrity, scalability, and ethical resources [7].

7.3 AI-Based Healthcare Using Smart Contracts (Blockchain)

Smart contracts are implemented using blockchain technology, which is technically called blockchain 2.0. A software protocol that allows the verification and execution of legal agreements between stakeholders is called a smart contract [8]. These are trustworthy as the data would be verified and validated. Smart contracts are being used more since the introduction of Ethereum, which is a blockchain-based smart contract solution released in 2015 [9]. Smart contracts provide the facility for storing health records of a patient and accessibility power will be given to patients. This ensures secure data transfer, avoids report counterfeiting, and prevents access by unauthorized people. However, blockchain also faces some challenges like scalability, transactions per second, latency, and development. So, the integration of AI with blockchain would overcome these challenges [10].

7.3.1 Opportunities That AI Bring to Smart Contracts

As we all know data that is stored in blockchain is diverse and greater volumes of data are being recorded. At some point, there will be a problem of scalability within the blockchain. There are some traces of recent advancement in AI which leads to the advancement in the field of big data analytics [1,3,4,6–15]. By integrating AI with blockchain we can also eliminate security and privacy vulnerabilities. Let us consider some opportunities that AI provides to blockchain:

1. *Assuring High-Quality Smart Contracts:* Smart contracts are software protocols with a programming code written concerning terms and conditions of involving parties. In some cases, smart contracts may contain bugs in the code that interrupt achieving high-quality smart contracts. In [16], they presented a platform called Oyente which runs by a symbolicexecution to detect potential bugs. Smart contracts usually run the program codes as per the execution costs. In Ethereum, the cost of executing smart contracts is

funded by gas. So [17] introduced a tool called GASPER which identifies and locates seven gas-costly patterns by checking byte codes of Ethereum. With the growing number smart contract users, we can automate the integration of multiple contracts. In some cases, the semantic definition and measurement criteria for quality-of-service (QoS) are lacking in smart contracts. We can eliminate these challenges by integrating smart contracts with AI, providing data-driven QoS evaluation.

2. *Maintaining Blockchain:*Blockchain records a great amount of heterogeneous information. Using this data, we can identify faults , predict failures, and also find the reason for low performance. Maintaining such an amount of data under different platforms is a hectic task. Integrating blockchain with AI would help us with maintenance, and data can be used by several AI algorithms to provide accurate predictive analysis.

3. *Detecting Malicious Behavior in Blockchain:* As we already know, blockchain transactions can be done anonymously, and they provide a decentralized system. This creates difficulty in auditing some malicious behaviors like phishing, money gambling, and other scams. All the transactions of the blockchain are recorded in a decentralized public ledger while displaying pseudo addresses. Nowadays AI is being developed in the field of big data analysis. This helps to identify malicious behavior in massive blockchain data. Some malicious users may also create multiple addresses to form a criminal gang to perform internet fraud. New machine learning algorithm approaches are well defined to address such issues. The advancement of AI and big data analytics would greatly help to address the challenges of blockchain.

7.3.2 AI Factors Need to Be Considered in Blockchain

Blockchain appears in several architectures. Being a blockchain architect, one can develop an architecture where they require permission to read information in the blockchain. They can also limit network users that transact on the blockchain, and can have access to validate the data or not. To improve smart contracts with AI, these are some factors need to be considered:

1. *Chain Data:* Multiple nodes present in the network duplicate the data stored in the blockchain. But as with personal health information, blockchain cannot contain data with large sizes. It will only guarantee transactions are transparent. They can store data by encrypting ahead of adding it to the chain and then decrypting it in the application interface whenever the data needs to be displayed. When there is a huge amount of data, they store it in offchain and refer to the hash address to locate the data. With the help of this process, many network users can participate and can validate the block in the consensus. The other method is using the permissioned blockchain with features like permissions, multiple copies, multiple input validation, and data transparency. The architect can give access to read and write only to a few parties. This is very useful in the case of medical health records as they containfeatures of the public chain but acts as a permissioned chain.

2. *Mining:* A smart contract gets executed by sharing transactions between multiple parties. Whenever a transaction occurs in the Ethereum smart contract, it costs you a specific amount of gas. A payment of gas is important as it prevents denial of service attacks from untrusted sources. It will avoid usage of the network over executing purposeless contracts and also prevents infinite loops or the halting problem.

3. *Network Actors:* Network users can be categorized as public, private, and permissioned. The public blockchain is a complete decentralized system where anyone can participate in the network and can access data. Private blockchain users are maintained by individual companies and they have the authority to decide the level of immutability and transparency. This specific area will be very beneficial to the pharmaceutical industry to track down the complete route during the supply chain cycle.

7.3.3 Smart Contracts Integration with AI

AI can help smart contracts by decreasing the use of paperwork and automating a code without any third party. Some rules and policies can be integrated into these smart contracts assigned to the chain. As we all know in AI the more data it has, the better predictions it can make. So with this integration, it can define the flaws present in the past contracts, identify the new clauses for development, and predict outcomes [11]. Here we can see some advantages of integrating AI with smart contract blockchain:

1. *Cost of Smart Contracts:* As we discussed earlier the execution of smart contracts would cost per gas in Ethereum; these gas costs are constructed on oracles and miners. When integrating AI, it needs to store and manipulate lots of data when compared with data stored in offchain. We can calculate the cost price:

$$Gas\,used * Gas\,price * USD\,/\,ETH$$

2. *Policies and Rules:* It is necessary to define some rules and policies while integrating AI. The blockchain data in AI can be used for various purposes like a recommendation for a search engine or data analytics. When the data is being used for the recommendation of a search engine, it needs historically huge data to predict precisely, so the rules and policies will be defined accordingly. We can also use the data stored in the offchain and can train a model with different scenarios in realtime. In this scenario, the data can be read-only public data with no cost. Some public data may also use sidechains for support.

When we are using data for analytics, the network users set rules and policies. Automation of tasks can be done by defining some sets of variants. Predefined rules can be formed by these variants, which are considered in smart contracts to automate the execution of tasks. In [9], they developed self-learned smart contracts with different approaches. At first, various application cases are considered to define rules in which all the rules do not trigger at the start, but they work in time. By using important factors in the chain, they set smarter rules based on data. New growth in the count of network users in the chain would help to define new rules. This is called analyzing the behavior of the network users and defining the new set of rules and policies to reach a target. A decentralized autonomous organization (DAO) is the last approach in which only an autonomous code will be present on the blockchain without any employee. When all the conditions in the smart contract are fulfilled, this autonomous code in the contract will be executed. This promises transparency, automation, and safety of the whole process.

7.3.4 Use Cases of Blockchain with AI in Healthcare

The blockchain integrated with AI is being used in managing changes in medical trials. At the initial trial, it sets an outcome and protocol. Accordingly, it records patient advancement in every phase of the protocol. Some group of patients can adapt a different path of medical chain that is personalized according to the patient based on a specific conditions of the smart contract [13]. Cortex is an AI platform-based decentralized application that supports smart contracts and AI execution. Cortex provides artificial intelligence models on the blockchain to improve the quality of smart contracts. All the logical connectives need not be hardcoded by developers. AI models do logical reasoning; they decentralize AI by providing a platform to give incentives to AI developers who share their models. This is built on the blockchain ecosystem. This cortex ecosystem will have mainly three stakeholders, namely smart contract developers, AI developers, and miners.

7.3.5 Future Scope of Smart Contracts in AI

Despite blockchain having features like immutability, transparency, and security, it does not have standards to check smart contract costs, storing data in offchain, and transaction fees. Off-chain data management would be a good option for real-time search engines and it would be costeffective. These cost-effective practices would be very useful in healthcare platforms like Cortex. This integration of blockchain with AI will lead to further developments in the field of blockchain systems. In the future, blockchain systems can monitor different performance metrics in realtime instead of performance monitoring and detection of faults with the help of AI. On the development of a single intelligent agent, we can develop collective intelligence in which all participants can engage in contributing their analysis and the reasoning for a better decision [11]. Finally, this integration will develop numerous machine learning algorithms for tracking and controlling blockchain data.

7.4 Decentralizing and Accelerating Biomedical Research and Healthcare by Integrating Blockchain Technology and Artificial Intelligence

A new wave of advancement in medicine is required in the healthcare industry. Improving accessibility of health records for health professionals, researchers, and patients would be very beneficial for electronic medical records and the digital healthcare system. The development of the facilities that are providing accessibility of the electronic health records to the patients improves the quality and efficiency of healthcare, and also most of the biomedical data is gathered from biomedical imaging and laboratory tests. The complexity and volume of data are increasing and would create new requirements and perspectives in the medical care industry. There will be a need for novel global healthcare approaches for treatments and disease control to meet the demand of the aging population. Different attempts were made on multiple kinds of data types to evaluate the healthcare data of patients, but they did not produce imperative results. The integration of blockchain with AI would scale up the outcomes in evaluating healthcare data of patients.

7.4.1 Advancements in AI

The integral analysis of data is hectic when health-related data and subsequent global projects are increasing gradually. Many conventional approaches are needed to preprocess and analyze heterogeneous high-quality biomedical data. Usually different healthcare fields follow various computational biology approaches and introduce these approaches to the pharmaceutical industry. These computational analyses are well led by machine learning techniques. The significant improvement in machine learning leads to an increase in computer processing power and advancement in algorithms. Nowadays machine learning algorithms are being used in the discovery of drugs and biomarker development. These techniques are being utilized in deep neural networks (DNN) for the healthcare data to find out major dependencies. Many machine learning approaches are in implementation and being developed even in the forecoming capsule networks, symbolic learning, recursive cortical networks, and natural language processing. Learning transfer and recurrent neural networks are gaining prominence in healthcare applications and these can be integrated with blockchain-enabled personal data [12].

7.4.2 Introduction to Highly Distributed Storage Systems

There is a great need for better data analysis and storage systems in this generation, which requires improved availability, scalability, and accessibility. Out of all the options for data storage, highly distributed storage systems (HDSS) are found to be a viable and useful option. HDSS stores data in multiple nodes and this data is replicated in all the nodes that make data quickly accessible. There are many failures in storage that have occurred in recent times. This challenge made HDSS popular as it allows data to be replicated in multiple nodes and protects it from failures.

There is some significant advancement in HDSS applications and optimization. To ensure consistency and affordability of data, HDSS introduced peer-to-peer (P2P) network nodes or data storage implemented in the blockchain. Blockchain can be defined as a decentralized database that stores data in blocks which are chained together using different cryptographic hashing mechanisms. The data stored in the blockchain maintain consistency and immutability. To store the data in the block, the data need to be verified and the block has to be validated by every node participating in the network with the help of consensus mechanisms. Each block contains a timestamp and parent block hash in the chain. As the data is immutable, the data cannot be changed unless it is changed by modifying the hash of the previous blocks and there is an agreement by the remaining participants in the network. Blockchain is an open distributed ledger that depicts integrity and immutability. In an extension of blockchain, we can create rules and policies in a code of blockchain that can be implemented as smart contracts. Smart contracts are very helpful especially for healthcare data where multiple restrictions and rules are to be met. These smart contracts are a kind of software protocol where the code gets executed once all the conditions or rules in the code are passed.

The blockchain contains mainly three types of users named maintainers, external auditors, and clients.

1. *Maintainers*: The decision of business logic and infrastructure of the blockchain are maintained by these. They store the full copy of the blockchain and have access to read data. They have the authority to decide the rules of the transactions and participate actively in the consensus.

2. *External Auditors:* External auditors are similar to maintainers, but they do not participate actively in the consensus. They verify the correctness of transactions by read access data of the blockchain. They also store a complete replica to declare the completeness of the transactions. Examples of regulators are law enforcement and nongovernment organizations.

3. *Clients:* Clients are the customers of the services of blockchain. They have accessibility to a minimum amount of data to verify its accuracy provided by the auditors and maintainers.

7.4.3 Exonum Framework for Blockchain Projects

In [12], they introduce a new open-source blockchain framework called Exonum. This is a permissioned blockchain that provides accessibility to read blockchain data. This has a service-oriented architecture (SOA) and is divided into three parts called clients, services, and middleware.

1. *Clients:*Transactions and read requests are initiated by clients. They are equipped with cryptographic data handling tools that are available to trigger transactions and check read request responses.

2. *Services:* Services usually contain the business logic of the application where the same service can be repeated in multiple blockchains at the same time. To receive accurate information from the blockchain state, read requests are processed to transactions and implemented at service endpoints.

3. *Middleware:* Middleware acts as a bridge between clients and services. It provides an ordering of transactions, interoperates between clients and services, manages the service lifecycle, and generates responses to read requests by controlling assistance.

Next are some advantages of Exonum when compared to other permissioned framework blockchains. Due to the design of Exonum, it creates an easy structure for clients and auditors for auditability service. As it uses SOA, the services developed can be reused for other Exonum applications. They can also add new services and reconfigure services used for the application. It can allow the involvement of third-party applications and provide interoperability with other Exonum-based applications. When compared to other permissioned blockchains, Exonum delivers higher power output and encoding capacity of complex transactional logic. The Exonumconsensus algorithm won't accept single-point faults. Blockchain storage in Exonum is carried out in the system of a key-value pair (Table 7.1).

TABLE 7.1

Characteristics of Exonum Service Endpoints

Characteristics	Transactions	Read Requests
Localness	Global	Local
Processing	Asynchronous	Synchronous
Initiation	Client	Client
REST service analogy	POST/PUT HTTP requests	GET HTTP requests
Example of the cryptocurrency service	Cryptocurrency transfer	Balance retrieval

7.4.3.1 Network

In Exonum, there are two kinds of interactions in which services interact with the external world. First, the blockchain state will be changed only by transactions. These transactions are executed according to their orders and the results follow a consensus algorithm. All the incoming transactions are broadcasted to every complete node that participates in the network. Secondly, every blockchain state provides information to the read requests along with the proof of existence. Any complete node present can also process such read requests internally.

7.4.3.2 Transport Layer

There is a chance of reconnecting to the same node by the clients once the transport layer verifies the transactions. In the presence of a malicious node, it may not modify the read requests, but it can delay the broadcasting of the information received from the client. The middleware layer has the task of abstracting transport layer capabilities from the application developers; the endpoints of the service could then be mapped according to local method invocations.

7.4.3.3 Authentication and Authorization

As the clients are the originators of the transactions, they ensure transaction integrity and real t-time verifiability by verifying public-key digital signatures. Authentication and authorization of reading requests can be done with the help of web signatures or by validating the communication channel as they are local. Service endpoints can be declared as private to improve safety, making it easy to control access and decrease the attack surface when we separate the private service endpoints.

7.4.3.4 Lightweight Client

In Exonum the lightweight client can provide the capability to communicate with the full node and verify the responses cryptographically. A lightweight client can authenticate requests with key management capabilities; it can also report non-repudiations and consistency among different read request responses.

7.4.3.5 Consensus

Exonum uses a leader-based Byzantine fault-tolerant consensus algorithm in order to maintain the transactions in the ledger and record results after the transaction execution. There would be no single point of failure as Exonum would continue to work even when one third of the validators are out of service or hacked. This is a completely decentralized process. The algorithm used by the Exonum has distinguishing factors like work split, unbounded rounds, and requests algorithm when compared to other BFT algorithms [12].

7.4.4 Health Data on Blockchain

Health data majorly faces the problem of data sharing and using that data for research and other commercial projects. Health data should maintain good standards of data privacy and security. The blockchain-based system allows users to access information directly in the system using a transparent price formula and provides the right to sell their data online. Data usage activity tracking is guaranteed. The users have complete ownership of the data and can sell directly to the clients. But purchasing through actual currency would be a problem. In [12], they introduce a crypto-token called Life Pound. When we place data for transactions on the blockchain-enabled marketplace, a Life Pound token will be created or mined. Marketplace clients aim to save the biomedical data of users for customers to buy. Data from users can be analyzed and validated by data validators. Life Pound contains major parts in its ecosystem called users, storage, and full nodes.

Users who sell their data can be anonymous and can secure their data privately by only providing data access to those who bought the data. Data validators initially buy the data and ensure customers by validating the data. No personal information is included in the interactions taking place in the marketplaces that are recorded in a blockchain.

The data is stored in cloud storage like Amazon Web Services (AWS). Storing data in the cloud would help to save data in the offchain, which would be very beneficial in the cases of data with huge sizes such as those usually used in CT scans and MRIs. Accessibility to this storage in the cloud is based on blockchain marketplace Public-Key Infrastructure (PKI). Data privacy and security are ensured by the use of a threshold encryption scheme on the user side [18–21]. Client workflow examples are explained in [12]. Blockchain full nodes contain full access to the data in the blockchain. There are three types:

- *Validators:* validate and chain new blocks to the chain
- *Auditors:* those who check the marketplace
- *Key Keepers:* Primary shares are protected by the threshold encryption scheme.

You can refer to Figure 7.1 for a workflow example for marketplace customers

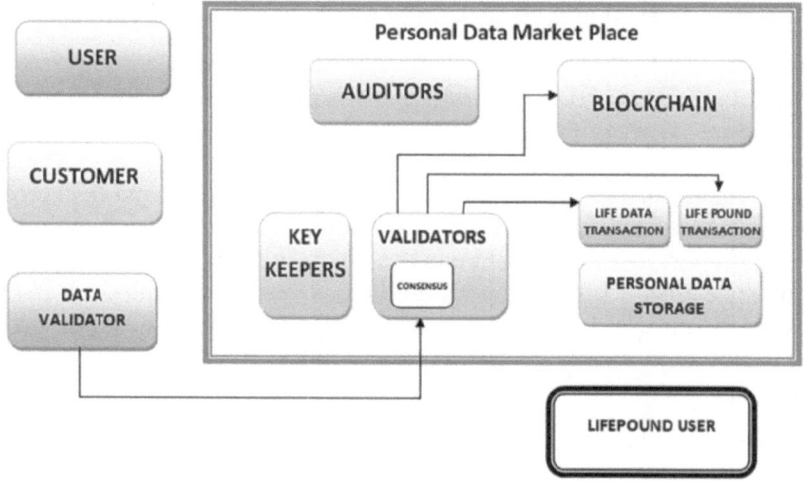

FIGURE 7.1
Workflow Examplefor Marketplace Customers (Section 7.3.4).

7.4.5 Ensuring the Quality of the Data and Consistency by Using Deep Learning Methods

The DNNs with exemplary generalization capabilities could be distinguished by the data they have trained on. Data-driven models prioritize data quality by applying data quality control. An unsupervised model that aims to detect anomalies from a data set is used for healthcare quality check initially. The data available from unsupervised methods that can be equipped to recreatedata is often used to identify anomalies. Data that is of poor quality would bring the highest reconstruction error. There are a different set of approaches for quality and consistency that are time-based models like RNNs. Initially, they learn from good quality samples, and then they are tested to adjust model behavior to save the data from being linked to poor quality samples. A collection of supervised methods can be applied to ensure data quality monitoring. As the detection of anomalies is unbalanced, we should note the fact that supervised models require labeled data sets.

7.5 Using AI and Blockchainto Personalize Treatment Plans and Diagnosis

Blockchain technology and AI give services to the healthcare industry individually where each of them has challenges. The integration of blockchain with AI would be the appropriate answer for almost all their challenges. The data stored in the blockchain is secure, immutable, private, and highly maintained. AI uses blockchain data for giving accurate predictions of the outcome by using its algorithms. The AI algorithms would read and analyze the data to capture the pattern, identify loopholes, predict failures, and more. Nowadays each user can store their data and provide access only to the desired parties. Using this personal data, they analyze health conditions and personalize treatment plans and diagnoses. The Patient Spherecompany recently employed AI and blockchain to customize treatment plans.

7.5.1 Patient Sphere Uses AI and Blockchain to Customize Patient Treatment Plans

Tatyana Kanzaveli, CEO of Silicon Valley started the innovative Patient Sphere where they combine different pipelines of data. The data given is patient healthcare data and is derived from required points. This regularly monitors the patient's progress and delivers personalized treatment plans using machine learning algorithms.

Patient Sphere collects data from various sources like electronic medical records, wearable devices, chatbots, health applications, and other programs like Google Fit. The collected data is gathered in a web dashboard to share with whoever requests access like physicians and specialists. This uses distributed blockchain as a database making it easy to retrieve old data by natural language queries. The patients can also track their progress and have the option of monetizing smart contracts [22].

7.5.2 Blockchain and AI Helping in Self-Testing

Blockchain technology integrated with AI would also help the healthcare industry by providing the facility of selftesting through wearable devices or mobile applications. There has been a novel coronavirus outbreak called COVID-19 all over the world where millions of people are infected. Due to the rapid spread of the virus, there has been a failure to find and report cases bygovernments. The testing rate is very poor in some countries that have limited public health infrastructure, inadequate surveillance, and weak health systems such as African countries [14]. Due to this limited surveillance of the infectious disease, it is difficult to prevent and contain the disease. In such situations, the rapid growth of point-of-care (POC) diagnostics would greatly help to stop the disease spread. We could achieve great solutions by combining blockchain and AI with POC diagnostics which can develop self-testing facilities for patients in isolation.

Blockchain and AI are already showing significant changes in the field of healthcare and providing the best solutions. They are being used in electronic health records, pharmaceutical supply chain management, drug anti-counterfeiting, education, biomedical research, health data analysis, and remote patient health monitoring. In [14], authors recommended using low-cost blockchain with AI-connected mobile health self-test facility. They provide patients with mobile or tablet application that request a patient's identification number before displaying pre-testing instructions. Accordingly, the patient can load the test results and access a local mobile health or ehealth system. The blockchain and AI systems will send a notification to the surveillance authorities and provide the data when there are positive test results. It also ensures that individuals who test positive are sent to a quaranting sire for isolated treatment. AI is very helpful in data collection, analysis, and security of medical data from blockchain databases. Data collecting from blockchain is verified and validated, which results in an outcome of high confidence and speed; we can also acquire deep insights using these immutable data sets of blockchain. This technology can be adapted by many other infectious diseases.

7.6 Conclusion

Blockchain and AI are conquering in their respective fields. The integration of blockchain and AI would bring us many benefits like using decentralized data storage in AI algorithms for better predictions. We could say that improvement in the availability, sharing, training of data in medical care is possible with the help of AI and blockchain. Until now, blockchain does not have the capacity to check on the quality of smart contracts, but AI would help in assuring the quality of smart contracts. We can solve the problem of latency by storing data in off-chain storage, which can be cost-effective. With AI, we can derive more sets of useful rules and policies in smart contracts. In this paper, we have also discussed advances in HDSS, AI, and their uses in the healthcare industry. We could provide the best personalization treatment plans and self-testing facilities using blockchain and AI by integrating them with POC.

References

[1] Russell, Stuart, and Peter Norvig. 2002. *Artificial Intelligence: A Modern Approach*. Prentice Hall.

[2] Makridakis, Spyros, Antonis Polemitis, George Giaglis, and Soula Louca. 2018. "Blockchain: The Next Breakthrough in the Rapid Progress of AI." *Artificial Intelligence-Emerging Trends and Applications*: 197–219.

[3] Salah, Khaled, M. Habib Ur Rehman, Nishara Nizamuddin, and Ala Al-Fuqaha. 2019. "Blockchain for AI: Review and Open Research Challenges." *IEEE Access* 7: 10127–10149.

[4] Taylor, Paul J., Tooska Dargahi, Ali Dehghantanha, Reza M. Parizi, and Kim-Kwang Raymond Choo. 2020. "A Systematic Literature Review of Blockchain Cyber Security." *Digital Communications and Networks* 6 (2): 147–156.

[5] Krittanawong, Chayakrit, Albert J. Rogers, Mehmet Aydar, Edward Choi, Kipp W. Johnson, Zhen Wang, and Sanjiv M. Narayan. 2020. "Integrating Blockchain Technology with Artificial Intelligence for Cardiovascular Medicine." *Nature Reviews Cardiology* 17 (1): 1–3.

[6] de Denus, Simon, Eileen O'Meara, Akshay S. Desai, Brian Claggett, Eldrin F. Lewis, Grégoire Leclair, Martin Jutras et al. 2017. "Spironolactone Metabolites in TOPCAT—New Insights into Regional Variation." *The New England Journal of Medicine* 376 (17): 1690.

[7] Siyal, Asad Ali, Aisha Zahid Junejo, Muhammad Zawish, Kainat Ahmed, Aiman Khalil, and Georgia Soursou. 2019. "Applications of Blockchain Technology in medicine and Healthcare: Challenges and Future Perspectives." *Cryptography* 3 (1): 3.

[8] Giordanengo, Alain. 2019. "Possible Usages of Smart Contracts (Blockchain) in Healthcare and Why No One Is Using Them." doi:10.3233/SHTI190292.

[9] Almasoud, Ahmed S., Maged M. Eljazzar, and Farookh Hussain. 2018. *"Toward a Self-Learned Smart Contracts."* In *2018 IEEE 15th International Conference on e-Business Engineering (ICEBE)*, pp. 269–273. IEEE.

[10] Zheng, Zibin, and Hong-Ning Dai. 2019. "Blockchain Intelligence: When Blockchain Meets Artificial Intelligence." arXiv preprint arXiv:1912.06485.

[11] Nguyen, Huu. 2018. Online: https://www.squirepattonboggs.com/en/insights/publications/2018/04/use-of-artificial-intelligence-for-smart-contracts-and-blockchains.

[12] Mamoshina, Polina, Lucy Ojomoko, Yury Yanovich, Alex Ostrovski, Alex Botezatu, Pavel Prikhodko, Eugene Izumchenko et al. 2018. "Converging Blockchain and Next-Generation Artificial Intelligence Technologies to Decentralize and Accelerate Biomedical Research and Healthcare." *Oncotarget* 9 (5): 5665.

[13] Ilinca, Dragos. 2020. "Applying Blockchain and Artificial Intelligence to Digital Health." In *Digital Health Entrepreneurship*, pp. 83–101. Cham: Springer.

[14] Mashamba-Thompson, Tivani P., and Ellen Debra Crayton. 2020. "Blockchain and Artificial Intelligence Technology for Novel Coronavirus Disease-19 Self-Testing." 198.

[15] Dai, Hong-Ning, Raymond Chi-Wing Wong, Hao Wang, Zibin Zheng, and Athanasios V. Vasilakos. 2019. "Big Data Analytics for Large-Scale Wireless Networks: Challenges and Opportunities." *ACM Computing Surveys (CSUR)* 52 (5): 1–36.

[16] Mavridou, Anastasia, and Aron Laszka. 2018. "Tool Demonstration: F Solid M for Designing Secure Ethereum Smart Contracts." In *International Conference on Principles of Security and Trust*, pp. 270–277. Cham: Springer.

[17] Chen, Ting, Xiaoqi Li, Xiapu Luo, and Xiaosong Zhang. 2017. *"Under-Optimized Smart Contracts Devour Your Money."* In *2017 IEEE 24th International Conference on Software Analysis, Evolution and Reengineering (SANER)*, pp. 442–446. IEEE.

[18] Shamir, Adi. 1979. "How to Share a Secret." *Communications of the ACM* 22 (11): 612–613.

[19] Al-Najjar, Hazem, and Nadia Al-Rousan. 2014. "SSDLP: Sharing Secret Data Between Leader and Participant." *Chinese Journal of Engineering*.

[20] Denning, Dorothy, and Elizabeth Robling. 1982. *Cryptography and Data Security*. Vol. 112. Reading: Addison-Wesley.

[21] Desmedt, Yvo. 1992. "Threshold Cryptosystems." In *International Workshop on the Theory and Application of Cryptographic Techniques*, pp. 1–14. Berlin, Heidelberg: Springer.

[22] Wiggers, K. 2018. "Patient Sphere Uses AI and Blockchain to Personalize Treatment Plans." *Venture Beat*. https://venturebeat.com/2018/10/25/patientsphere-uses-ai-and-blockchain-to-personalize-treatmentplans/.

8

Fusion of IoT, Blockchain and Artificial Intelligence for Developing Smart Cities

M. Kiruthika

Jansons Institute of Technology, Coimbatore, Tamil Nadu, India

P. Priya Ponnuswamy

PSG Institute of Technology and Applied Research, Coimbatore, Tamil Nadu, India

CONTENTS

8.1 Introduction

According to the published by the United Nations, the world population will reach 9.8 billion by the end of 2050. It is deduced that almost 70 percent of that population will be an urban population with many cities accommodating over 10 million inhabitants. As the number grows, we will encounter challenges regarding making a provision for resources and energy for all inhabitants at the same time while avoiding environmental deterioration. Other critical challenges are administration and management to prevent sanitation issues, mitigate traffic congestion, and thwart crime. Many of these problems can be controlled by converging some advanced technologies such as artificial intelligence (AI), Internet of Things (IoT), and blockchain. Using technological advancements to facilitate inhabitants can make their day-to-day living more comfortable and secure. This has given rise to the concept of smart cities. A smart city is a city that makes use of technological advancements to enhance the quality of urban services (like energy and transportation), thereby reducing the consumption of resources, and preventing wastage and overall costs [1]. Smart cities thereby aim to provide a sustainable environment for individuals with improved quality of life at minimal cost of living. The most popular technology that has been used to build smart cities is IoT.

IoT is a trending communication technology where objects used daily such as home appliances, vehicles, and cameras are equipped with microcontrollers, transceivers, and necessary protocols to enable them to communicate with one another [2]. IoT technology can give a great impact on every human's day-to-day life. The Internet is used by IoT to integrate heterogeneous devices with one another. Many applications based on IoT can be used for building smart cities. A few of them are smart homes, smart energy management, smart transportation, smart healthcare, smart agriculture and farming, smart surveillance, and smart environmental control. Figure 8.1 depicts these IoT applications for building smart city. Though numerous applications can be provided by IoT, there are a few issues that must be considered when building smart cities. IoT disseminates data among connected systems, so some strong technology is needed to provide security and privacy to the data. When billions of devices start to communicate with each other, certain large amounts of data will be transferred. Hence it is important to adapt a technology that can store, retrieve, and analyze a huge volume of data [3].

IoT can be integrated with blockchain technology to provide secure data access and storage. Blockchain stores data in a distributed pattern and it can be accessed by authorized persons without modification. That is, the data stored in blockchain is immutable. A smart city interconnects different domains. This makes a connected system where a

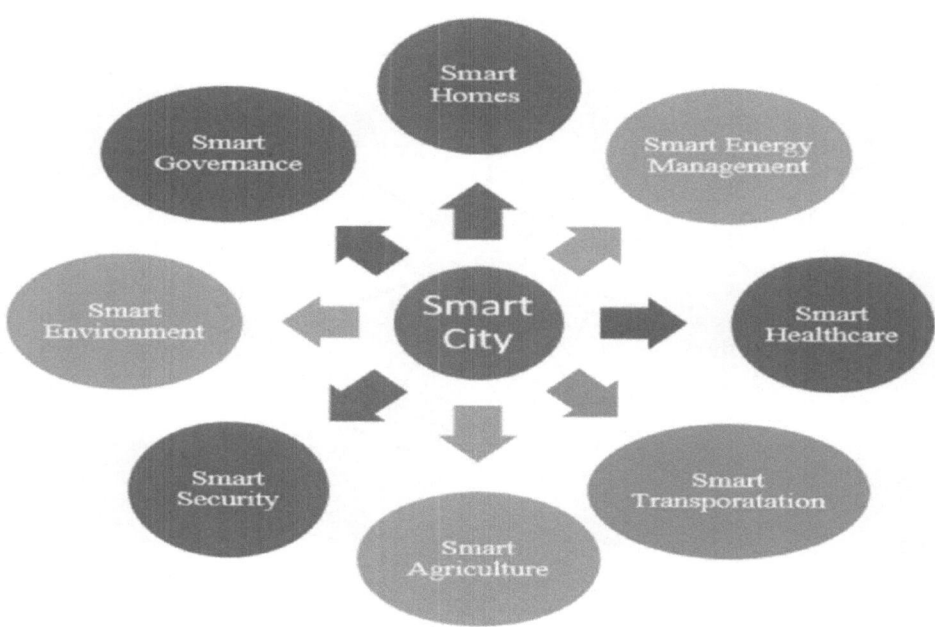

FIGURE 8.1
IoT Applications in Smart City.

large volume of data is involved. The technologies in IoT—Radio-Frequency Identification (RFID), Sensors, etc.—will produce digital measurements that can be used for investigation and research purposes. Similarly few data must be available to everyone and at the same time the data should be immutable. So data storage and secured access can be implemented using blockchain technology [4]. Numerous applications can be built by integrating IoT with blockchain technology. One such application is an intelligent water management system (IWMS) [5] where IoT is unified with blockchain for providing a secure and reliable solution. There are many potential use cases defined by IWMS like managing water during floods, checking water quality, collecting water tax using secure e-payment, and more.

In [6] a solution is proposed by combining IoT and blockchain technology in order to provide trust. Blockchain ensures robustness and reliability of data. In a smart city environment, a zone known as IoT-zone is defined where the user's interaction is stored in a blockchain as a transaction. Series of these transactions is called the IoT-trail. The user's interaction is allowed only if the user has a unique digital crypto-token. These tokens are pre-generated using a prediction model known as Variable Order Markov Model. Data generated using IoT technology in smart cities are in very large volume and many enterprises store these data in the cloud. The drawback of cloud technology is that it relies more on central servers and is prone to a single point of failure. The data used in smart cities may pertain to social media, smart homes, security surveillance, and more as depicted in Figure 8.2.

Storage of data becomes a significant trait in developing smart cities. An advanced storage technology is needed to store and access data. A decentralized blockchain-based dynamic data storage protocol known as Provable Data Possession (PDP) is proposed in [7] that supports data updation and user verification. The fairness of all the participating

FIGURE 8.2
Types of Data Used in Smart City.

entities is verified using smart contracts. Data is replicated at multiple blocks in order to ensure availability and robustness. The main advantage of blockchain is that the data stored in blockchain is tamper-proof. All the security issues pertaining to IoT can be solved using blockchain. In [8], the authors have developed two systems: an IoT system with and without blockchain. The protocol used for communication among IoT devices is message queuing telemetry transport (MQTT). The blockchain platform used is Ethereum along with a smart contract. Both the IoT systems are simulated and the results prove that the IoT system with blockchain technology is more secure than the IoT system without blockchain. Blockchain is sure to strengthen IoT by providing secured access to data and devices. Several companies such as IBM are taking initiatives to integrate blockchain into their production and supply chains.

IoT devices usually generate huge volumes of data. These voluminous data can be used to perform analytics using several machine learning and deep learning techniques that have given way to AI. When the working of AI is based on a centralized architecture, then there will be a risk of data modification. When the data is manipulated by hackers, then the analysis/decision outcomes will yield highly erroneous results. Also, there is a risk of data provenance and authenticity of sources generating data. In order to overcome this issue, AI can be integrated with blockchain technology. Data handled by AI will be distributed and blockchain will ensure secure storage and access to the data. A huge volume of data can be stored in a distributed way using blockchain and so the decision outcome can be trusted by all the parties across the network [9].

By integrating IoT, AI, and blockchain technologies, decentralized smart city applications can be developed. All involved parties can access an identical data source that is secure enough. There are numerous benefits gained by integrating blockchain and AI. A few of the benefits are enhanced data security, improved decision making, and disturbed intelligence [10]. Integration of AI and IoT architectures with blockchain technology can enable convenience and comfort for the user. Assume a refrigerator is equipped with IoT sensors to monitor all the items' availability. Blockchain assures data integrity from devices to trusted decentralized systems. The sensors in IoT sense the availability of an item in the refrigerator and inform a chat bot with AI when the item becomes unavailable. Even an order can be placed with the store supplier using blockchain technology, ensuring a secure transaction. So, all the emerging technologies like IoT, blockchain, and AI can be clubbed together to efficiently build a smart city that can ensure comfortability, security, efficiency, and reliability.

8.2 Consent Management Based on Blockchain and AI in IoT Ecosystem

Today, the internet is used widely in all areas of day-to-day life. IoT has evolved in various sectors notably in transportation, energy management, and agriculture. Multiple resources or devices can be connected in IoT. These devices can collect personal information about the user. Eventually, this may evolve as a threat to the user's privacy. In a smart city environment, this threat is even high because many citizen's data are tied to IoT. When information goes to a hacker, then the risks are countless.

When opening a website, we might notice the website asks for consent from users. The General Data Protection Regulation (GDPR) [11] article, passed in the year 2016, states that personal data processing is strictly prohibited. Personal data can be processed only when allowed by law or the user has consented to use their data. A focus on consumer consent is looked at after this, more seriously. It forces the companies to give significance to the personal data of the customer. So, consent management refers to a process where a website or any source collects data from the user only after obtaining their consent/permission. We might have noticed cookies that are displayed to receive consent from the user. Enforcement of GDPR allows users to regulate their personal data.

ADvoCATE architecture proposed in [12], has a user-centric approach that allows the user to give consent to access their personal information in the IoT system. ADvoCATE uses a blockchain infrastructure to ensure personal data integrity and security. The architecture works in line with the requirements defined by GDPR. An intelligence service is used for analyzing the user's policy conflicts and for providing recommendations to users to protect their personal data from accidental exposure. The ADvoCATE approach employs a series of sensors in devices used in the IoT environment to gather data about the user. As discussed earlier, IoT can be used in all activities related to a smart city such as agriculture, education, entertainment, healthcare, and smart homes. The data controllers used in ADvoCATE architecture communicate with the user and collect the user's consent. ADvoCATE architecture comprises of functional components such as the consent management component, consent notary component, and intelligence component. Figure 8.3 shows the components of ADvoCATE architecture.

8.2.1 Consent Management Component (CMC)

This component manages the user's personal data disposal policies and data consent. The policies defined by the user may be generic or domain specific. The data controllers may have collected this consent-related data from the user while using a particular IoT device. The same consent policy may not be applicable to other devices in the IoT system. Another

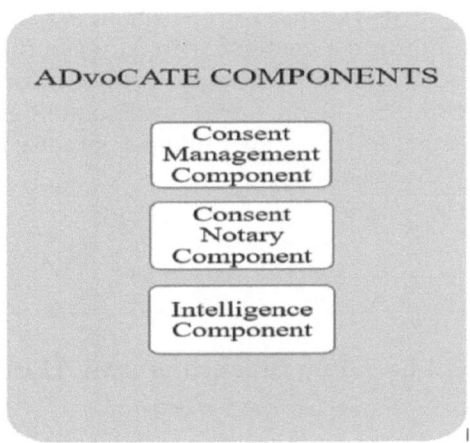

FIGURE 8.3
Components in ADvoCATE Architecture.

main issue deals with the security of information exchanged between the users and data controllers. Also, the data controllers have to make sure that the privacy policies defined are displayed to the users in a clear way using data privacy ontologies. ADvoCATE architecture follows the eXtensible Access Control Markup Language (XACML) for defining policies based on semantics. ADvoCATE allows not only a single ontology, but it supports similar other ontologies also. These ontologies may even be designed based on a machine learning technique that ensures correctness and reliability.

8.2.2 Consent Notary Component

ADvoCATE architecture employs digital signatures and blockchain to ensure non-repudiation, correctness, integrity, reliability, and validity of data consent policy, both at the user's end and the data controller's end. CNC acts as an intermediary between the CMC and blockchain architecture. CNC checks if the consents are up-to-date and they are protected from unauthorized access. ADvoCATE architecture uses smart contracts to define rules for the consent agreement and the penalties that can be imposed upon misuse of the user's data. When blockchain is used, all the consents are made publicly visible, but the contents are immutable. ADvoCATE additionally obtains digital signatures in the consent policies from both parties to ensure non-repudiation. Also, the hashed version of the consent is deployed in the blockchain structure. The working of CNC is depicted in Figure 8.4 and the steps can be illustrated as follows:

1. CNC receives the latest version of the agreed consent from CMC.
2. User and the data controller are requested to sign the consent policy independently.
3. Hash values of both digital signatures are stored in blockchain using a smart contract.

The smart contract deployed in the blockchain is liable for all updates and withdrawals of consent. Smart contracts are executed and controlled automatically. It takes care of the events and actions given in the consent. Smart contracts verify the latest versioning and integrity of the consent. Finally the CNC component returns the latest version of consent

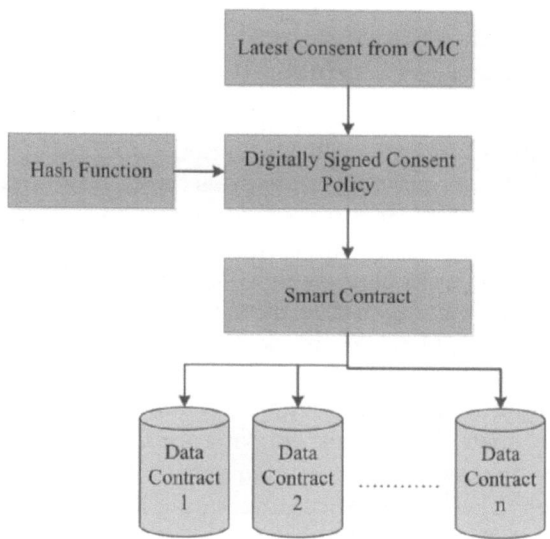

FIGURE 8.4
Working of CNC.

with the digital signatures and address of the smart contract on blockchain to the CMC component. Later if needed, the user and the data controller can verify the validity of the consent by verifying the digital signatures and by retrieving the latest version of consent from the blockchain through a smart contract. Blockchain plays a major role in securely saving the consents in a distributed way.

8.2.3 Intelligence Component

ADvoCATE architecture has implemented the intelligence component (INC) as a fusion of two mechanisms: Intelligent Policies Analysis Mechanism (IPAM) and intelligent recommendation mechanism (IRM). IPAM is used to identify if there are any contradictory policies related to the exposure of user's private data and it also makes sure that the collected data cannot be used for profiling the user. Many traps can be included in the consent policy that can ask the user permission for encrypting data while transferring and displaying fields in the submission forms that are not needed. The information that is least bothered by the users may turn out to be a big security breach. Hence utmost care must be taken to monitor the consent policies periodically. AI can be used to find underlying knowledge and unwanted activities in the consent policy. Fuzzy Cognitive Maps (FCM) are used to find out the relations among elements in the consent policy using learning algorithms and decisions that are taken based on knowledge learnt. Learning algorithms are implemented by training and updating FCMs based on artificial neural networks (ANNs). ANNs reduces human involvement in the learning process. IPAM uses FCM to monitor for violation of user's privacy policy periodically with a consent agreement.

IRM is used for providing real-time personalized information for protecting a user's privacy by using a machine learning method known as cognitive filtering (CF). CF works by collecting non-private information from the user. Some rules are framed by CF based on information and the user's consent policy. This way, personalized intelligent rules can be framed for individual users and IRM assures a strong security strategy.

8.3 Blockchain to Secure IoT against Attacks

With the increasing popularity of blockchain, there is no doubt that it has been adopted for commercial applications. Many IoT applications need blockchain to secure the users and their data. Devices in IoT are vulnerable to various attacks and there is always a threat to privacy of data [13]. Few intrinsic features of blockchain such as immutability, data integrity, distributed nature, transparency, and confidentiality, make it suitable for IoT applications. Integrating blockchain technology with IoT can bring a revolution in security aspects.

Huge amounts of data are generated by the sensors in IoT devices and this data can be used for an analysis purpose. Hence more focus has to be given to securing data, and the smart city architecture has to ensure that the data transmitted and stored is threat-free. Various types of data are handled in a smart city environment including e-governance information, electricity metering, vehicle tracking information, and healthcare data that are more susceptible to attacks. The data used in a smart city environment have to be transmitted and stored safely. The framework of integrating IoT and blockchain for a smart city environment can be portrayed as in Figure 8.5.

Multiple smart devices such as smart information desks, smart meters, smart traffic signals, and automatic ticket vendors are used in a smart city environment that are prone to external attacks [14]. When these devices are attacked by attackers, personal and financial data may be leaked, which can cause more damage to citizens and the city. Threats that occur in a smart city environment that uses IoT devices are threats on availability, integrity, confidentiality, authenticity, and accountability [15]. IoT devices usually have a limited battery life and are prone to attacks where the limited power is misused by attacker nodes, directing to availability threats. The popular attack on availability is distributed denial of service (DDoS) attack. Also a threat to availability occurs when data are stored in a centralized configuration. The data that are collected by IoT devices are used for various analysis and improvement. An integrity threat leads to making modifications and corruptions to the data. Attacks that affect the integrity of data are a modification attack, a byzantine attack on routing information, and an injection attack. Many sensitive data are handled in the smart city environment, and a confidentiality threat includes disclosing these sensitive data to unauthorized third-parties. Attacks on confidentiality include identity spoofing

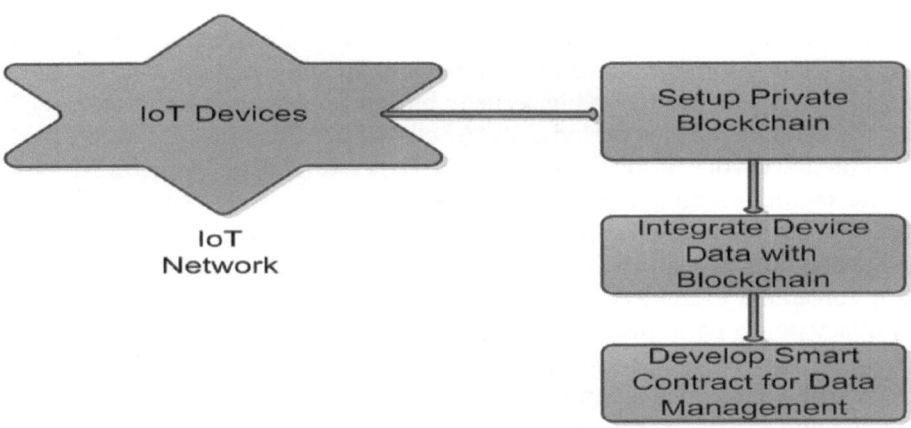

FIGURE 8.5
IoT and Blockchain in Smart City.

and traffic analysis. Multiple IoT devices/nodes work together in a smart city and an attacker can enforce an authenticity threat by gaining unauthorized access to these devices or their data. Also, the nodes can send or receive information from another device, but later deny the transmission or reception of data leading to accountability threat.

8.3.1 Role of Blockchain

Blockchain is blooming as global technology in protecting and optimizing users of IoT. Blockchain relies on four techniques to provide security [16]. They are consensus, distributed ledger, cryptography, and smart contract. These techniques easily handle the threats against IoT devices. As we have seen before, consensus provides proof of work and prevents the webpages or third parties from stealing a user's private data by providing certain protocols for consent management. Distributed ledgers ensure availability of data by storing the complete details about transactions of the users and devices in a distributed manner. Hash functions and cryptography ensure authentication and integrity of data by restricting unauthorized access. These functions are applied on the data that are stored in distributed ledgers. Finally, smart contracts are used to validate and verify the users of the network. Another important aspect of blockchain is that it does not allow a particular transaction to be linked to a particular user. Figure 8.6 depicts the role of blockchain in IoT. Blockchain can be integrated with IoT to enforce security policies and to maintain publicly viewable IoT records. When blockchain is used for security, IoT does not want to rely on third party software to provide security, which can turn out to be economical and beneficial [17].

8.3.1.1 Blockchains against Availability Threat

Blockchain follows a distributed architecture wherein the nodes and data are stored in a distributed nature. Since centralized architecture is not there, blockchain naturally acts as an opponent of a DDoS attack. In other words, a DDoS attack cannot be launched on blockchain. ProvChain architecture [18] is proposed to secure the IoT devices against availability threat using a blockchain concept. Here, provenance data is embedded into blockchain transactions for data that are stored in the cloud. The provenance data is first collected, stored, and validated using a public and private key pair. ProvChain architecture is implemented in three layers as a data storage layer, blockchain layer, and provenance layer, and these layers perform data collection, data storage, and data validation

FIGURE 8.6
Role of Blockchain in IoT.

respectively. A blockchain-based auditing architecture [19] uses hierarchical cryptographic techniques with blockchain to ensure availability and accountability of data. In this method, consent is received from each data owner to ensure privacy of data using smart contract methods. This architecture facilitates a regulatory framework to enforce laws based on consortium blockchain.

8.3.1.2 Blockchain against Integrity Threat

Modification attack is a famous attack that alters the content of the stored data in the blockchain. The modification can be done in many ways such as the following: manipulating data content in the block, adding new data in a block, or deleting data in a block. With the decentralized architecture in blockchain, it is easy to prevent the modification attack. In [20], integrity of data is maintained by using a reference integrity metric (RIM) for data in blockchain. The value of RIM is verified each time that data is downloaded. A central hub is used to store metadata information, and the original data are stored in a distributed manner. The data about the owner of data, policy of data sharing, and address of the data are maintained in one blockchain and another blockchain holds the RIM information. This way integrity of data can be maintained using blockchain.

8.3.1.3 Blockchain against Confidentiality Threat

Cryptography is the common approach to ensure confidentiality of data. Blockchains use the public/private key pair to protect sensitive data from intruders. The data is not directly stored in blockchain, instead the hashes of data are only stored to protect data from the intruders. The major problem occurs in the distribution of private key/secret key when using asymmetric cryptography/symmetric key cryptography, respectively. Many algorithms such as the Diffie Hellman algorithm are used for key distribution in a secure way. A decentralized outsourcing computation is proposed in [21] and here the data owners may check if the servers are reliable by requesting the servers to perform homomorphic computations on encrypted data. No plaintext data are given from the data owner. The same scheme is applied to IoT that results in a confidential blockchain-enabled IoT known as BeeKeeper 2.0. Another method is suggested in [22] that uses symmetric key encryption (only one secret key is used) and claims that symmetric key encryption is much faster than public key encryption. This method also devises a way of securely transmitting the secret key between IoT devices and the key manager.

8.3.1.4 Blockchain against Authenticity Threat

Authenticity plays a vital role in ensuring privacy and security. Most of the blockchains are controlled by users. For example, when blockchain is used in the healthcare industry, patients play a major role in deciding who will access their medical records and who will not access their medical records. So, authenticating users is necessary for deploying blockchains. In [23], the authors have proposed a technique called digital fingerprints to authenticate IoT devices. A unique unclonable function is generated for each device. The manufacturers of the device generate and store a cryptographic hash of the device ID in a universally accessible public blockchain. When the particular IoT device logs into the blockchain, the end user can verify the authenticity of the device by calculating the hash of the device ID and verifying that hash is present in the global register. In order to prevent cloning, device authentication is done periodically.

8.3.1.5 Blockchain against Accountability Threat

Nonrepudiation is a popular threat where the user later denies that they have done a particular event. That is the users will not take responsibility for a particular occurrence. Smart contracts in blockchain technology will definitely help in the process of injecting accountability. Smart contracts are automatically executed documents that can record an event or action performed by the users. These contracts can speak up legally as it is executed after the consent of the users at both ends and it generates an agreement between users. In this case, it becomes impossible for a user to refuse that a particular event is not performed as the contract is generated automatically.

8.4 Artificial Intelligence for IoT Enabled Smart Cities

There are statistics that state over 600 urban areas use computerized innovation. This consists of various emerging technologies like blockchain, AI, and IoT that could be a part that drives the world market to increase economic progress. In the next ten years AI, IoT, machine learning, and blockchain will create new innovations in all fields, not only in smart cities. A number of projects are being developed by blockchain and are utilized by Google, Uber, Amazon, IBM, and other corporate giants. Distributed databases stored in blocks in different locations are combined and form a chain like blocks. In this, AI is used in making appropriate and exact decisions about the stored data.

When IoT devices are used for various applications in a smart city, there is no doubt that it generates a huge volume of data. So, a powerful technology such as AI is needed for processing and analyzing these smart data. This analysis may turn out to be very important for making improvisations in most cases like healthcare, e-governance, electrical metering, and vehicle tracking. In all these places, a huge amount of data is involved, and maintaining data turns out to be a complex task. AI algorithms are increasingly useful in making decisions about whether the user should be blocked or investigated.

For instance, in smart transportation, blockchain can store details about the vehicle, traffic congestion places, parking areas, traffic lights, location of the vehicle, and many more. AI can be applied to derive analysis of transportation data and can be helpful when drivers need to make decisions on finding the shortest path to reach a destination, or the driver needs to make a decision on which parking lot the vehicle should be parked in. By gathering the vehicle details from the distributed ledger, AI can be efficiently applied to manage traffic and to make clear decisions. In order to enhance and modernize traffic management, AI plays a vital role. The ability to maintain and monitor huge data volumes stored in ledger and make intelligent decisions can have a major impact on congested cities.

New policies and decisions can be imparted in the smart city environment based on data analysis. However, data should be analyzed correctly; failing can give a negative input to the result. Many organizations have moved on to cloud platforms for satisfying storage needs. Storing data in the cloud has few drawbacks. Integrity of data stored in the cloud cannot be assured. Similarly, there are security issues and threats to the data stored in a cloud platform. These issues can lead to serious effects in a smart city environment. In order to ensure the integrity and security of data, blockchain technology can be integrated into a cloud platform, thereby providing a distributed and reliable data access. No third parties have to be relied upon for providing integrity and security when blockchain is used.

FIGURE 8.7
Data Audit Using DAB.

Blockchain-based data integrity auditing is introduced in [24]. The proposed method proves to be effective and reliable. Data auditing blockchain (DAB) is used to collect proofs about auditing. Data owners/data users store the data and perform auditing of the data in a cloud platform. An AI concept is used to analyze the data to derive fruitful results. The data owner first sends a request to the cloud provider to perform an audit. After approval from the cloud provider, the data owner audits the data. After auditing, the cloud provider generates an audit proof that is included in the block in DAB. Now the blocks in DAB will have the audit records that are tamper proof and secure. This procedure of storing data and audit records is shown in Figure 8.7.

Algorithms are used for representative selection, key generation, tag generation, challenge/ response, consensus management, and data verification. Blockchain nodes hold data owners/ users and cloud providers. These nodes use the representative selection algorithm to select a master node and few slave nodes. The data owner generates a public key-private key pair using the key generation algorithm. Homomorphic tags are generated by data owners for all the data stored in the cloud, and these tags are generated using atag generation algorithm. Auditing is performed by data owners/users by sending a challenge to the cloud provider, and these providers generate a response for the challenge that generates an audit proof. A consensus algorithm is used by the slave nodes to publish new data blocks. The data owners use the audit proofs stored in DAB for cross verifying it with the audit results using the verification algorithm.

8.5 Optimizing Risk Minimization in Smart Cities Using Blockchain

Citizen participation in making public decisions has the power of improving the scope of smart and digital cities, especially with new technology called blockchain. Blockchain technology has the power to reorganize modern complicated systems. Blockchain technology enables smart cities to operate more efficiently by increasing productivity and economic growth. Blockchain has an impact in all fields especially transport systems, smart energy metering and power plants, healthcare, education, civil registration, E-Governance, agriculture, and defense as shown in Figure 8.8.

Blockchain reduces human and social resources, digital identity for billing and track data for energy prosumers, and track vehicles in logistics. Third-party involvement is mainly reduced in blockchain technology. More security is provided by cryptographic algorithms. Mainly, investment risk in all the public sectors gets reduced by implementing blockchain [25,26]. Types of blockchain risks are:

- Type 1: Blockchain allows all users to access the network without security check—permissionless

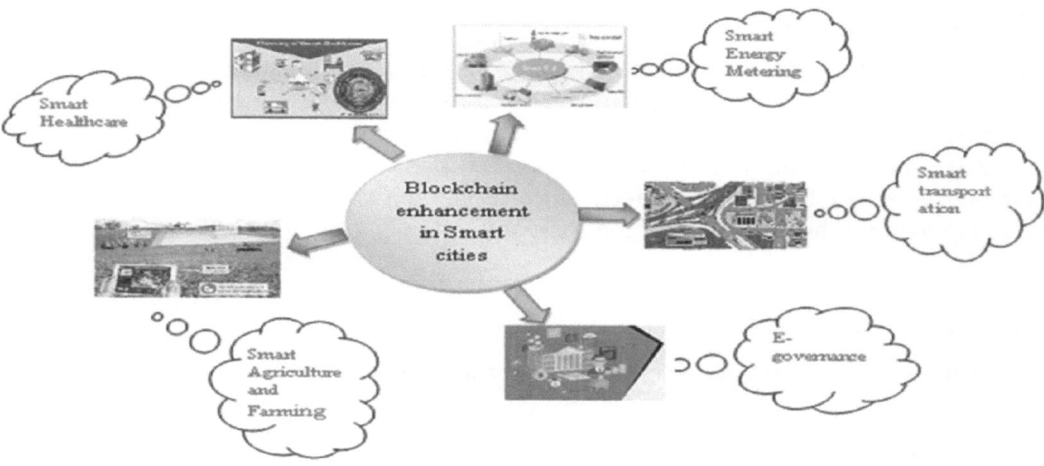

FIGURE 8.8
Risk Minimization Needed Areas.

- Type 2: Administrator validates the details of the user to participate in the network—permission

8.5.1 Different Risks in Blockchain

8.5.1.1 Business Continuity Risk

Blockchain technologies are generally flexible due to the distributed nature of the technology. However, the business procedure prepared for blockchains may be at risk in using technology and operational failures as well as attacks and hackers. Organizations need to have a strong business continuity plan and a governance framework architecture to overcome such risks. Moreover, blockchain solutions minimize the time duration and investment cost of many businesses.

8.5.1.2 Information Security Risk

While blockchain technology is focused on transaction security, it does not provide security in all the areas of a transaction like individual account holder accessing and wallet security. The non-centralized database and the secured algorithm sealed ledger can be hacked by any attacker. Additionally, there are more cyber security risk protocols and algorithms implemented for the blockchain network to overcome.

8.5.1.3 Contract Risk

Service-level agreements (SLAs) are done between the customers and the administrator. This SLA is used to monitor the raised compliance from the customer end.

8.5.1.4 Supplier Risks

An organization or industry may experience third-party risks since most of the technology implemented is from external vendors.

8.5.1.5 Reputational Risk

Blockchain technology is implemented with the help of basic infrastructure. Time taken to implement the new legacy infrastructure takes more time. While using the available structure and same methods, the new technology may result in implementing issues and client experience. Reputational risk comes in when companies fail to integrate blockchain to their legacy system. If not done correctly, it can result in poor customer experience — and can easily hamper the reputation of the company.

8.5.1.6 Consensus Protocol Risk

In a distributed ledger, each peer acts as both a host and a server. The information is exchanged between peers' needs to reach a consensus node. Sometimes some block nodes will not be working and there will also be some malicious attacker, which will destroy the process and working of consensus. Therefore, an excellent consensus protocol is needed to tolerate the occurrence of these phenomena and reduce the harness. Ripple, delegate proof of stake, proof of stake, and proof of work are some of the consensus protocols used in blockchain.

8.5.1.7 Key Management and Data Confidentiality Risk

The blockchain ledger should be protected and no corruption of past transactions is possible. It's still vulnerable to private keys that associate with public keys since all participants can view the metadata of the ledger. So the ledger is maintained with a strong encryption algorithm to protect the data or ledger with more lengthy private keys and public keys.

The main objectives of risk minimization using blockchain in smart cities are reducing involvement with a third party, improving security in the transaction process, and reducing the operational time and investment cost. Considering all the above points with the risk mentioned prior, some of the use cases are discussed next [27].

8.5.2 Risk Minimization in Smart Energy Usage in Blockchain

New technology, a new scheme, and a new policy are to be planned for the increasing population. New developments, new ideas, and new technology are emerging in different fields to be incorporate in cities. With the growth of innovation and advancement in the digital era, different challenges need to be faced. Information technology has the power to develop a relationship with the transformation of cities' systems into a green country.

One of the modern technologies used by smart cities is blockchain technology. Blockchain technology provides a solution to control and manage decentralized power systems and generate electricity through microgrids. The generation of electricity through an array of networked sensors and digital meters are placed inside the electricity generation plant to monitor the system data. The decentralization of generation of electricity through blockchain provides a solution in the energy-trading platform by improving the distribution of energy process, maintaining the losses, facilitating renewable energy sources, and providing security in power supply.

8.5.2.1 Impact on Energy Sector Using Blockchain

Blockchain technologies may be useful in different scenarios associated with the operations and commercial enterprise methods of energy organization. Peer-to-peer (P2P) energy trading is the buying and selling of excess energy produced in solar energy. The customer can buy and sell the excess kilowatts or megawatts of power generated in their resident's solar systems with a neighbor's or industry via Ethereum blockchain technology. This energy-trading process is done in three steps as shown in Figure 8.9.

- The first step is to generate electricity based on the customer demand or predict the demand of the electricity, and information is passed for initiating the transactions.
- The second step is when the generated excess electricity is sent to the microgrid operator. From the microgrid power plant operator, the customers will buy energy. Based on the real-time monitoring data of the microgrid operator, excess electricity is sent to the main power plant. Microgrid supports energy production and consumption and the same will be used for distribution and transmission. Microgrid provides network adaptability; supports basic services like frequency and voltage; and provides energy services in terms of failure in power generators, transmission lines, and transformers.
- The third step is checking the constraints of the energy balance based on the orders sent to the microgrid and fixing the selling and buying price. The energy transaction process is done by blockchain technology to improve the revenue of the distributed network.

Blockchain technology faces challenges in balancing the supply and demand of electricity, synchronizing the transmission and distribution of excess power, fault tolerance in microgrids, and improving the usage of distributed resources. Maintaining the microgrid, distribution, and transmission need to be connected to the blockchain. The new data set produced from the blockchain needs to be carefully analyzed for further processing of real time grid management.

8.5.2.2 Research Projects Using Blockchain in Energy Sectors

The research project named filament develops a blockchain technology such as smart energy metering and monitoring of real time data. This project allows the electronic devices to be connected using blockchain technology.

- **Slock.it** concentrates on developing an Ethereum-based technology and energy web blockchain.
- **Dajie** proposes a method to save excess energy in coins and stores them in a digital case to use by peers for energy trading in microgrids.
- **ElectricCChain** industry gathers more than a million solar power plant data and is stored in a single blockchain. All the solar power plants data is connected with the solarcoin blockchain for monitoring and rewarding the solar producer.

FIGURE 8.9
Energy Trading Process between Customers.

- **Fortum** recommends a solution that facilitates consumers to control home appliances using IoT. This company mainly focuses on optimizing energy requirements, reducing the electricity bill, predicting energy demand, and current electricity price.

- **GreenRunning** a startup company developing an artificial intelligence-based solution for predicting the energy demand and market price of an electricity producer and consumer of electricity generation. This AI technology is very useful in energy trading between peers.

- **Tavrida** electric is a leading electric equipment supplier using blockchain technology for energy transactions. The energy transaction details are stored in blockchain and it is transparent to all other energy companies.

- **Wanxiang** automotive component manufacturer in China invests billions of dollars in developing a smart city project.

- **Oli** is focused on optimizing less, and a more complex energy system. More energy cells are present in an energy system and all energy cells need to work together to get the desired result. These energy cells are networked together by blockchain technology.

A large number of energy industries explored blockchain technology and gain transparency in smart energy meters, trading the transaction of generated power and improving the power supply. Blockchain offers a new solution for empowering consumers and renewable energy generators in the energy market [26].

8.5.3 Risk Minimization in Transportation Systems Using Blockchain

Blockchain technology is experiencing fast growth and a transformative impact on smart transportation systems. With the growing population in the world, a vision of modern and innovative technology is needed to provide a higher quality of infrastructure. The smart traffic system is incorporated with blockchain technology to manage the traffic congestion during peak working hours and busy places, and road safety reduces the number of accidents. A smart transportation system collects all the current traffic information using onboard units and different sensors in the vehicles. The collected traffic information is stored in a centralized server to analyze traffic issues [27]. As shown in Figure 8.10, the risks in road transport system are reduced by using blockchain for:

- Efficient monitoring of current traffic data in the busiest places at peak working hours near schools and industrious areas.
- Identifying accident locations and emergency areas.
- Monitoring whether vehicles are following the lanes during travel and if they parked their vehicles in proper places.
- Traffic signal automation during ambulance arrival.

Vehicular ad hoc networks are used to monitor traffic, reduce accidents, and alert busy roads (traffic congestion places). This vehicular ad hoc network system is integrated with blockchain technology to focus more on a smart transport system. The blockchain technology will assist and help to regularize the public transport system on time. Using vehicular ad hoc network and blockchain technology, details about the vehicles (i.e., current location of the vehicle, speed of the vehicle, and time the vehicle passed in the location) are stored

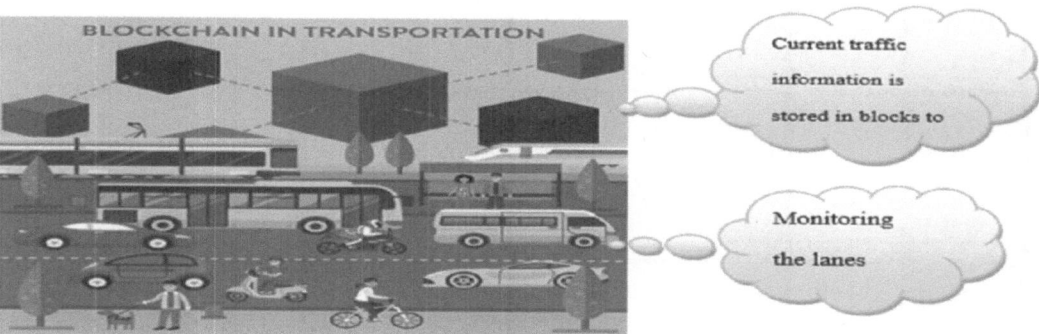

FIGURE 8.10
Simple Blockchain Usage in a Smart Transportation System.

in database ledger. This report is sent to a state transportation system and also a distance relative to other vehicles. The stored information is helpful for other vehicles to know about traffic. Blockchain usage in smart transportation is shown in Figure 8.11.

How blockchain is used in smart transportation is given in these four steps:

1. A Wi-Fi or IoT device is used to send and receive information about the vehicle
2. Checking the authenticity of information about the vehicle
3. True information is stored in blocks. Each block contains information about a greater number of vehicles. Information is stored in blocks shared with other vehicles. Similar to that each block is added with other blocks and forms a blockchain.
4. The chained information is shared so that all the other vehicles know the traffic congestion places, emergency needs, and possible routes.

The vehicle and infrastructure information stored in blockchain is very useful to show the current condition of the road in the form of pictures. In Figure 8.12 each line represents the path. This pictorial representation helps travelers to change their path to reach their destination based on traffic and is shown in Figure 8.12. Each color shows the traffic information (green color shows low traffic, yellow color shows medium traffic, and red color shows heavy traffic). This will benefit the public by allowing them to choose a path during their drive. Some of the other risks in the transport industry are overcome by blockchain technology.

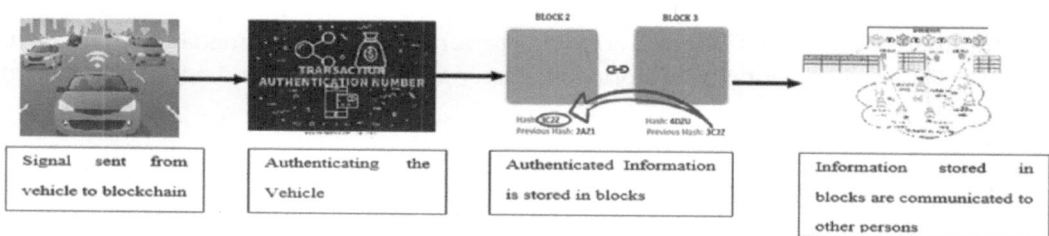

FIGURE 8.11
Four-Step Process of Blockchain in Smart Transportation.

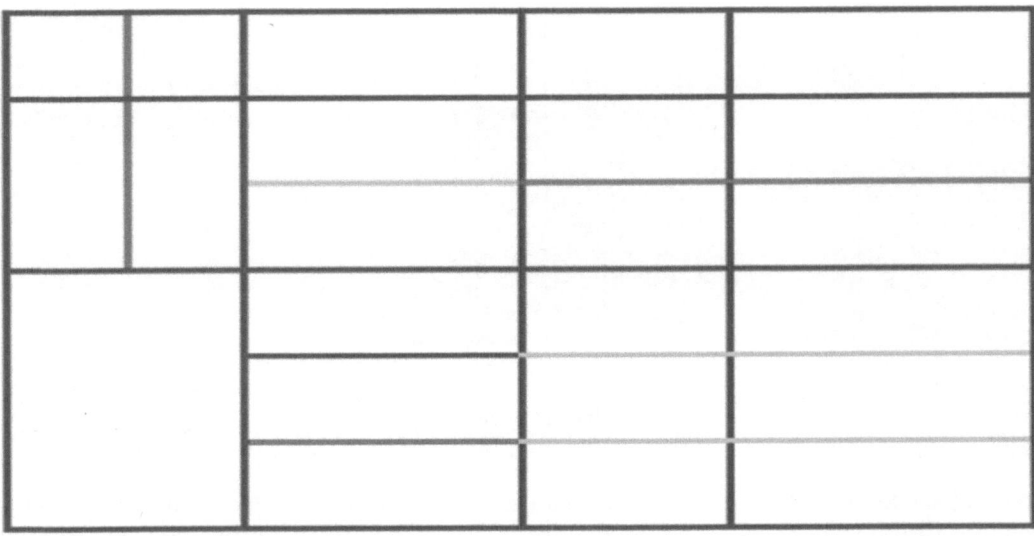

FIGURE 8.12
Path to Different Cities Are Shown with Lines.

8.5.3.1 Useful in Tracking

In a digital era, most of the public uses an ATM for money transaction and online product purchasing. Everyday some cores of money are transferred through vehicles in and around cities. In order to track the vehicle and to know the status of the vehicle, a blockchain tracking system is very efficient (risk involved in tracking gets reduced). With blockchain technology data validation and authentication is done in the entire network.

8.5.3.2 Advantages of Smart Contract

When the online purchased product is delivered to the customer, payment is received on the other end. Therefore, resources for billing and collecting money are very easy by validating information stored in the blockchain such as that the payment is completed without human intervention.

8.5.3.3 Reduce Paper-Based Ticketing System

Most of the transport industry is moving to the digital era; Metro in the year 2013 had stopped selling ticket books. A digital system is used for generating tickets, producing receipts for confirmation. The world is moving towards paperless maintenance. Blockchain-based distributed ledger helps in storing all passenger information and retrieving it whenever necessary.

8.5.3.4 Decentralized Public Transportation System

Traffic is very heavy in larger population cities. In these places, bike and car sharing services can lead to reduce the traffic and pollution. Blockchain could make it easy by giving tokens, tickets, and other prizes for shared public transportation.

8.5.4 Risk Minimization in E-Governance

Blockchain technology is expected to drive the next modern era of digital infrastructure. Online software helps the public to store and retrieve their details at anytime, anywhere, and on any device. Governance models such as the Neo Blockchain are an example of digital assets and democracy [27]. Most states like Andhra Pradesh, Gujarat, Maharashtra, and Karnataka started exploring the blockchain technology for E-governance.

The government has already recorded civil supplies and land registration records in order to protect documentation of subsidies and land ownership from scam artists and fraudsters. The state believes the blockchain can provide adequate protection against cyber-attacks. 'All our sensitive data is in digital format. With blockchain, even if someone hacks, our data will be safe,' said Chowdary IT adviser of CM in AP.

8.5.4.1 Online Registration of Birth and Death Certificate

The number of online registrations has increased daily over the last two decades. Birth registration has gone up from 58% percent to 88.8 percent% between 2001 and 2014; similarly, incidents of registered deaths have gone up from 54% to 74.3% percent in the same period [28,29]. While these are positive growths, many risks still exist with regard to accuracy, checking uniqueness, complete information about the person, and timeliness.

Most of the villagers in India are not utilizing the online birth and death registration; instead they use supplementary documents. Only a few states have moved to the digital era to register the details. The important information is sometimes stored incorrectly. This incorrect information makes the government need to find the mortality rate and death rates. This will create an impact in planning future resources. These issues are overcome by blockchain-based technology.

Simple digital storage ensures the ease of birth and death registration for citizens. Minimal interactions with the government representatives are required to get approval for getting birth and death certificates. The availability of documents in a digital format helps citizens to carry them wherever they go. Current birth and death registrations are easily obtained with increases in transparency and reduction in the storage of redundant data. This technology is very helpful to find a double entry within the interstate. The birth and death registration details are maintained in a smart contract and the system uses a protected private distributed ledger to store details, and it is more effective in information exchange.

8.5.4.2 Real Estate

Nowadays human interaction is more in land and property registration so it is very time-consuming, inefficient, and more maintenance is needed. Hacking of data is also possible with the existing system. Blockchain would provide an efficient solution where all land and property registration data can be stored permanently. No third party would be required to get real estate data or change ownership of a piece of land. Because blockchain is immutable, a transaction is completely transparent with an unchangeable ledger of data.

8.5.5 Risk Minimization in Agriculture Using Blockchain

Blockchain technology has an impact in agriculture also. It allows one to trace and track a product produced in farms. Blockchain solutions lead to the elimination of third-party intermediates thereby leading to improved fair pricing and decreased transaction fees, thus eliminating issues of hoarding. The Table 8.1 shows the advantages of using a blockchain in agriculture. The blockchain application is deployed to work with a simple mobile phone interface system for data capture.

The following steps describe the user cases of a blockchain from farm industry to processing plant:

- A farmer takes a geo-tagged picture of their farm during harvest. The harvest is packed into gunny bags and takes photos of the packed gunny bags. The farmer may also share the advantages and move to the nearest buying agent to register.
- The agent authenticates the details of the farmer, a QR code is generated, and the same code is printed on their gunny bags.
- The buying agent then combines multiple registered farmers products and creates a new consignment. Each consignment box is secured, and the weight of the consignment is recorded and placed on top of the box before it is shipped. This weight is then inputted into the blockchain by the buying agent, and a consignment QR code is generated, which is linked with the farmer QR code.
- During transportation, with the help of the consignment QR code the vehicle and conditions of the consignment can be tracked.
- Any deviation in the journey or consignment will be dealt with by a smart contract on the blockchain. For example, if frozen vegetable, fruits, or milk products are being transported and the temperature maintained for the consignment is likely to be below certain degree Celsius till reaching the processing plant, a deviation shall be recorded by the smart contract at any point of time, and the same is intimated.
- After reaching the processing plant, the weight and QR code of the consignment is verified with the data stored in the blockchain.
- For each and every process the farmer gets an instant message about their harvest. At the primary/secondary processing plant, an employee will be able to read the QR code of each gunny bag to trace the produce to the geo-tagged farm.

TABLE 8.1

Advantages of Blockchain in Smart Cities

Areas	Risk Reduced-Blockchain
Smart energy meters	Energy generation data from prosumers, energy consumer and supplier data records, smart energy meters for billing, supply on demand, tracking of resources.
Transport and logistics, business and distribution	Transport records, good delivery and shipping data, logistics service identifiers, toll data maintenance, vehicle tracking, shipping container tracking.
Smart city	Smart service offerings, energy management data, water management data, pollution control data, digital data, enabling digital transactions, smart data maintenance, smart transactions.
Agriculture	Soil data, processing records related to agriculture data, shipping of agro-products, sales and marketing data of agro-seeds, yields, growth.

8.6 Securing Smart City Using Blockchain Technology

The world has experienced unexpected growth due to an increase in population and resources becoming scarce. Recent research shows that more people live in cities (54%) than rural and urban areas (46%) and this number will increase to 66% by 2050 [30]. Because of this growth, cities are focusing on recent technologies as well as aiming to reduce investment costs, use resources efficiently, and increase the quality of life in urban environments. The significant advancements in new technologies have made it easy to interconnect a range of devices and enable them to transmit data ubiquitously even from remote locations [30].

Chances of device spoofing, vulnerability, insusceptibility to tampering, false authentication, integrity, or less reliability in data sharing could happen while storing data in a centralized point. To address such security and privacy issues, a central server concept is eliminated and blockchain technology is introduced. Blockchain allows P2P messaging transfer in a faster way with the help of distributed ledger. The distributed ledger is very protective and does not allow any misinterpretation or wrong authentications in data.

8.6.1 Digital Signature

In blockchain-based technology, the distributed ledger database acts as a brain. To protect the database from active attacks, a digital signature-based crypto hash algorithm-1 (SHA-1, SHA-2, and SHA-256) are used. Public-key cryptography creates a secure digital identity of a user in blockchain. Secure digital identity is about who is who, and who owns what; these are the basis for P2P transactions.

Public-key cryptography allows checking one's unique identity with a set of cryptographic keys called a private key and a public key. The combination of both keys creates a digital signature. This digital signature proves the content with an individual's private key and is confirmed by the individual's public key. The digitally signed document offers benefit in:

Authentication	Before accessing the blockchain, each and every user should be authenticated. Once authenticated, a QR code or unique is generated as token
Non-repudiation	A corresponding person should not claim that they are not involved in the transaction.
Integrity	Verifying the originality
Confidentiality	Ledger data is hidden from unauthorized people

8.6.2 Homomorphic Encryption

Without changing the blockchain properties, one can use the homomorphic encryption techniques to store data over the blockchain. The homomorphic encryption technique allows complete access to encrypted data over the public blockchain for auditing and managing fund transactions [31,32].

8.7 Conclusion

A smart city is the current project under development that enhances the lifestyle of the citizens living in it. Major cities are tightly populated and hence a complex technological integration is needed to implement smart cities. Developing smart cities needs to consider crucial aspects such as infrastructure, energy management, transportation, healthcare, waste management, and more. Smart cities are implemented using the network of smart IoT devices. Smart cities must develop an architecture for storing data generated by IoT devices. It should also ensure that data should be handled in a secure manner with integrity in consideration. Hence blockchain can be used for securely storing smart data, and there is proof that blockchain can combine well with IoT and AI technologies to meet up challenges in building smart cities. This chapter specified the importance of consent management to secure users and has given the overview of how consent management can be performed. A clear idea has been devised to secure data generated by IoT devices using blockchain. This chapter specified the needs of artificial intelligence techniques to manage smart city data. Finally the risks undergone by smart cities are elaborated with measures of how to minimize risks.

References

[1] Gharaibeh, A., M. A. Salahuddin, S. J. Hussini, A. Khreishah, I. Khalil, M. Guizani, and A. Al-Fuqaha. 2017. "Smart Cities: A Survey on Data Management, Security, and Enabling Technologies." *IEEE Communications Surveys & Tutorials* 19(4): 2456–2501.

[2] Zanella, A., N. Bui, A. Castellani, L. Vangelista, and M. Zorzi. 2014. "Internet of Things for Smart Cities." *IEEE Internet of Things Journal* 1(1): 22–32.

[3] Arasteh, H., V. Hosseinnezhad, V. Loia, A. Tommasetti, O. Troisi, M. Shafie-khah, and P. Siano. 2016. *"IoT-Based Smart Cities: A Survey."* In *2016 IEEE 16th International Conference on Environment and Electrical Engineering (EEEIC)*, pp. 1–6. IEEE.

[4] Ibba, S., A. Pinna, M. Seu, and F. E. Pani. 2017. "CitySense: Blockchain-Oriented Smart Cities." In *Proceedings of the XP 2017 Scientific Workshops*, pp. 1–5.

[5] Dogo, E. M., A. F. Salami, N. I. Nwulu, and C. O. Aigbavboa. 2019. "Blockchain and Internet of Things-Based Technologies for Intelligent Water Management System." In *Artificial Intelligence in IoT*, pp. 129–150. Cham: Springer.

[6] Agrawal, R., P. Verma, R. Sonanis, U. Goel, A. De, S. A. Kondaveeti, and S. Shekhar. 2018. *"Continuous Security in IoT Using Blockchain."* In *2018 IEEE International Conference on Acoustics, Speech and Signal Processing (ICASSP)*, pp. 6423–6427. IEEE.

[7] Chen, R., Y. Li, Y. Yu, H. Li, X. Chen, and W. Susilo. 2020. "Blockchain-Based Dynamic Provable Data Possession for Smart Cities." *IEEE Internet of Things Journal* 7(5): 4143–4154.

[8] Fakhri, D., and K. Mutijarsa. 2018. *"Secure IoT Communication Using Blockchain Technology."* In *2018 International Symposium on Electronics and Smart Devices (ISESD)*, pp. 1–6. IEEE.

[9] Salah, K., M. H. U. Rehman, N. Nizamuddin, and A. Al-Fuqaha. 2019. "Blockchain for AI: Review and Open Research Challenges." *IEEE Access* 7: 10127–10149.

[10] Daniels, J., S. Sargolzaei, A. Sargolzaei, T. Ahram, P. A. Laplante, and B. Amaba. 2018. "The Internet of Things, Artificial Intelligence, Blockchain, and Professionalism." *IT Professional* 20(6): 15–19.

[11] General Data Protection Regulation (GDPR)—Official Legal Text. Available online: https://gdpr-info.eu.

[12] Rantos, K., G. Drosatos, K. Demertzis, C. Ilioudis, A. Papanikolaou, and A. Kritsas. 2018. *"ADvoCATE: A Consent Management Platform for Personal Data Processing in the IoT Using Blockchain Technology."* In *International Conference on Security for Information Technology and Communications*, pp. 300–313. Cham: Springer.

[13] Panarello, A., N. Tapas, G. Merlino, F. Longo, and A. Puliafito. 2018. "Blockchain and IoT Integration: A Systematic Survey." *Sensors* 18(8): 2575.

[14] Mistry, I., S. Tanwar, S. Tyagi, and N. Kumar. 2020. "Blockchain for 5G-Enabled IoT for Industrial Automation: A Systematic Review, Solutions, and Challenges." *Mechanical Systems and Signal Processing* 135: 106382.

[15] Biswas, K., and V. Muthukkumarasamy. 2016. *"Securing Smart Cities Using Blockchain Technology."* In *2016 IEEE 18th International Conference on High Performance Computing and Communications; IEEE 14th International Conference on Smart City; IEEE 2nd International Conference on Data Science and Systems (HPCC/SmartCity/DSS)*, pp. 1392–1393. IEEE.

[16] Singh, M., A. Singh, and S. Kim. 2018. *"Blockchain: A Game Changer for Securing IoT Data."* In *2018 IEEE 4th World Forum on Internet of Things (WF-IoT)*, pp. 51–55. IEEE.

[17] Ali, M. S., M. Vecchio, M. Pincheira, K. Dolui, F. Antonelli, and M. H. Rehmani. 2018. "Applications of Blockchains in the Internet of Things: A Comprehensive Survey." *IEEE Communications Surveys & Tutorials* 21(2): 1676–1717.

[18] Liang, X., S. Shetty, D. Tosh, C. Kamhoua, K. Kwiat, and L. Njilla. 2017. *"Provchain: A Blockchain-Based Data Provenance Architecture in Cloud Environment with Enhanced Privacy and Availability."* In *2017 17th IEEE/ACM International Symposium on Cluster, Cloud and Grid Computing (CCGRID)*, pp. 468–477. IEEE.

[19] Kaaniche, N., and M. Laurent. 2017. *"A Blockchain-Based Data Usage Auditing Architecture with Enhanced Privacy and Availability."* In *2017 IEEE 16th International Symposium on Network Computing and Applications (NCA)*, pp. 1–5. IEEE.

[20] Banerjee, M., J. Lee, and K. K. R. Choo. 2018. "A Blockchain Future for Internet of Things Security: A Position Paper." *Digital Communications and Networks* 4(3): 149–160.

[21] Zhou, L., L. Wang, T. Ai, and Y. Sun. 2018. "BeeKeeper 2.0: Confidential Blockchain-Enabled IoT System with Fully Homomorphic Computation." *Sensors* 18(11): 3785.

[22] Huang, J., L. Kong, G. Chen, M. Y. Wu, X. Liu, and P. Zeng. 2019. "Towards Secure Industrial IoT: Blockchain System with Credit-Based Consensus Mechanism." *IEEE Transactions on Industrial Informatics* 15(6): 3680–3689.

[23] Guin, U., P. Cui, and A. Skjellum. 2018. *"Ensuring Proof-of-Authenticity of IoT Edge Devices Using Blockchain Technology."* In *2018 IEEE International Conference on Internet of Things (iThings) and IEEE Green Computing and Communications (GreenCom) and IEEE Cyber, Physical and Social Computing (CPSCom) and IEEE Smart Data (SmartData)*, pp. 1042–1049. IEEE.

[24] Yu, H., Z. Yang, and R. O. Sinnott. 2018. "Decentralized Big Data Auditing for Smart City Environments Leveraging Blockchain Technology." *IEEE Access* 7: 6288–6296.

[25] Kodym, O., L. Kubáč, and L. Kavka. 2020. "Risks Associated with Logistics 4.0 and Their Minimization Using Blockchain." *Open Engineering* 10(1): 74–85.

[26] Office of the Registrar General & Census Commissioner, India. n.d. *Civil Registration System Division*. Ministry of Home Affairs, Government of India. Retrieved from http://www.censusindia.gov.in/vital_statistics/CRS/CRS_Division.html.

[27] Kumar, N. M., and P. K. Mallick. 2018. "Blockchain Technology for Security Issues and Challenges in IoT." *Procedia Computer Science* 132: 1815–1823.

[28] Oliveira, T. A., M. Oliver, and H. Ramalhinho. 2020. "Challenges for Connecting Citizens and Smart Cities: ICT, E-Governance and Blockchain." *Sustainability* 12(7): 2926.

[29] Andoni, M., V. Robu, D. Flynn, S. Abram, D. Geach, D. Jenkins, P. McCallum, and A. Peacock. 2019. "Blockchain Technology in the Energy Sector: A Systematic Review of Challenges and Opportunities." *Renewable and Sustainable Energy Reviews* 100: 143–174.

[30] Soto Villacampa, J. A. 2019. "Towards a Blockchain-Based Private Road Traffic Management Implementation."

[31] Zhang, R., R. Xue, and L. Liu. 2019. "Security and Privacy on Blockchain." *ACM Computing Surveys (CSUR)* 52(3): 1–34.

[32] Camboim, G. F., P. A. Zawislak, and N. A. Pufal. 2019. "Driving Elements to Make Cities Smarter: Evidences from European Projects." *Technological Forecasting and Social Change* 142: 154–167.

9

Impact of AI, BC and IoT for Smart Cities

Geethu Mary George and L. S. Jayashree

PSG College of Technology, Coimbatore, Tamil Nadu, India

CONTENTS

9.1 Introduction

Several modern-day applications of the Internet of Things (IoT) domain include a smart grid, e-health, and supply chain; the insurance industry is trusting millions of interconnected digital devices scattered across people, homes, and communities. At present, the internet holds not only typical computers and quantum computers associated, but also an important heterogeneity of utensils such as TVs, laptops, fridges, stoves, electrical appliances, cars, and smartphones. Even though the IoT devices seems to be stimulating and cracking various problems in real time, achieving security and privacy is puzzling due to its characteristics like low processing power, dispersed nature, and lack of standardization. At the same time, IoT also provides some valuable benefits to the field of information technology to save time and make money, increase productivity, improve the customer experience, and make better business decisions. However, the existing IoT devices are prone to provide stable services due to a singleton point of failure, malicious attacks, and threats. Formerly the onset of the information leakage (manipulation and modification) and denial of service remained the most serious security threats testified. So in considering all these concerns, like the need to prevent integrity and privacy aspects, the effective use of blockchain must be considered.

Blockchain technology truffled with IoT, artificial intelligence (AI), and big data are the core three pillars in computing knowledge for future a generation of financial business. To realize the impact of blockchain, it is compulsory to know what blockchain exactly provides. Blockchain is an exposed, scattered database that accounts for all transactions in a supportable and stable means that occurred in a network. A blockchain is a series of connected blocks where each block associates with the previous one in chronological order. This public distributed ledger influences cryptographic techniques that can timestamp information in a system since all the transactions to nodes then to block are signed initially using the user's private key (digital signature). It is anticipated [1] that the yearly returns of blockchain-based innovative enterprise tenders worldwide will range around $19.9 billion by 2025 with a yearly progression rate of 26.2% from about $2.5 billion in 2016.

Blockchain is a refined counterpart to IoT with enhanced interoperability, privacy, security, consistency, and scalability. The interoperability is the capability to interrelate with the physical world and the transferring of information between IoT platforms. Privacy is boosted by the blockchain technology in computing over encrypted data and is scattered across the entire network as one of the holy grails of computer science. The security concern is that data cannot be tampered with. Data protection and privacy is another important phase of data security. Consistency ensures the quality of IoT data is trustworthy. Scalability issues arise in terms of storage capacity deferment due to fractional block size and existing consensus algorithms like the proof of stake (PoS) mechanism, where each single node in the complex network begins validating each transaction node in the network published in the blockchain. We scrutinize a new paradigm of incorporating blockchain with IoT to eliminate a singleton point of failure attack, and improve data transparency and immutability will convert a modern commercial organization and governance [2]. Blockchain also facilitates several transactions and progressions as a soothing background for a system of sprinkled grids [3]. With blockchain, yet, many industrial and environmental monitoring data is plunged in a dispersed cloud which allows us to trust engineers and protect the sustainable developments with transparent data.

Blockchain technology generation 2.0 is Ethereum, and has become the de-facto standard platform for dApps. This system was launched in 2015 to support smart contracts. A smart contract runs on top of the Ethereum platform. In the Ethereum web network when a node coalmines a block, a grant is provided in the form of a crypto-token called Ether. Ether is the legal currency cast to arrange smart contracts and pay the transaction fees linked with each transaction. Ethereum unlocks a novel aspect of sequentially decentralized presentations on the blockchain web, spending a core enable program code called smart contracts. There are two major disputes with incorporating blockchain into IoT deals with speed and computational complexity in terms of cost and space. In the current scenario, Ethereum can handle 15 transactions per second. These speeds are not fast enough to be useful for IoT networks with hundreds or even thousands of connected devices all working and transacting concurrently. But Ethereum is developing with an expectation that there is a high chance that the number of transactions will increase shortly after. IoT devices are frequently built with higher interconnectivity, no computation calculation in mind, and average processing power. IoT networks cannot handle computationally complex consensus algorithms. Proof of work strains far too much for it to be meritoriously used in IoT.

A smart contract enables programmable interfaces installed in a scattered network that can attain arbitrary information from outside and modernize the interior state routinely. This Ethereum platform can produce any contractual-transaction type and adversaries over the 'smart contract' methodology. Smart contracts can analyze the computerization, intelligence, sufficiency, and genuineness of transactions without reversing the chain. The vital goal of smart contracts is to make use of computers to enhance the operational activities and to make the contractcode less vulnerable so it creates sufficient and perfect transaction efficiency. Smart contracts are a great way to safeguard transactions and interactions. Christie, who was sold the first Block 2, considers relating the blockchain and smart contract technique to the IoT that can crack the problem of simplifying, visioning, and automating the services lecturing the password verification. Distribution of data to the entire system can protect an unlimited amount of time and costs. Smart contracts are remarkable assets in IoT networks, permitting a high grade of coordination and authority. IoT is always built on the idea of being able to take the right volume of action at the right time. The combination of smart contract in IoT grows and scales new revenue streams.

The enhanced IoT can be measured as the Internet of Everything (IoE) since the widespread real time applications in the fields of agriculture, healthcare, and public safety range from industrial to commercial sectors. IoT consists of various devices that interconnect collectively with each other in the form of things, objects, devices, and machines without any human intervention. IoT enabled devices to lead operating enhancements in terms of performance, efficiency, and wellbeing. Over the longer term, as both technologies advance, companies will use ledger technology with IoT to improve and scale new revenue streams. We need to be more concerned about securing data in IoT devices using Ethereum blockchain. Much data produced by IoT is greatly personal—for example, smart home devices have admitted to warm details about our lives and daily practices. This is data that wishes to be collective with further machines and amenities to be useful to us.

Major IoT device security issues are classified into three. At first tampering data from client to server; second, modifying results at the server side; and at last, disconnecting the connection between the IoT device and server. One such cyberattack related with blockchain is the stalker attack. The stalker attack is a selfish mining attack. A user can extract his funds to pay himself the entire contents of the contract and then eradicating the funds by executing the proposal. Selfish coalmining in blockchains was recognized by Eyal and Sirer [4] who proved that mining practices are not stimulus comparable and selfish coalminers may negotiate the system and attain more advanced booties than their owed shares. If the miners invest properties and do not collect their reward incurably, they will vacate. In addition to the logic of reward service, fees are offered to protect against Denial of Service (DoS) attack, when the evil attacker tries to slow down the whole network by requesting time-consuming calculations. Permitting contact to have data from IoT devices be accomplished over blockchain means a supplementary layer in the network offers security that would be defended by most robust encryption standards. Furthermore, there are many more openings for hackers to hypothetically attack users. Businesses and governments at a large scale have invested in IoT technology and also have to struggle with this enlarged scope for a data breach by criminals, rivals, or foreign enemies.

The scope of this chapter is to include the perceptions nearby and process of blockchain analyzing, and how blockchain technology can be recycled to provide security and privacy in IoT. The entire content of the chapter is planned as follows: In Section 9.2 we introduce the working mechanism of blockchain along with its structure, the importance of hash function and encryption, digital signatures, and transactions with mining. In Section 9.3, we discuss blockchain security along with the comparison on standard cyber security versus blockchain security, security, and privacy properties of blockchain along with new technologies. Section 9.4 presents the integration of blockchain and the Internet of Things (IoT). Section 9.5 deals with the stalker miner. Section 9.6 presents the conclusion followed by the references.

9.2 How Does Blockchain Work?

Blockchain has become one of the most feature-rich technologies in the internet. Blockchain has exposed its incredible capability in upsetting how digital transactions are supported more securely and transparently. Blockchain consist of several blocks. In simple terms, a blockchain is a form of a shared database that is simulated over a peer-to-peer (P2P) network and will be able to modify safely and securely even if users do not trust each other. Some additional four key features are that blockchain allows it to stand as the mechanism for cryptographic stigmergy. First among these is it is arranged as a chain of blocks of verified transactions ad infinitum, giving it an *infinite memory*. The blockchain holds the potential for providing more informative signals. The second main feature is the *decentralized nature* or a *distributed ledger*. No central authority will support the transactions-trades or fix truthful rules to have transactions accepted. Every node agrees to, stores, and validates all transactions and conserves a local copy of the complete state of the account, balances, contracts, and storage. In a blockchain, data is shared between different nodes in the network without any central point so it cannot be stolen. Blockchain providing *enhanced security* is the third main feature. A database can only be enlarged, and previous records cannot be altered and are designed to achieve a reliable agreement over the record of events between independent participants. The most distinctive fourth feature of blockchain technology is that it is *transparent*—anyone can join the network. If someone creates a new block that will be forwarded to everyone present in the network and records can be tracked by anyone, which means there is no chance for a fraud type of transaction.

9.2.1 Structure of Blockchain

Blockchain is an expert technology that contains a collective list of registers, called blocks, connected using cryptography. It consents digital data to be dispersed but not to lack its originality. Blockchain technology ensures trust with working proof of decentralized trust. It is fashioning the strength of a new type of internet. The main advantage of storing in block says with immutability that once it is stored it is permanently saved.

The major five key components of blockchain include the following:

Cryptography: Many cryptographic techniques have been used including hashing and employing SHA-256, a public key infrastructure scheme in timestamp, and Merkle tree techniques. A single document is diced and deposited in blockchain straightforwardly, since it is difficult to store large amounts of data as the original. Each data has to be hashed, which provides limited space and less transaction cost to help increase performance. On one-time hashing, data will guarantee the same output. The main objective of hashing is to produce a hash value that always has a static length no matter how long the original input data is. Any change in hash value instantly indicates data has been tampered.

The hash values ought to be exclusive. An example would be for any two inputs E and F, the output-yield messes of H (E), H (F) would not be identical. For every output 'Z', it is impossible to find an input 'X', represented in (1) as such

$$H(k|X) = Y \qquad (9.1)$$

where k is an arbitrary value since the circulation with high min-entropy.

Numerous hash functions occur out of these; the succeeding one is SHA-256, which are the most commonly used hash functions in blockchain technology (SHA-256 in Bitcoin and Keccak-256 in the Ethereum platform).

Public Ledger: Anyone can join the network. The public ledger is accessible to everyone as each node stores and processes all transactions and preserves a duplicate of the complete state of the user's address and transaction details like account balances and contract details. By this ledger perception, we can forget about the tampering issues since it requires a large amount of computation power to make alterations.

Consensus Mechanism: Consensus mechanism holds much more promise in various events in recording transactions and logging important events. These mechanisms are standard protocols that give a guarantee that all nodes are synchronized and checked along with the miner to confirm the transaction is included in blocks. Miners will get the reward and transaction fee from the sender. The consensus protocols differ with different blockchain networks. There are three core consensus techniques including Proof of Work (PoW) which is the initial blockchain consensus mechanism and was first used by Bitcoin. Second is the Proof of Stake (PoS) which solves the lag of computing power in PoW. This consensus protocol is more environmentally friendly and is a randomized process used to determine who will produce the next block. Next comes traditional Byzantine fault tolerant (BFT) protocols that instrument a three-phase commit scheme for blockchain enlargement.

Point-to-Point Network: Point to point provides a dedicated link between two nodes in a network. Blockchain is free from a singleton point of failure, meaning no central authority. It is a distributed ledger technology that agrees that data be stored globally on thousands of servers. Peer-end to Peer-end file-sharing protocols are broadly recycled and aimed at allotting enormous data above the internet [5]. This feature allows everyone to see every other entry in the next real time block. Many blockchain systems are performed by whole users in a peer-to-peer (P2P) network environment, which marks it hard to explain whether users are controllers or processors.

Legality Rules: A lot of controversy regarding legality exists still to do with other things. Rules all need to be reviewed since the disruptive evolution of technologies like blockchain. The evolution of new laws will bring some standard certification on security and user privacy features. This way helps to bring a more trusted IoT network in combining with a blockchain network. At best, governance rules can normalize users of the blockchain to esteem privacy laws when they load the personal data to the blockchain [6].

Figure 9.1 describes a simple blockchain structure that focuses on cryptographic hashes, timestamp, and transaction details. The opening block in the blockchain is termed a genesis block, the root of the chaining. Subsequent blocks are numbered as Block 1, Block 2, and Block N. Each block is equipped with a cryptographic hash. In Bitcoin, it is followed by the secure hash algorithm-256(SHA-256) technique. SHA-256 is the cryptographic hash function most commonly used in the blockchain which has a digest length of 256-bits. It is a keyless hash function and hash is similar to a remark signature in a data set. In Ethereum we are using a not merely common hash named Keccak 256. As the Ethereum platform

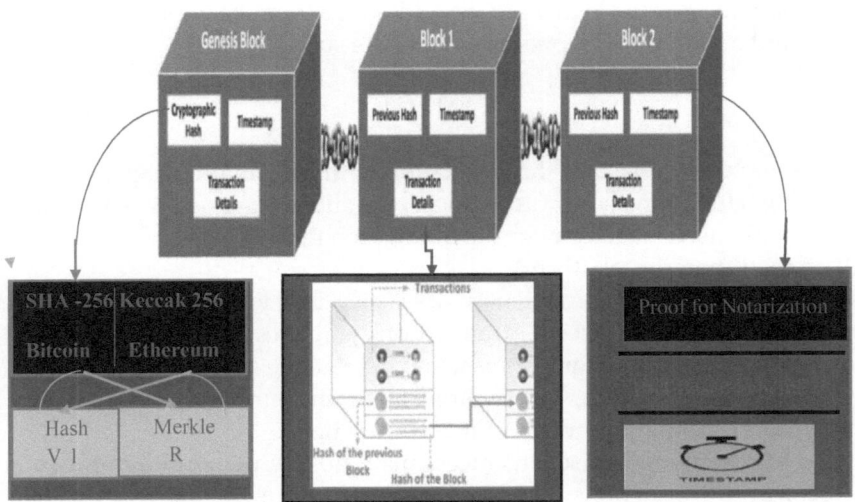

FIGURE 9.1
Simple Blockchain Structure That Focuses on Cryptographic Hashes, Timestamp, and Transaction Details. (Adapted from ResearchGate-Typical Blockchain Structure, Blockchain-Wikipedia).

relies on the *Keccak 256* hash *algorithm,* it has an infinite input space [7]. A small change in hash leads to a drastic change in the whole stage. Each curve of Keccak is a quadratic mapping to elegantly catch a pair of messages which track a great possibility of differential characteristic natures. Keccak is widely suited for protection against side-channel attacks. Merkle root is a binary hash tree for summarizing efficient and secure verification the integrity. This is the reason why blockchain grows day today. Next is a transaction which mainly embraces the source and receiver address, and the transfer amount. The transaction mainly focuses on carrying instructions such as querying, storing, and operating data. Each block transaction props on the hash value of the previous block. The actual transaction cost calculated as

$$\text{Ether} = \text{Gas owned} * \text{Gas tariff} \tag{9.2}$$

In the above Eq. (9.2)gas owned is the real gas used for a particular block in executing all transactions on it. Gas tariff refers to the price per gas. Each action in the Ethereum virtual machine has a specific depletion, which is calculated in terms of gas. Timestamp in the blockchain can be used as proof of survival and it keeps proof for notarization. This action can prove that an assured document exists, or not for a while. Any unauthenticated modification can certainly be acknowledged under a timestamp from when the manuscript was in existence.

9.2.2 Hash Function and Encryption

Hashing is more important in the context of blockchain technology. It is practically impossible with very large-sized invariable inputs data to get hash deposited in the blockchain so that each of the data is hashed and kept in the blockchain. Any solo document can be hashed and deposited in blockchain effortlessly with a timestamp. The process is accomplished by a hash function with the support of a hashing algorithm.

Hash Functions: Hash function is a method involving the input of variable interval, and produces a fixed-sized output. The output of a hash function is known as hash. SHA-256 is considered the standard hashing mechanism in blockchain technology. In practical implications, the SHA-256 algorithm is used in Bitcoin hashing. If we use the SHA-256 algorithm for hashing it will always produce an output of a 256-bits length. This is considered one of the successor hash functions to SHA-1, which is one of the strongest hash functions. For SHA-256 there are 2^{256} possible combinations. SHA-256 is not more complex to code than SHA-1 and has not yet been negotiated with in any way. The 256-bit key makes it a good partner function for Advanced Encryption Standard (AES). In Ethereum, the Keccak hashing technique is used. The keccak hash function uses multi-rate padding: with a message it appends a single one bit followed by a minimum number of zeros [8]. The sponge construction works on the state of r bits which is again divided into two halves. The first half works on an $r0$ bit called an outer state. The second half contains the last $c = r - r0$ bits of the state, which is known to be the inner state. After the overall processing of bits, the first r bits are returned as output, and then the permutation is applied, consisting of 24 rounds. This is continued until n output bits are produced.

One-Way Hash Function: A hash function method is used along with a magical-math algorithm that picks variable input length and yields stable-sized output. One-way function is one of the most important characteristics of hash algorithms. In one-way function, it is computationally very difficult to find input from hashes. For example, a mathematical transformation ensures the reversion of hashing, which meansthe fixed string back into the text message is difficult. One-way hash function can be used for message integrity and authentication.

Encryption: Without encryption nothing can proceed to make a date secure. It is an indispensable part of the interior working of blockchain. The blockchain platform ensures data is encrypted by the trait of immutability meaning modification of data is a difficult task. Asymmetric key encryption is more common in blockchain technology. Encryption systems labor goes as follows: Set a message in the form of plain text and a pair of keys. Either of the two keys, one reserved for encryption and other keys, will be used for decryption or vice versa. Encryption produces a cipher text to be transmitted over unprotected channels. The blockchain encryption security works through a mining network. By encryption, we can secure the privacy and confidentiality of data stored in the blockchain database.

9.2.3 Transactions and Mining

Structure of Block Transactions: Transaction needs to be an asset since users require transactions to be private and identity is not linked to a transaction. Transactions generated are protected by asymmetric encryption, digital signatures, and cryptographic hash functions (SHA-256 or Keccak). Blockchain permits transactions to be grouped into blocks and then recorded. For a transaction operation, blockchain exploits asymmetric keys, both private and public, to protect identities and hash functions to brand the blockchain immutable.

How Do Blockchain Transaction Occurs? Each block maintains a pack of transactions. Figure 9.2 explains transactions in a detailed manner. When someone requests a transaction, transaction content details are passed to a hash function termed SHA-256 lectures with the amount of bits it picks up into the memory. The hash function brings in variable-input data and yields an alpha-numeric output of 64 characters. Secure hash algorithm (SHA-256) has a one-way hash function significance; it is tough to decode the original value utilizing the hash value. The output of SHA-256 hash function is termed as hash

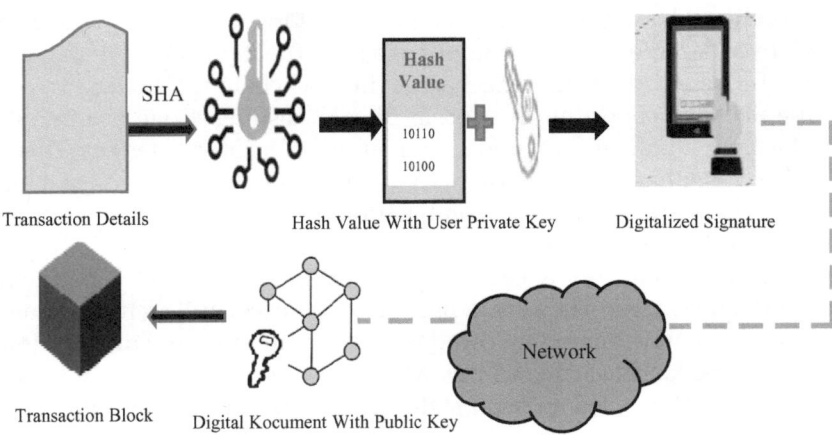

Transaction Details   Hash Value With User Private Key   Digitalized Signature

Transaction Block   Digital Kocument With Public Key   Network

FIGURE 9.2
Blockchain Transaction Model. (Based on Pininterest-Pulling the Blockchain Apart – The Transaction Lifecycle Apart | Blockchain).

value and it is unique. This hash value is traversed over a digitally signed signature algorithm (encryption process) along with the user's private key and it produces a signed document. This document is then floated through the public network and distributed over the blockchain network beside the user's public key. Miners use a digitally signed document and the public key to verify the transaction. Each miner tries to solve a complex mathematical puzzle. The miners use the consensus algorithm to bring trust over the transmitted network. The first miner who cracks the puzzle gets rewarded because they spent a large amount of energy in the form of computation power to perform the consensus mechanisms like Proof of Work (PoW), Proof of Stake (PoS), Practical Byzantine Fault Tolerance (PBFT). PoW is challenging to solve but easy to validate, but PoS works like a voting scheme where one node acts as a polling officer that allows this officer node to validate the next blocks. PBFT consensus algorithm adopts a new procedure to check if the nodes are honest or not. After checking the validity of transactions it is placed into a new block in the chaining network.

Miners: Miners are resource-profound blockchain nodes. The main task of a miner is to find out the *Nonce* value, which is an arbitrary number to produce a hash value that should be less than the target, which is to have some leading zeros. In the Bitcoin network, the penalty area is accustomed to every 2016 blocks. Regularly a block is coal mined every 10 minutes. To identify nonce, miners have to try about 26 quadrillion nonce values to get one valid hash.

Miners will validate the unproved transactions with consensus protocol and add them to the blockchain. Miners strive to crack a problematic mathematical puzzle built on cryptographic hash function well-known as Proof of Work (PoW). The strain of solving puzzles is sustained in a manner that never diverges greatly from the typical average time occupied to dig a block. Figure 9.3 shows how a mining activity is carried over. The first miner who pops the hard math puzzle gets rewarded. Unfortunately, someone else that has mined the block has to wait for constructing a candidate block by assembling the unproven transactions from the transaction mining pool. For this purpose, miners invest a huge expenditure over hardware in terms of computing power and energy in which the mining machine is consuming to acquire the remuneration. When contained transactions are weight confirmed,

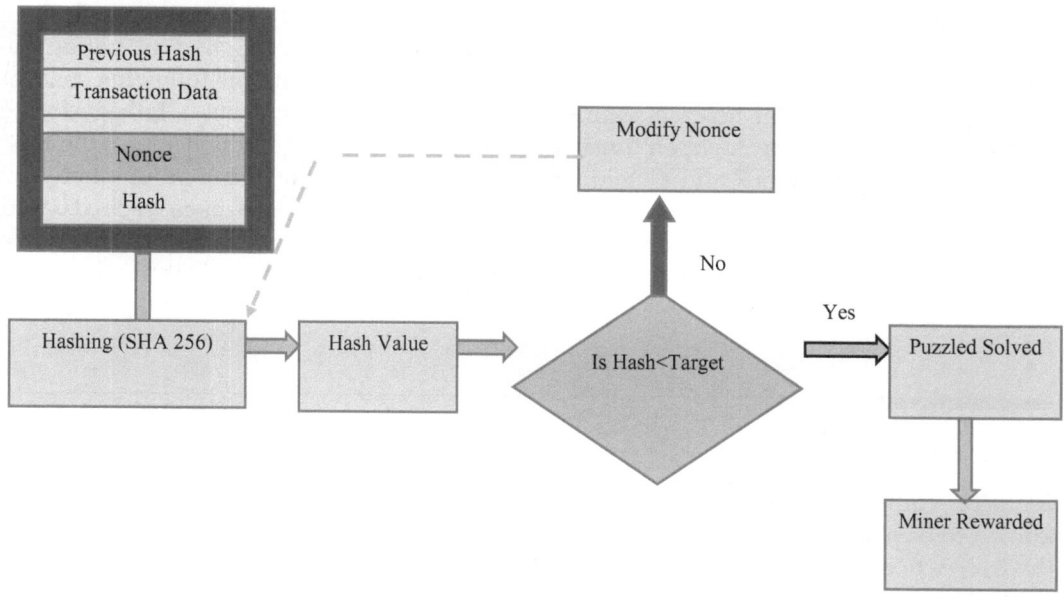

FIGURE 9.3
Mining Phases. (U.S. GAO-Science & Tech Spotlight: Blockchain and Distributed Ledger Technologies; Carlos Olin Montalvo III, How a Bitcoin Transaction Is Processed).

a fresh block is supplemented to the current blockchain for mining purpose and can use the open-source tool Prom. The mining technique process is a novel tool for extracting information from event logs that connect the areas of data science and business process management (BPR). In terms of business, it asks customers first to apply, verify, and check its validity then approve and send it back. If it is not valid due to some practical or technical issues, reject and intimate with notification.

Process Mining: Mining is the activity carried out by the networking neighbors that include certain consensus algorithms like PoW, PoS, and many more algorithms like Byzantine fault algorithms and the results generating in the form of coins will be rewarded for the miners who solve the difficult mathematical puzzle at first for each block. The major intention is to safely exchange the transaction message with other participants. The mining action starts in this manner: at first mining nodes itself assemble and aggregate new transaction data. Upon receiving data each node, it independently verifies the transactions against a long list of criteria that benefits in conserving the reliability of data.

1. It can track the source of the amount being transferred to prove the identity securely.
2. Checks whether they have undergone double spending or attack spending the same money twice.
3. Checks whether the volume of money has not crossed the maximum limit of 21 billion dollars.

Nodes in the network also perform some checks for balance. After verifying transactions are assembled into pools called *mining pools* or termed as *mem-pools*. The transaction waits until they are incorporated in the block. Miners strive for each to come up former with a valid block also called a candidate block and to win the miner reward. They will make sure the most important thing, too, as transactions included in the mem-pools have not previously been included in the previous block. After collecting and arranging the verified transactions, miners need to construct the block header with the following components. Each block header holds a summary of all transactions and has a linkage to the earlier block in the chain. An open timestamp standard is included which shows the time of the creation of the block and a valid consensus algorithm. The entire mining network performs a function call and bounces a contract depending on the consensus reached based on consensus protocol. The final result is replicated through blockchain and provides a commission in the form of reward for miner transactions with established interest rates.

9.3 Blockchain Security

Security requirements during online transactions are discussed and each requirement is known to have any type of vulnerabilities. Blockchain gives complete security in terms of the transaction. Since a block must record all transactions, it imitates similar to a record book. After a finalized transaction joins the blockchain, it is an everlasting database catalog. If a block completes a transaction, a new block is formulated [9]. We need a lot of security roles for blockchain-like vigorous peer-to-peer end network functionalities meaning the total of right nodes including CPU, RAM, storage, and the right bandwidth sandwiched between the nodes. We actually have the right circulated delivery utilities for transactions and new blocks and need a right state of the art of crypto algorithm and crooked hash function to accomplish the security requirements. A most vital part of the security concern is that the blockchain application is free from malware. By combining the security and transparency of many new applications, business prototypes and ecosystems can be established stating we have a trusted automatic transaction store that helps to save money and time and also offers a strong trusted mechanism for ensuring the authenticity of data.

9.3.1 Standard Cybersecurity versus Block Chain Security

Cybersecurity: Cybersecurity is considered one of the key national security issues of all time. Cyberattacks are estimating a global business of as much as $500 billion per year. When the world is moving with advanced technology, new cyber threats are also evolving. Cybersecurity is a constantly evolving, continuously active process just like the threats it targets to prevent. Achieving any kind of growth for a cybersecurity industry will require additional support of the government, private sector, and academia. The two main pillars of cybersecurity include detection and recovery. When a breach occurs, the quicker it is to be detected and answered, the smaller the adequacy of a great loss. Here comes the role of a security officer who cleverly and quickly can respond to the loss whether it may be financial loss or any reputational idea or otherwise. Recovery deals with how quickly you can regain the corrupted data and the preservation and analysis of log helps identify how the breach occurred and to stop them as a part of the recovery process. This should be a seed for preventing similar breaches.

Blockchain Security: In the blockchain, blocks are included after proper validation and verification. The blocks can be copied, deleted, or updated which means holding property of immutability. People can only read the information from the blockchain. An alternative kind of security article is a series of blocks, each block is associated with the previous hash value. Secondly, data undergoes strong encryption when added to the chaining network. So it is very difficult for an attacker to perform corruption which in turn affects all other nodes in the chain. The third case transactions are authorized with a private key that is unique to the individual. This will upturn the protection of sensitive data or information.

Blockchains are secured through a variety of means like advanced cryptographic techniques, hashing techniques, using mixing protocols, wallet scheme, and by the mathematical model of behavior and decision making. A new concept called crypto economics also plays a vibrant role in controlling the security of blockchain networks. The most important information to be protected includes user information, operational network, and process control systems.

9.3.1.1 Security Analysis

For the sake of simplicity, we undertake every protocol and certain clouds are secured by using encryption techniques. Cloud holds a large number of informative systems. The way the cloud can monitor the blockchain network will respond to the request from terminals. To maintain privacy and security, the system should ensure the anonymity of terminals.

Moreover, by security analysis, it is observed high granularity using access control systems and is more flexible and more informative it can outfit for. Revocation is another important feature that simplifies the capability to withdraw an agreement to admit the resource and to assure that the user cannot access the revoked resources anymore [10]. By the need for security in the blockchain network node, the head chooses too many attributes, then the total of qualified miners will be reduced, which leads to some malicious attack. To avoid this problem a new blockchain technique will initially verify a few miners. However, after verification, the blockchain network can donate to mining a new block.

Figure 9.4 explains the remedies provided by blockchain in overcoming common cyber threats. Each cyber threat is tackled differently by blockchain. Note that some countermeasures may state more than one issue so that we can easily find a cure for the associated problem. The proposed consortium on Ethereum blockchain with a smart contract technique[11] makes it necessary to attach a majority of valid mechanism in the form of control policies and several protocols make a block valid. In the blockchain, data is collected and dispersed to everyone who trusts the blockchain to award accessibility and reliability. In outdated cyberdata, it is organized and sold with owner-controlled limited access, which lays the accessibility, reliability, and confidentiality of the data user data [12]. Privileges on the blockchain itself are restricted to only authorized parties. Data on the blockchain holds only information about transactions and not any sensitive content like health data or shipment details. Both environments are vulnerable to intended abuse attacks. Modern networks are virtually compacted, and their cybersecurity concern mainly determined by permissions protocols. Availability and integrity are delivered by affording its design in blockchain technology. But the organization of a modern environment is constructed on reliability and confidentiality using numerous access control policies and smart contract techniques.

9.3.2 Security and Privacy Properties of Blockchain

In simple terms, blockchain technology is a decentralized application with a public ledger of distributed transactions. Therefore, all transactions occurring can be viewed, validated, and verified by the peer-to-peer (P2P) networks [13]. Security and privacy assets of

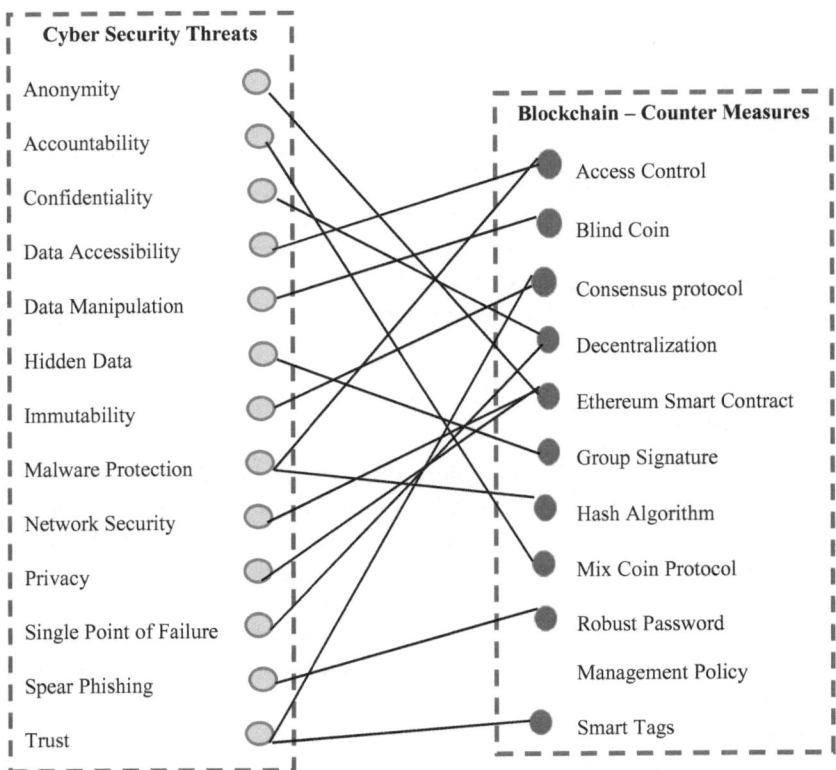

FIGURE 9.4
Relationship on Cyber Threats with Blockchain Countermeasures. (ResearchGate: An Example of Relationship between Cybersecurity Threats and Countermeasures; MDPI JSAN, a Comprehensive Study of Security and Privacy Guidelines, Threats, and Countermeasures: An IoT Perspective).

blockchain are discussed because of the more severe cryptographic attacks that will be disclosed in the next paragraph.

Data Security in Blockchain: Security needs to be explained as the safety of transaction evidence and data included in the blocks against internal, external, and malware threats. Here comes the need for securing the blockchain in securing the personnel data from storage or to prevent stolen information. The need to secure the database and appropriate access control policies have to be taken to prevent threats.

Process Integrity: Users can believe completely that the transaction will be implemented accurately depending on the protocol commands eliminating the necessity for a trusted third party. During online transactions for investment and asset, transfers have to depend on many intermediaries spending a lot of amount of transaction fees on sustaining middlemen. By using blockchain assurance on the reliability of transactions, users can avoid transactions from being altered [14].

Trustless Exchange: Financial balances are secured on the blockchain. Truly no third-party involvement is required for performing any transactions so it can act as a service provider and not collect any extra charge from users [15]. This security feature is useful in many areas in the fields of industries, practicing academics, and more common in financial sectors. It was estimated that 52% of transaction fees are in the hands of the intermediary during turmeric production in the Erode district Coimbatore. So blockchain helped to save many farmers from the avoiding suicide.

Durability, Reliability and Longevity: The prime benefit of blockchain is the property of decentralization. No single authority is needed to favor the transactions or establish explicit rules to have transactions recognized. Every node in the blockchain stores and routes the transactions need to separately maintain a copy of the entire state of account balances, contract details, and data storage.

Un-relievality of Transaction Details: Users in the blockchain do not have to disclose all their details. A certain consensus technique known as zero-knowledge proof (ZKP) has developed to protect transparency. In ZKP protocol each user can prove and verify they are a verified user. This protocol helps to improve the privacy and security of personal data through blockchain.

Data Privacy in Blockchain: The key feature of privacy in blockchain points to the usage of secret keys. More commonly blockchain systems use multi-party asymmetric cryptography to make an individuality in their transactions. The keys generation is mathematically difficult, making it challenging for a stranger or attacker to conjecture another user's secret key from the public key. Secret keys are commonly used to protect user identity and security through the digital signature concept.

Encryption Methods: In blockchain ledgers [16] the data secrecy is guaranteed by strong encryption methods like Advanced Encryption Standard (AES) and an attribute-based encryption (ABE) technique and allowing well-ordered access to ledgers. After encryption using two keys, anyone can see the cipher text information but they do not see the address of related users and cannot copy the encrypted file from interplanetary file systems (IPFS). In IPFS, large files are fragmented and stored in different storage nodes, so we can ensure fine-grained access control over the data. As long as the Ethereum blockchain network and encryption are both with ABE and AES scheme, the proposed scheme is safe and strong privacy is guaranteed.

Decentralized Storage Infrastructure: This infrastructure facilitates by cryptographically securing data across participating networks [17]. It also helps to maintain data in a highly relevant manner so that accurate and complete data or information can be collected from reliable data cradles. In a decentralized storage level, each node network preserves a client-centric application to be deployed at node levels to safeguard data availability for anticipated clients. This helps blockchain to maintain privacy and data provenance.

Group Signature Scheme: This scheme is an essential component in anonymity revocation which helps in privacy-preserving blockchain-based smart contracts. This technique holds multiple inputs and outputs that are more commonly used in having a high chance of malicious attack. By this, the malicious center and signature recipient can verify and locate the original signer identity and threaten the anonymity of the program. The main property of this scheme is that owners can hide the amount, especially during the transaction and fast validation of data/asset/value transfers depending upon discrete logarithm problems and bilinear mapping property.

Data Transparency and Access Control: Each user has given entire transparency consisting of knowing who knows what about them, how the data will be used and for what purpose or for any future reference with whom it will be shared in the network, and all that is possessed by an individual and how they access those possessions. Any change to the public blockchain can be observed by all parties creating transparency among all in the chaining network. Access control should conserve its user privacy to gain trust since it no longer depends on user credentials only; it also depends on the system and its environment.

The most crucial attacks reported over the last three years are mentioned in the below Table 9.1. Blockchains themselves have relevance to their specific definite set of security

TABLE 9.1

Top Five Blockchain Security Issues Report over the Last Three Years. (Data from [18–22])

Attack	Description	Security Risk	Revenue Loss	Counter Measures
51 % attack[18]	Group of Sybil nodes achieve the mainstream of the network's hash rate to employ the blockchain	High	$20 million (in 2019 alone)[19]	Utilize higher hashrate consensus mechanism or joint consensus
Double Spending attack[20]	The attack especially refers to what was spent twice without account balance update	High	Revenue of an attack is greater than the cost of launching it[21]	Use of strong encryption techniques
Cryptojacking[19]	It is an unauthorized way of mining cryptocurrency which leads to the performance issue to increase electricity usage and opens the door for other hostile codes	Very High Internet Security Threat Report (ISTR) reveals cryptojacking attacks on websites have enlarged by 8500%during 2017 [22]	Over $3 million (June 2018)	A software tool called "MineGuard" to perceive and stop hiding mining operations in cloudjacking
Phishing attack [18,19]	Common software engineering attack occurs by clicking on a malicious link	High	$3 million vanished due to social engineering in 2018[19]	Neversend anyone your login credentials and private key
Software flaws[19]	Software bugs include wallets and decentralized applications	Very High	$63 million Nice Hash-A cryptocurrency company in Dec 2017[18]	Use a lattice-based algorithm to generate private ECDSA keys.

issues and if not accounted for, they can be injurious to industry and business. There were five main blockchain security issues in 2019. This includes a 51% attack termed as one of the major attacks. The 51% attack made the consequence of blocking the transactions from being verified and reverting the transactions, preventing miners from discovering blocks for a short period. The second issue deals with the double spending attack. In this attack data is being transferred more than once and there are more chances of undergoing alterations without any intimation. Cryptojacking is one type of malware that can affect the entire system. This issues range from malevolent crypto mining software to code that could even seal down a company server completely. Another blockchain security issue is phishing. The goal is continually the same: to obtain the private keys and other login details straightly which puts sudden breaks in user confidentiality. The last security issue deals with software flaws in the form of wallet theft and decentralized apps (DApps). In the case of Bitcoin by default, data is stored in the form of an unencrypted manner. Even when wallet data is protected safely it has more of a chance of a malware attack which leads to stealing wallet contents.

9.3.3 How the Infusion of Blockchain Solves Vulnerable Issues over Three Zones: MLCL-FoG

Blockchain will light up many industries by adopting digital transformations in a new technology to stay competitive and also aids with a unique touch in organization growth. Here we will be discussing how the impact of blockchain in Figure 9.5 helped in preventing the security challenges in three wide research areas including machine learning, cloud computing, and IoT. As world digitization progress, different infrastructures will have to compete with each others to achieve the fastest economic growth. The winner will be the infrastructure that manages complexity to its best. Infrastructure needs to be smart enough to maintain a competitive advantage.

Machine Learning: Today both blockchain and machine learning are getting strong support and trust along with the whole world; combining the two we can form a disruptive nature. In machine learning, data are placid from a repository and stored in a secured and trusted area whereas blockchain assists as a distributed ledger with its replicated storage and decentralized management, offer storage with high integrity, and provide integrity that can be used in a bigger transaction [16]. The integration of both techniques helps in creating a secure, immutable, decentralized system for storing highly sensitive data which will be useful in various fields like medical, banking, supply chain management, personnel, financial, and legal data [23].

Cloud Computing: Cloud computing has a very close relationship with a blockchain network by incorporating cloud in blockchain with security notions between the user and data managing in the cloud. From the user point outlook, data are hidden and stored securely meaning the cloud service permits its customers to have control over data and where it is being stored. But in terms of blockchain technology, it holds an immutable property stating none of the data stored in blockchain undergoes alteration, and by a

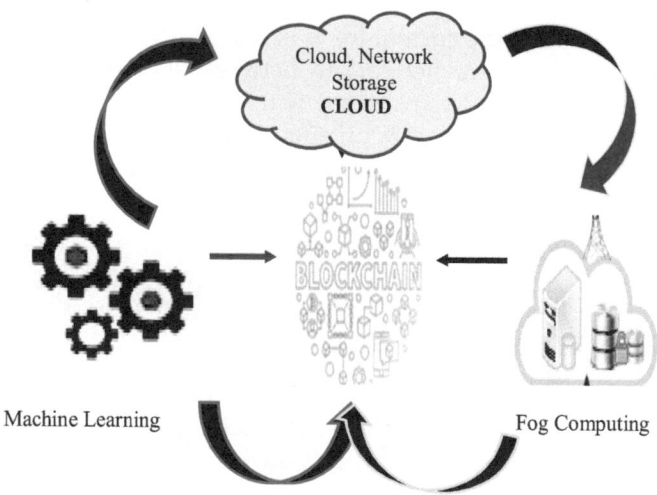

FIGURE 9.5

Blockchain on Integration with MLCL-FOG: Hybrid Approach. (MDPI-Applied Science Blockchain Based Resource Allocation Model in Fog Computing; ScienceDirect, https://doi.org/10.1016/j.iot.2019.100118).

TABLE 9.2

How Blockchain Technology Benefits Three Zones, MLCL, FoG with Countermeasures. (Data from Papers [16,23–27])

Machine Learning	Cloud Computing	Fog Computing
Byzantine attack: Byzantinefault tolerance consensus protocol	Centralized concept: decentralization	Access control role-based access control
High efficiency: intelligent decentralized autonomous agents(DAOs)	Data availability: blockchain	Data privacy: smart contract
Privacy leakage Learning blockchain	Data security: mixing potocols in blockchain	Location privacy: lightweight mixing protocols
Probabilistic data: immutability feature	Network security: Ethereumsmart contract	Security challenges: strong cryptographic hashing
Trusted third party: decentralized stochastic gradient descent	Privacy management: smart contract	Tampering threat: *group* signature scheme
Searching technique: fair access control policy	Storage systems: blockchain –Storj, Swarm, and Filecoin	Trust: No trusted third party

strong signature, the technique is used to hide the real data contents. This method helps in reducing the risk of fake knowledge [24]. The most important issue addressed in cloud computing is that it takes more time for data access to be solved in a few seconds by blockchain configuring all mathematical models. Blockchain's public nature helps cloud computing services to guarantee data provenance, reviewing, running of digital properties, and distributed consensus [25]. These features all are achieved by the P2P ledger system and by the consensus mechanism.

Fog Computing: Blockchain as a decentralized framework is used to establish security and trust for computing fog familiarized by Cisco with IoE in which a vast quantity of information is produced. Fog is a follow up from the cloud. Cloud services face some problems in data handling due to the current trend in IoT devices. Fog is a virtualized platform benefit in computing, storage, and networking between cloud data centers and end devices [26]. In fog computing, each node creates a new transaction encrypted with the user's private key and broadcasts the encrypted contents along with a public key to the blockchain network [27]. Table 9.2 describesthe benefits of the three zones.

9.4 Integration of IoT over Blockchain

The IoT revolution shaped to profoundly upgrade and build new systems from start to increase productivity and value of products. The main goal of integrating IoT with blockchain is concerned with the security and privacy of IoT data. In 2015, the U.S. company Veracode found serious security issues. The preliminary issue deals with the execution and safety of the communication procedures recycled in IoT devices.

The team analyzed poor insecurity in terms of weak passwords on front end connections between the user and other high-end platforms like the cloud or edge devices. This vulnerability led to man-in-the-middle network attacks. The same squad investigated a lack of proper encryption scheme in backend connections which led to replay attacks. The company Ubi from Unified Computer Intelligence Corporation botched to suitably shelter the sensitive data. Mirai software reported having a friendly base program that even untrained hackers would be able to use to remove distributed denial of service (DDoS) attacks.

Figure 9.6 explains general discussions on many aspects such as vacillating the impression experience, joined constructions to application domains, and research experiments; we understood there is a strong need for integrating IoT with blockchain. We decided to fragment into three layers: the bottom is the IoT layer; second is the core layer where block-to-blockchain transformation concentrates more on data accessing and data sharing, transferring data to block, how it is verified, and finally added to the entire blockchain; the third layer is the IoT-BC application layer.

1. **IoT Layer**: IoT devices are the fastest ones in terms of security; subsequently it can work without the help of the internet. Each device can communicate with each other with the help of any routing mechanisms [29]. IoT devices certainly need to harvest data from local environments and forward the data into the neighboring gateways as

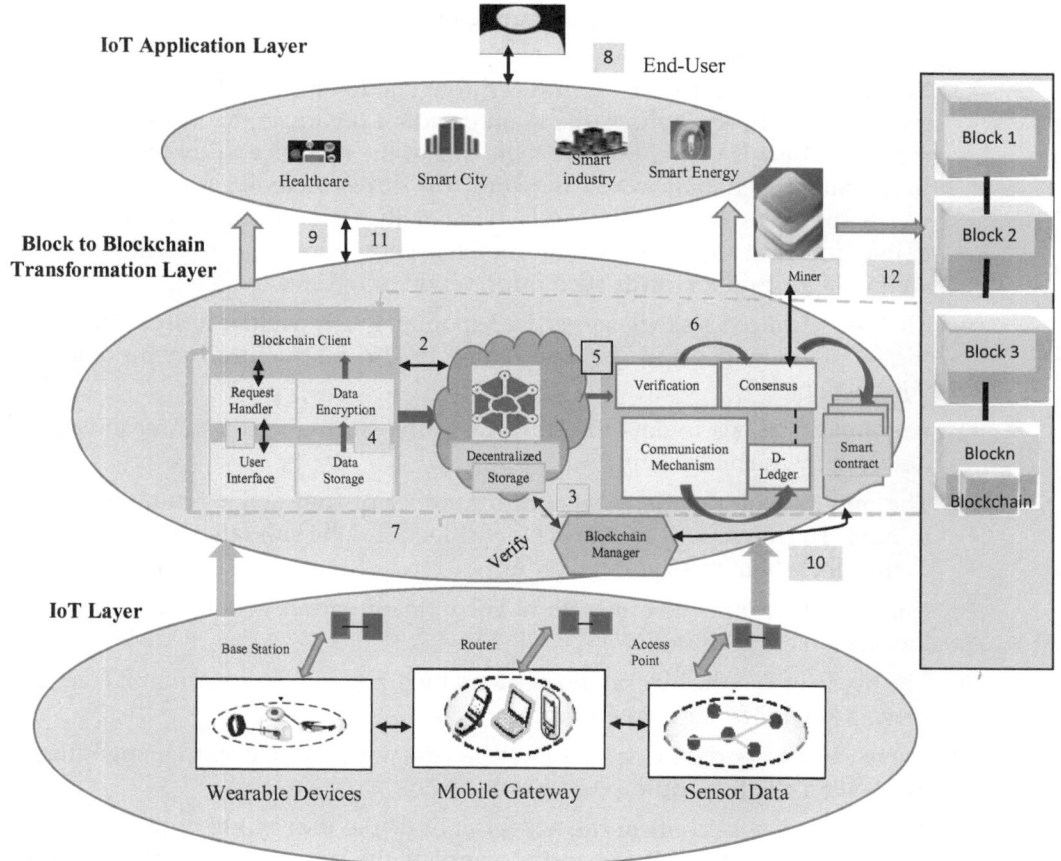

FIGURE 9.6
Blockchain IoT Three-Layer Architecture. (Adapted from IEEE and ScienceDirect Papers [28–31]).

base station, router, or wireless access point. Especially, each resource-limited to IoT device can contribute to blockchain consensus from the side-by-side confirmation on hash values without holding the total blockchain data, thus helping to work with low resources and interactions that take place with low latency. IoT devices can cooperate with each other over IoT gateways to reach a strong-join announcement. Here the communication taking place is a device-to-device interaction.

2. **Block-to-Block Transformation Layer**: In this layer, all interactions happen through blockchain. This approach ensures all the connections are noticeable and can be included in the blockchain itself. On the other hand, IoT data related to all transactions need to be deposited in the blockchain. In this layer, we more concentrated on the data collection, uploading, and sharing, including those data in the blockchain. A seamless arrangement of this approach will be the finest way to incorporate blockchain with real-time IoT interactions [28]. For instance, when high performance is mandatory, unaided blockchain may not be the right solution. In this layer, we collect the following.

   1. Data from various IoT devices is collected upon user requests with the help of a handler. The handler transfers the request to the blockchain client module, which outfits a complete function to contribute to the blockchain network.

   2. The blockchain client module is liable to encrypt the transactions and data requirements, sign digitally on transactions, and interconnect with blockchain for contract tracing.

   3. The request from the user will travel through a decentralized storage device to a blockchain manager, who is the main deciding organizer, to check whether the requested data is valid or not. It is the role of the manager to control all transactions that happen in the whole network. The duties include verifying the right to use with access control approach along with a smart contract, they decrypt demanded data expending an asymmetric encryption algorithm like the lightweight algorithm forwards such information to retrieve the data.

   4. The decentralized storage server will spontaneously yield a hash of the loaded file to the manager and the hash value is also restructured in the board for verification.

   5. Any modifications to data files that are found during verification can easily be detected by the blockchain manager.

   6. The miners will include a maximum number of transactions into the pool for mining. The miner will verify each data block and the validated block is attached to the blockchain in sequential order.

   7. Miners will update the transaction. The uploading transaction is updated to the blockchain client module.

   8. The user can join the blockchain network by preparing a transaction ID using the private key for faster data entrance.

   9. Users can access data from the interplanetary file storage space expending the blockchain client module.

   10. The manager will confirm the access right of the user along with a fair access control policy. Once the demand is complete the requested content of data will be forwarded.

The previously suggested model avoids perils of data leakage and thus confirms great data privacy. To reduce the gap between IoT and blockchain a large number of devices with combined capabilities are available on the market. One of them is Eth Embedded [31] which states the connection of Ethereum on embedded devices such as Raspberry Pi and Android.

3. **IoT Application Layer**: Various industrial applications gain from blockchain IoT integration. The key revolution includes where the IoT information can be toughly shared among many participants after this incorporation. It is especially beneficial in smart healthcare, smart transportation, smart city, smart energy, and smart industry. For example, in smart healthcare this can assist healthcare providers in finding the most intelligent technique for a patient to receive better medical care. Food traceability [29] helps to avoid any type of attack identified by a potential system vulnerability. This method helps to increase the quality of food and thus can save many human lives.

From our point of view, there is a great need for integrating IoT with blockchain that helps in developing current IoT technologies. IoT provides an extensive variety of events envisioned to reinforce security. Numerous start-ups have initiated integrating blockchain to encourage trust in the supply chain management that is planned to be distributed ledger agnostic. Supply chain management has been recognized as one of the main applications in merging blockchain with IoT since it helps vendors to share, exchange, and collect products within a trustless environment [30]. A collection of technology and financial companies declared to agree to a new customary for safeguarding IoT applications spending blockchain. The group aimed to introduce a blockchain protocol as a mutual platform to figure IoT devices, applications, and networks. Using protocols, users can register and bind with strong cryptographic techniques that are immutably related to physical and digital worlds.

9.4.1 Need for Securing IoT Using Blockchain

IoT states that an extensive system of internet-connected devices are skilled in collecting and transferring information; it could be consumer goods, any industrial equipment, or any healthcare devices, and more. Data is a chunk of the IoT so it is very important to store and protect it. There are many worse complexities and problems working in this field that may face security as well as a functional point of view. There includes a lot of IoT protocols for holding the security of data. IoT protocols include ZigBee, Z-Wave, Radio Frequency, and 4G. These all provide multiple standards but it they are not good enough to secure important data. However, many architectures are ready to collect and store information in a rich stack. One is the cloud with support from Amazon Web Services (AWS), Google Cloud Platform (GCP), and many private clouds. In the cloud all data are processed, copied, and presented in a way that is useful for customers and businesses.

The problem with IoT especially in India is that it is broadly done by small- and medium-sized businesses. Many companies reported saying we believe in suppliers for hardware, software, and security and it is their prime responsibility for any security bug. The need for security in IoT is mainly to protect confidential data. Security is not just about confidentiality; it focuses more on integrity and availability which is the prime essence of Ccyptography and network security. Unfortunately, attacks on IoT are actively happening. The consequences of such attacks lead to data loss, hardware failure, and damage to online reputation. The cost of each attack is estimated at US$135 which is not big in terms of a high-value attack [32].

A challenge in securing IoT devices is a basket with huge cells. It is impossible in the current era to make everything using a manufacturer. Instead, buy each physical device like hardware from one country, and software from Europe, and assemble. The second challenge is with the existing software provided; the security development life cycle cannot be edited depending on our users' cases. The third challenge is meeting with the open-source software; the rampant and ultimate challenge emphasis is on software size restrictions. This view is known as 'Linus's Law.' The cyberattack is not simply based on a single element or component, and should have to boost the detective and preventive control mechanisms.

How does blockchain help in securing IoT is explained in Figure 9.7. It is like a software protocol that does not work without any internet connection can be integrated into multiple areas. Blockchain technology uses many techniques to archive IoT security during transactions for storage, utilization of data, and network security because of the propagation of IoT in our daily needs, military and healthcare, and the constant growing claim for IoT solutions [33].

Time Reduction: It helps in settling the trade-transaction quickly. Since the copy of data is shared among all participants, it will not take much time in ordering, settling, transferring, and clearing all regarding a transaction into a block.

Unchangeable Transactions: This property is called a standard label immutability. Blockchain includes all transactions in chronological order. When a new block is included that ranges through the entire network with the rest of the participants verify it and affix it to its blockchain copy.

Reliability: Blockchain certifies and confirms the individuals of each interested party. This property is attained because of the consensus that can be achieved quickly. The most popular consensus is PoS. These protocols work defining big stakeholders in the sense of who carries more coins on the network and therefore who is most suitable to carry a huge responsibility for protecting the system. In this consensus mechanism if they don't validate

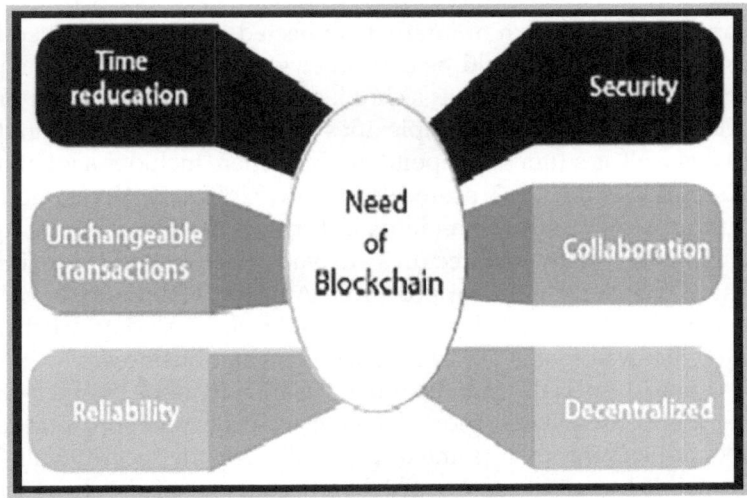

FIGURE 9.7
Primary Need for Blockchain in Securing IoT. (Based on Blockchain IoT Stock Illustrations; OpenMind: A Secure Model of IoT with Blockchain).

the transaction, there is nothing to lose on the blockchain but it creates a delay in creating a new block.

Security: By using very advanced cryptography techniques and the hashing mechanism to make definite that the information is sealed inside the blockchain by property immutability. The SHA-256 is used as a hashing algorithm. The idea behind this is that blockchain bids restricted space, and it is very expensive to have a huge galaxy and transaction cost. So all the data is to be hashed and deposited in a blockchain network.

Collaboration: It permits each party to perform straightly without demanding a third-party intermediary. This is the most unique property of blockchain. Since a middleman is not included, a lot of savings are reported in all areas especially in the agriculture-farming industry. It is reported by farmers from the Modakurachi taluk in Erode District that previously they were unaware of these techniques and experienced a huge loss in terms of monetary benefit. It was estimated around 52% of the amount will be in the hands of a middle-man or third-party brokers and a smaller amount of money will be in the proper hands.

Decentralized: The decentralized setting of blockchain does not depend on a central socket of the controller to manage transactions. This exceptional property brings talented benefits, abolishing the singleton point of failure hazards to the interruption of central authority, authenticating operational costs, and augmenting fidelity. This decentralized infrastructure facilitates in cryptographically securing the data storage through the whole participating network.

9.4.2 Advantages of Integrating IoT with Ethereum Blockchain

Blockchain is a peer-end to peer-end network. Growth in IoT provides vast opportunities to access and share information [29]. The need for integrating IoT is to increase the security in sharing the data between many participants representing a key revolution. The cradle of data can be recognized at any time and data will be secured for the whole by the property of immutability.

Access Control Policy: These policies are expressed by the scripting language. The policy behaves like a set of conditions to access a resource from User A to User B through blockchain without the intervention of the resource owner. The rules are combined, and they must be satisfied with the right accessgranted [31]. By using such access control policies fairness improves because nobody other than the owner has the right to their data. The use of the smart contract technique empowers coarse access control policies over the framework. This helps users in constructing our access control framework to attain the needed security level. Without appropriate policies, attackers might gain control over communal necessities for all IoT applications. Certain access control policies include a token mechanism to access a protected resource that helps in reducing communication cost, end-to-end security will be achieved, and no further authentication mechanism is required. The token can be used for many access control operations like updating, receiving, and delegating in a better manner. One application of this policy helps in improving the existing permissions dialog in mobile applications.

Consensus Mechanism: These algorithms are used to preserve the reliability of transactional data in the organization, substituting the necessity for an intermediary. In public blockchain like Ethereum, a consensus is mandatory for selecting a miner to create a new block to the prevailing blockchain network. By this mechanism, all participants agree in the transaction. In our design, we opted for PoS, which can conduct hundred and thousands of transactions every few seconds. PoS indirectly solves the security problem and does not lose anything on the blockchain, but it may cause a delay in creating blocks. Blockchain that holds any invalid transactions could be revealed immediately[16].

Fault Tolerance: Since in public blockchain networks like Ethereum the ledger is replicated and shared, failure to any one or more component will not affect the entire system. All users in the network hold a complete record of transactions. Blockchain features make it a more attractive option in conserving trust in systems of record and the exchange of value between parties. To assure the continuous availability and maintain the same level of resiliency, blockchains commonly exhibit an advanced need for fault tolerance on a per-node basis.

Decentralized Nature: In centralized storage units like the cloud, data centers, and edge computing become a foremost bottleneck for developing highly secure privacy-preserving submissions. This is a unique feature of blockchain that benefits to acquire no singleton point of vulnerability or failure. There is no fundamental agency in the blockchain network, a small transaction fee is incurred, and transactions happensooner. Consensus algorithms uphold data consistency in a decentralized, scattered network. In conventionally large-scale systems data from the repository undergoes scaling and capacity related issues, whereas in blockchain a more secure decentralized data is stored across the entire network. A few decentralized data storage examples includeinterplanetary file systems (IPFS), Storj, Swarm, and Filecoin.

Immutability: In traditional systems, databases are vulnerable to both accidental and malicious manipulation. Computers or nodes on a blockchain are scattered and mined, and computing power is required to bring more belief and integrity to the data making it immutable and reversible. But since blockchain will verify the blocks using consensus algorithms, verified blocks are immutable and resilient to all types of attacks. All the transaction details will be stored in a cryptographic form. By this feature, our confidential and sensitive data from IoT devices will be more secure.

System Scalability: IoT merely deals with the scalability concern. In today's blockchain platform scalability is one major factor; if the block size is increased, potential threats are created leading to double-spend attacks. It is a misbelief that every transaction must be validated by every other nodeor else it leads to a scalability issue. The system can achieve greater scalability with different consensus mechanisms by reducing the size of block parameters. When blockchain is integrated with IoT this issue is solved by 80%, allowing service providers to convey and then spread uninterruptedly their marketplaces [34]. The blockchain offers a large amount of security but confines scalability to not being able to process extra transactions, and particularly guarantee the robustness of the scalable application.

Traceability: With the speedy growth of recent detecting, cooperating, computing technologies, current years have shown fabulous progress in intelligent transportation systems (ITS). Traceability in several food products is a key aspect to safeguard food safety. Tracing is done for a lot of participants in terms of manufacturing, feeding, distribution, quality, and quantity testing. All of these affect human lives. Therefore, the usage of blockchain can be a counterpart to IoT with reliable and secure information.

Smart Contract: A smart contract handles a programmable applications that turns on a blockchain network. The initial smart contract platform for Ethereum was released in 2015, which is termed as blockchain 2.0. Users can directly interrelate with smart contracts by the contractual-agreement address and the application binary interface (ABI). Smart contracts can diagnose and authorize the request, and award access permissions for handlers by activating transactions or messages. A smart contract helps to verify and guarantee correction operations essential for them to widely and safely adopt for clients and providers. Smart contracts offer immutability, deterministic execution, and transparency that is mandatory in untrusted environments.

9.5 Stalker Miner

The security of blockchain is noteworthy for the appropriate tolerability by latent users [35]. A stalker attack is considered one of the major attacks. This attack can be demoralized when an attacker, a crowd of sybil nodes, or a mining pool in the network achieves the popular network's hash rate to deploy the blockchain. There are two types of miners: an honest miner and a stalker miner. An honest miner is likely to miss its block through a fork and there is no warranty that the honest miner will succeed under race conditions. All honest mining trails protocols will disclose a block instantaneously after mining. They admit the supreme protracted chain and coalmine on top of it. The main goal of a stalker miner is to deny a particular target, not disturbing about profit. This attack's duty is to approve, interval, and publish heuristics. A block is disposed of cheating if a small portion of hashing power is used. In the context of selfish coalmining, the miners hold the mined blocks without spreading to the network and will telecast after some conditions are met. This results are a huge loss for the honest miners by worsening a slice of time and resources, whereas the private chain is excavated by selfish miners to acquire an extension rather than the public blockchain. These events are prime to a contest among the permissioned or private blockchain and the permissionless or public blockchain. Without any consideration, selfish miners issue their blocks to label block rewards. The selfish miners will have extra power to win the mining race. Many effective techniques were invented to spot selfish mining behavior but they are not always reasonable and there are unpredicted deferrals in the dissemination of honest blocks that may courtesy the selfish miner. Solat et al. [36] familiarizes a zero block wherever miners are allotted a time slot for releasing blocks in that particular timestamp. Certain miners play a dummy game by not releasing in time, create mock blocks, and append along with the original blockchain. Selfish miners affect the entire system and acquire higher rewards than their owed shares. Propagation delay also leads to winning the selfish miner. There are mainly two types of miners in selfish mining and they include: the single selfish miner and the multiple selfish miner. In the circumstance of a single, merely one block is issued at a time. The blocks that the opponent keeps in secret are stated as a private chain. In the case of numerous selfish miners, countless blocks can be suddenly issued at a time. If a block is issued in the public chain, other selfish miners will not be able to disclose their secret blocks. Almost att the mining (at least 78.7%) is done on a pool basis and not on a single selfish manner basis. Since mining pools are public, miners are free to link and vacate the pools. Miners in a similar mining pool are enforced to share some valid information due to the organizational nature of the pool.

To summarize, a detective mining along with a strong lightweight height algorithm is very effective against the selfish mining pool. Further, it gives a more advanced revenue to the miners than the honest strategy. But if there are adequate detective miners, the selfish mining pool acquires revenue, then it practices an honest approach.

9.5.1 Defense against Stalker Miner

The authors of [36] introduced a zero block solution, a strategy to prevent selfish mining using one of the most important components in blockchain termed as the timestamp. This can be recycled as a proof exists and it preserves proof for notarization. This method will help to prove that a certain document exists for a certain period so any unauthorized modification can be detected easily. The stalker, mentioned in Algorithm 1, is a different way of selfish mining. The goal of the stalker is not financial but to prevent a fairly specified node from being capable of having its block printed in the chain; so whenever that goal node places its block in the challenge of a conceivable publication, the attacking nodewill also put its block in the clash. Many cryptographic techniques in the form of algorithms were included to prevent the stalker attack. It is observed that the frequency of stalker attacks is much high and is expected to be around 3.35% for every ten minutes. This algorithm is called the lightweight hight algorithm, and with it, the detection of an attack is easy. HIGHT (new block cipher) algorithm consist of a 64-bit block and 128-bit key length. HIGHT was measured to be appropriate for execution in the low resource situation such as radio-frequency identification (RFID) tag or little IoT pervasive devices. From a security study, we assure you that HIGHT has plenty of security.

9.6 Conclusion

In this chapter, we discussed the major role in preserving security and privacy by using blockchain using smart contract technology. We sketched the nature of cybersecurity attacks and its displayed profit restrictions in blockchain was discussed along with its countermeasures. We examined the prior work and discussed their methodology and limitations. We endeavored to highlight the need for security and privacy issues of the IoT to be solved using blockchain technology. Although some restrictions occur in blockchains and several inventive applications are crucial to be executed, blockchain is expected to develop with smart contract technology so that everyone moves in the direction with its maturity, requiring adaptations and optimizations. The grouping of blockchain and IoT appears powerful, as blockchain offers resilience to attacks and the capability to interact with peers in a consistent and auditable mode. Our proposed algorithm effectively frightens selfish mining and cheers fair mining. By charting these attacks and measuring their countermeasures, we highlight new investigation guidelines that must be dashed concerning more secure and active usage of blockchain.

References

[1] Coin desk. 2017. "State of Blockchain - Q4 2017." November 27. Accessed February 4 2020. https://www.coindesk.com/wp-content/uploads/research/state-of-blockchain/2017/q4/sob2017q4-2018.pdf.

[2] Davidson, S.; P. De Filippi, and J. Potts. 2016. "Economics of Blockchain." Accessed May 10 2017. http://dx.doi.org/10. 2139/ssrn.2744751.

[3] IBM Corporation. 2015. "Device Democracy: Saving the Future of the Internet of Things." Accessed April 29 2017. http://www01.ibm.com/common/ssi/cgibin/ssialias?htmlfid=GBE03620USEN.

[4] I. Eyal, and E. G. Sirer. 2014. "Majority Is Not Enough: Bitcoin Mining Is Vulnerable." In *Financial Cryptography and Data Security*, 436–454. Springer.

[5] Z. Zhou et al. 2014. *"EEP2P: An Energy-Efficient and Economy-Efficient P2P Network Protocol."* In *Proc. Int. Green Computer. Conf.* Dallas, TX, U.S.A., November.

[6] "Blockchain and Associated Legal Issues for Emerging Markets." 2019. Accessed February 17 2020. www.ifc.org/thought leadership.

[7] "How does Keccak 256 hash function work?" Accessed February 17 2020.https://ethereum. stackexchange.com/.

[8] Dinur, I., O. Dunkelman, and A. Shamir. 2012. "New Attacks on Keccak-224 and Keccak-256." In: Canteaut, A. (eds) *Fast Software Encryption. FSE 2012. Lecture Notes in Computer Science.* Vol. 7549. Berlin, Heidelberg: Springer. doi:10.1007/978-3-642-34047-5_25.

[9] Stephen, Reyma, and Annena Alex. 2018. "A Review on Blockchain Security." *IOP Conf. Series: Materials Science and Engineering* 396: 012030. doi:10.1088/1757-899X/396/1/012030.

[10] Ouaddah, Aafaf, and Anas Abou Elkalam. 2017. "Fair Access: A New Blockchain-Based Access Control Framework for the Internet of Things." *Security And Communication Networks Security Comm. Networks.* doi:10.1002/sec.1748.

[11] Emanuel Ferreira Jesus, Vanessa, and R. L. Chicarino. 2018. "A Survey of How to Use Blockchain to Secure Internet of Things and the Stalker Attack." *Hindawi Security and Communication Networks* 2018: doi:10.1155/2018/9675050.

[12] Staples, M., S. Chen, S. Falamaki, A. Ponomarev, P. Rimba, A. B. Tran, I. Weber, X. Xu, and J. Zhu. 2017. "Risks and Opportunities for Systems Using Blockchain and Smart Contracts." doi: 10.4225/08/596e5ab7917bc.

[13] Lee, Jae Hyung. n.d. "Systematic Approach to Analysing Security and Vulnerabilities of Blockchain Systems." Accessed March 2 2020. https://web.mit.edu/smadnick/www/wp/2019-05.pdf.

[14] Zhang, R., R. Xue, and L. Liu. 2019. "Security and Privacy on Blockchain." *ACM Computing Surveys* 1 (1). doi:10.1145/3316481.

[15] Wang, Qi, Xiangxue Li. 2018. "Anonymity for Bitcoin From Secure Escrow Address." *IEEE* 6: 12336–12341. doi 10.1109/ACCESS.2017.2787563.

[16] Salah, Khaled, and M. Habib Ur Rehman. 2019. "Blockchain for AI: Review and Open Research Challenges." *IEEE Access:* 10127–10149.

[17] McConaghy, et al. 2018. "Bigchain DB: A Scalable Blockchain Database." *Big Chain DB*, GmbH, Berlin, Germany. Accessed January 10 2019. https://www.bigchaindb.com/whitepaper/ bigchaindb-whitepaper.pdf.

[18] Saad, Muhammad, and Jeffrey Spaulding. 2019. "Exploring the Attack Surface of Blockchain: A Systematic Overview." *arXiv*: 1904.03487v1 [cs.CR].

[19] "Top Five Blockchain Security Issues." Accessed March 2 2020. https://ledgerops.com/ blog/2019/03/28/top-five-blockchain-security-issues-in-2019.

[20] Frankenfield, J. n.d.. "DoubleSpending." *Investopedia.* Accessed February 10 2020. https:// www.investopedia.com/terms/d/doublespending.asp.

[21] Jang, Jehyuk, andHeung-No Lee. 2019. "Profitable Double-Spending Attacks." https://www. researchgate.net/publication/331543701.

[22] Singh, D. 2018. "Crypto-Jacking Attacks Rose by 8,500% Globally in 2017: Report." Accessed February 15 2020. https://goo.gl/qpGcZy.

[23] Marwala, Tshilidzi and B. Xing. 2018. "Blockchain and Artificial Intelligence." https://arxiv. org/abs/1802.04451.

[24] Zhi, Li, Xinlai Liu, Ali Vatankhah. 2018. Cloud-based Manufacturing Block chain: Secure knowledge-sharing blockchain for injection mould redesign. *Procedia CIRP* 72: 961–966.

[25] Tosh, Deepak K., and Sachin Shetty. 2017. *"Security Implications of Blockchain Cloud with Analysis of Block Withholding Attack."* In *17th IEEE/ACM International Symposium on Cluster, Cloud and Grid Computing*, pp. 458–467. doi:10.1109/CCGRID.2017.111.

[26] Bonomi, Flavio, Rodolfo Milito, Jiang Zhu, and Sateesh Addepalli. 2012. *"Fog Computing and its Role in the Internet of Things."* Proceedings of the First Edition of the MCC Workshop on Mobile Cloud Computing. ACM.

[27] Jeong, Jun Woo, and Bo Youn Kim. 2018. *"Security and Device Control Method for Fog Computer using Blockchain."* In*ICISS'18: Proceedings of the 2018 International Conference on Information Science and System*, pp. 234–238. doi:10.1145/3209914.3209917.

[28] Gauhar, Ali, Ahmed Naveed, and Muhammad Asif Yue Cao. 2019. "Blockchain-Based Permission Delegation and Access Control in Internet of Things (BACI)." *Computers and Security* 86: 318–384.

[29] Reyna, Ana, Cristian Martin, and Jaime Chen. 2018. "On Block Chain and its Integration with IoT. Challenges and Opportunities." *Future Generation Computer Systems* 88: 173–190.

[30] Ali, M. S., M. Vecchio, M. Pincheira, K. Dolui, and M. H. Rehmani. "Applications of Blockchain in the Internet of Things: A Comprehensive Survey." *IEEE Communication Surveys Tuts* (to be published).

[31] "Ethembedded." 2017. Accessed March 11 2020. https://ethembedded.com/.

[32] Article from "Electronics for you" January 2020 Edition Security: The Perils of Trivialising the IoT Security, pp. 46–47.

[33] Gu, J., B. Sun, X. Du, J. Wang, Y. Zhuang, and Z. Wang, 2018. "Consortium Blockchain-Based Malware Detection in Mobile Devices." *IEEE Access* 6: 1211812128.

[34] Nguyen, Dinh C., Pubudu N. Pathirana, and Ming Ding. 2019. "Integration of Blockchain and Cloud of Things: Architecture, Applications, and Challenges." *IEEE Communications Surveys & Tutorials*. arXiv: 1908.09058v1 [cs.CR].

[35] Pilkington, M. 2016. "Blockchain Technology: Principles and Applications. In *Research Handbook on Digital Transformations*, p. 225.

[36] Solat, S. and M. Potop-Butucaru. 2016. "Zeroblock: Preventing Selfish Mining in Bitcoin." *CoRR*. http://arxiv.org/abs/1605.02435.

10

Influence of AI, IoT and BC for Healthcare in Smart Cities - I

R. Sujatha

Vellore Institute of Technology, Vellore, Tamil Nadu, India

CONTENTS

10.1 Introduction to Evolution of Healthcare Industry

10.1.1 Healthcare Industry

The healthcare industry is the key factor that decides the wellbeing of people in the country. Each and every day moves with a great challenge for the medical field due to many new diseases that are entering the world. When a disease is contagious in nature, the effect and countermeasure that needs to be taken is very high. In the initial stages of many diseases, home remedies have been the solution, but due to the fast-growing world and medical technology, the way of treating is modernized. There is no query that the growth of healthcare is exponential in nature. In some countries, the entire healthcare industry is taken care of by the government and in some countries, it is the combination of federal and private agencies. The growth of the industry is marked with the beginning of digitalization of paper files, providing quality treatment at the earliest by diagnosing the disease at the initial stages, interacting with insurance agencies, drug regulations, and so on. The diagnosing methodology adopted is greatly increased, and intervention of the human has been reduced. The discussion that hefty data in healthcare is followed by the information drone environment is an evolution in strategy or revolution in manner [1].

The interest of work in the healthcare industry along with artificial intelligence (AI), Internet of Things (IoT), and blockchain (BC) is very impressive. The impact of this field is more in the way of diagnosing and treating the affected person. AI is in the process of using software and perpetual complex algorithms based on the requirement. Many devices work on the basis of the AI perspective. IoT is highly used hardware with software-related jargon in the healthcare industry. The sensors are used to monitor and track people's health. Highly developed countries completely integrate people in a centralized manner with the help of components, and great data is generated that requires careful analysis. BC is the quiet new addition in this field, and this is sort of ledger management makes the patient's access to records easier with this advent in healthcare turning into a smart healthcare environment [2–4]. The dominant trends in healthcare are value-based care, data-driven personalization, social determinants, patient engagement, and in a nutshell, coordinated care. Keywords and phrases that run the healthcare industry are cloud environment meant to store data; virtual private network; sensors; robots to treat patients in an unmanned manner; image processing for analyzing the X-Ray, CT, MRI images; data processing in EEG, ECG, and EMG data points; machine learning and deep learning methods to predict disease in advance stages; and treatments based on severity with the help of devices. The list is growing based on the nonstop research in the interdisciplinary context. Making decisions for patients with the assistance of the computer is an interactive manner that gathers the patient data, flows over the inference engine, and has the knowledge base that provides the diagnosis and treatment plan [5–9].

The interest of this AI, IoT, and BC could be easily found in Google trend that are represented in Figure 10.1. For the past 10 years, the graph was increasing in order. Another interesting inference is the difference in the varied role of the field. Figures 10.2, 10.3, and 10.4 represent 'AI, IoT and BC in healthcare' web searches across the globe. Inferences from Figure 10.2 indicate that the United Arab Emirates tops the AI in healthcare interest followed by Singapore, Australia, the Philippines, and so on. Similarly, in Figure 10.3, inference

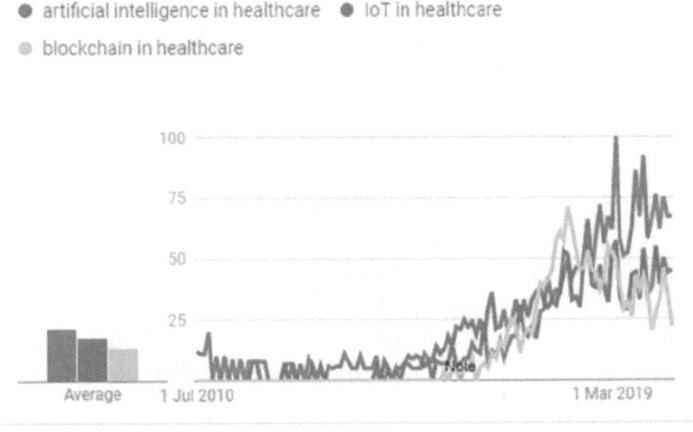

Worldwide. 20/06/2010 - 20/06/2020. Web Search.

FIGURE 10.1
AI, IoT and BC: Web Search.

FIGURE 10.2
AI in Healthcare: Google Trends.

FIGURE 10.3
IoT in Healthcare: Google Trends.

in India tops the IoT in healthcare interest followed by the United States, Canada, the United Kingdom, and so on. Figure 10.4 indicates the BC interest leads in India followed by the United States, the United Kingdom, and so on. It is very evident that the evolution of the healthcare industry is thought provoking.

10.1.2 Health Information Management (HIM)

HIM is the way of organizing information pertaining to all the components of the health-care industry in a comprehensive manner. In 1920, the usage of medical records started,

FIGURE 10.4
BC in Healthcare: Google Trends.

and in this process, practitioners began to note down individual patient information about complications and outcomes that soothed the patient. After 40 years, the computer made a change in the HIM. It begins the era of proving the platform for standardization and the need-based distribution of health records. Slowly the term healthcare information technology (HIT) bloomed by the way of integrating the computer with other technologies on the way. Various challenges faced are the unstructured data, and automation of the system-required time for syncing with the existing system and infrastructure requirement. The growth of primary healthcare information management is the required component in developing nations to ensure the safe health of all people even in the rural areas, and the work mentioned in this shows more privilege towards primary health information management [10]. HIM is a hefty area involving various roles but the outset is effective maintenance of hospital data in all viewpoints. Record preservation is a very vital area that requires a higher level of attention. HIM comprises data like clinical, demographic, financial, epidemiological, and reference information like the treatment history of patient and health databases.

10.1.3 Electronic Health Record

An electronic health record (EHR) is the digital format that helps in tracking patient's health condition from birth throughout their entire life. An HER is maintained through the medical facility, and may include all important administrative clinical data, including demographics, notes on medical progression, medical history, vital signs, medications, past immunizations, laboratory data, and radiology reports. The healthcare service depends on the EHR that holds complete information pertaining to diagnosis followed by a method of treatment to the patients. Healthcare intelligence gets useful information with the cumulative stored data from various patients and healthcare centers. Data provides insight into a prevailing health condition and also information to make decisions in times of emergency [11].

10.1.4 Electronic Medical Record

An electronic medical record (EMR) is a digital form of the paper files in the clinics. In the respective clinical practice, the patient's record is available in the EMR format that provides information about the individual. EMRs have advantages over paper records. Based on the visits the EMR is maintained by multiple hospitals and healthcare facilities. Primary access of data is given to patients and past data accessing is not easy. In a real-life situation, EMRs assist practitioners to keep track of a patient's data. It also reminds us about screenings or checkups. It analyzes important health parameters like blood pressure regularly. However, EMR and EHR serve different purposes, in spite of sharing certain characteristics.

10.2 Smart Healthcare

Data is the main component that runs the entire smart city and obviously in smart healthcare, the hefty, unstructured, highly velocity data keeps on flowing making the area so competitive and tough in the case of decision making. Fast, accurate, cheap, reliable healthcare provisions are the ultimate aim of the smart system. Data generated at the backend over time acts as a great aid in the case of a critical situation [12–14]. According to Deloitte [15] Global Healthcare Outlook, rather than data management, various key needs to be addressed are medical devices and wearable security and more work on telemedicine security. They have given the digital health ecosystem by considering various ingredients that are multidisciplinary in manner. Figure 10.5 indicates interoperability between all the success factors for the flawless, efficient working of the medical sector. Cloud computing solutions meant for storage of data in a virtual and secured way, robotics for an accurate and highly inflectionless environment for the patients, sensors that provide fast reporting for quick treatment purpose, AI that is the combination of data mining, machine learning, deep learning, statistics, data analysis, and enterprise applications mend the organizational capability for smooth functioning of all the departments. Blockchain meant for secured ledger maintenance and natural language processing acts as a background for the chatbot applications that work in the remote area and provide the input on the expert knowledge point of view. 5G along with data generated from various devices that is internet of medical things acts as a backbone and provides the great platform for the highly interoperable health ecosystem. Real-time challenges are the higher ratio of the aged population that require medical attention, chronic diseases are rising, advancement in technologies makes the investment in infrastructure and models increase exponentially, particularly in developing places the shortage of workforce and labor cost is high. Many leading organizations are working hard by considering many factors and it is varied because each landscape across the globe is not uniform in the way of culture, socio-demographics, people, lifestyle and so on. Smart cities are the dream of many developing countries to ensure the best lifestyle for the future generation and the government is taking steps to fulfill the requirements of the citizens [16].

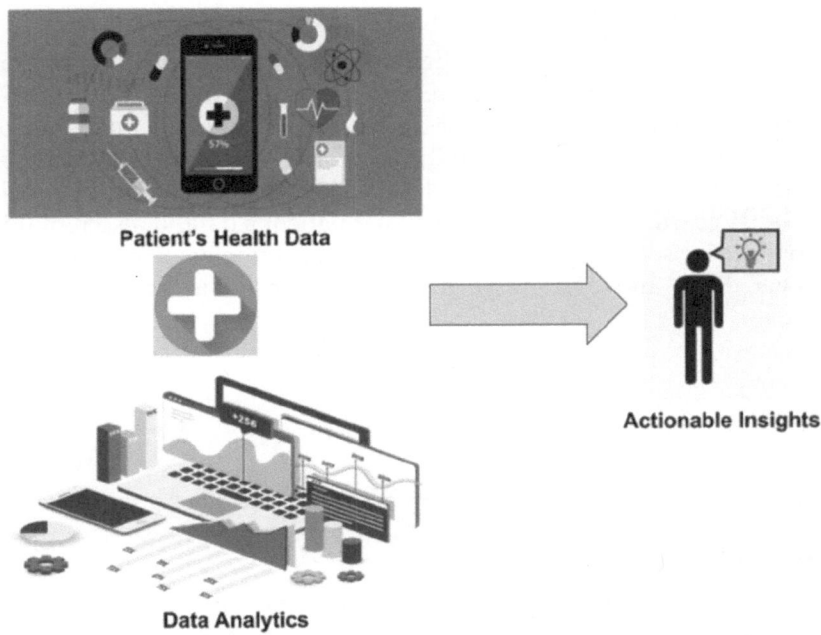

FIGURE 10.5
IoT – Healthcare System.

10.2.1 Artificial Intelligence in Healthcare

Various tools have been developed and are being developed to provide an apt diagnosis followed by prediction and treatment. The health field is moving toward deep medicine in a way of interpretation with the help of deep learning algorithms. A strategy to be considered in creating AI supporting the healthcare should include more vigilance in the governance of data usage of statistics and decision-making algorithms in a meticulous manner, efficient data visualization techniques that provide great insight in a glance, and proper inference of changeover in the clinical process by merging AI [17,18].

Computer information systems play a vital role in the present medical field and help in improving the caretaking processes. The predictive statistical model with data from physicians will make the system a more informative way to support patients and practitioners. Providing the benefits to stakeholders along with cost reduction is the success factor in the healthcare industry [19–23].

The change in AI healthcare is enormous and the magnitude of it is very impactful. Recently *Harvard Business Review* provided an interesting chart that shows the top 10 AI applications in healthcare from the viewpoint of potential annual value by 2026. The same is given in Table 10.1. Avenues mainly show that the need to change human intervention is more to provide the best medical service revealed in the first two applications, namely robot for surgery and virtual nursing.

Builtin.com mentioned in an article that in 2015 the U.S. had more misdiagnoses and a 10% death due to error. Humans face some issues in working with huge datasets or previous histories of sickness to make a perfect diagnose [24]. AI was used in rescue for this and in their article, they have presented impressive recent activity. Let's explore it.

TABLE 10.1

AI Application in Healthcare

Application	Potential Annual Value by 2026
Robot-assisted surgery	$40B
Virtual nursing assistants	$20B
Administrative workflow	$18B
Fraud detection	$17B
Dosage error detection	$16B
Connected machines	$14B
Clinical trial participation	$13B
Preliminary diagnosis	$5B
Automated image diagnosis	$3B
Cybersecurity	$2B

- PathAI – ML to assist pathologist – Accurate cancer detection
- Buoy Health – AI – Symptom checker
- Enlitic – AI DL – Radiology diagnoses
- Freenome – AI – Earlier cancer detection
- Beth Israel Deaconess Medical Center – AI – Early detection of deadly blood disease
- Zebra Medical Vision – AI – Clinical findings
- BioXcel Therapeutics – AI – Developing new medicines in immuno-oncology and neuroscience
- Berg Health – AI-based biotech area – accelerate medicine finding
- XtalPi – AI, cloud, and quantum physics – Drug discovery
- Atomwise – AI – Clinical trials
- Deep Genomics – AI – Neuro-related drugs
- BenevolentAI – AI, DL – Targeted treatment
- Olive – AI – Handle repetitive task in an optimal manner
- Qventus – AI – Automation for optimizing healthcare service
- Tempus – AI – Massive data collection for personalized health
- Proscia – Image Processing, AI – Detecting cancer cells
- Google Deepmind – AI – Alerting practitioners based on patient's health
- Vicarious Surgical – AI, Robots – Virtual enabled surgery
- Auris Health – AI, Robot – Endoscopy procedure
- Microsure – Robots – Surgical precision
- Mazor Robotics – AI, Robots – Spinal surgeries

Change in healthcare because of AI is very thoughtful. Research in this domain is seamless and the way of interpretation makes the birth of new invention possible. Various field experts meeting in the middle for discussions will make this a highly remarkable field.

Natural language processing along with text mining helps in extracting data from faxes at OneMedical, and Google is joining hands to build prediction models that give signals during the critical condition of sickness [25]. The Philips future health index actively participates in an innovative way of technology and has given many good healthcare products that benefit several stakeholders. Recent work done by Philips mentioned six prominent niche areas, namely precision diagnosis, cancer care, computational pathology, acute care, image-guided therapy, and access to care [26–28]. Interestingly, survey results about the ration of healthcare professionals comfortable with AI is given in Table 10.2. The output of healthcare professionals that are comfortable with AI provides the way to research to make it ergonomically comfortable.

10.2.2 IoT in Healthcare

A smart hospital is the output of managing the entire healthcare with the help of IoT. In this process, data is generated from each activity via sensors. Sensor technologies join hands to make the system highly automated and personalized but the handling of data is cumbersome and proper knowledge about the domain is mandatory [29–31].

Very commonly used IoT devices are headsets that measure brainwaves, blood pressure monitors, EEG, ECG, EMG monitors, glucose monitors, pulse oximeters, and more are added frequently. A preliminary service to the remote at a fast phase is called TeleHealth; obviously part of the smart component. TeleHealth is the need of an hour in this 21st century and various services are the following:

- TeleMedicine
- TeleMonitoring
- TeleSurgery
- Remote Medical Education

Advantages of TeleHealth are immediate and remote services, and it is very handy during disaster and emergency times, waiting time in a queue is nil, distance is not a matter and service is at your doorstep, cost-effective, with reduced documentation, and better communication. Figure 10.5 precisely shows an IoT-based healthcare system. The data is gathered in various source views and with effective data analytics, the actionable decision is generated. Many organizations work on this and the big boom is anticipated. Frost and Sullivan in their analysis stated that the internet of medical things will expand exponentially to show a 26.2% growth in annual rate. Also, by the end of 2021, there is a possibility to

TABLE 10.2

Percentage of Healthcare Professionals Comfortable with AI

Clinical	Actioning treatment plans	45%
	Diagnosis	47%
	Recommending treatment plans	47%
	Flagging anomalies	59%
Workflow	Patient monitoring	63%
	Staffing and patient scheduling	64%

reach US$72,000 million. It's not only in the part of economic growth but also the standard of treatment will be optimized in the manner of collecting data at the earliest and processing effectively.

Since huge data is evolving, the challenge prevails in all the phases. It begins with a security breach where care is taken to ensure nonauthorized persons are not intruding, when a person is having multiple ailments for each different device required and then the need to integrate them, and decision making from hefty data is tough. Effective working in real time with thorough knowledge is needed to make it a more interesting and informative platform.

10.2.3 Blockchain in Healthcare

The healthcare industry has been at the bleeding edge of innovation in recent years. Hardware, programming, medicine, surgeries, and the nature of care accessible to patients in 2020 is better than ever before. However, the organizational effectiveness and information that the executives are supporting that care are seriously lagging behind. The answer to this is blockchain technology. A more competent and secure way of keeping electronic health records that gives the innovative environment for health information exchange (HIE) is the great achievement of blockchain technology.

Blockchain is quite the latest in the field to provide a distributed platform for storing and retrieving electronic health records in the healthcare system. Ledger technology in the blockchain ensures secure handling of patients records and drug supply chains. Major characteristics like transparency with private access and incorruptibility along with the decentralized nature make it too unique. In a nutshell, it can be illustrated as the component reviving healthcare. Blockchain-based medical records helps in streamlining healthcare and costly mistakes are avoided. The medical supply chain is greatly optimized in their level of working by initiating the procurement of drugs once the requirements for the patient are seen. The handling of drugs from the origin until it reaches the end-user is transparent and trackable. Counterfeiting of drugs is drastically reduced by intervening in the blockchain concept in the pharma industry that links the healthcare platform, pharma industry, and end-user. The timestamp component in the blockchain and peer-to-peer (P2P) way of communication in the blockchain is highly commendable during disease outbreak based on patterns generated by the accumulated data that help get clarity about the origin and severity of disease across various locations. The features also make the path in the genomic industry, too.

Healthcare is facing hurdles and losing US$300 billion each year in the process of maintaining huge data, and security is a great threat intertwined with it. With the help of this intricate blockchain system, the health industry can produce a secured and trusted environment. Once data is created it is peer checked and then block is created. Once updating happens again it passes to the intended users and then a new block is generated. Only a person who is holding the matching key can visualize the data. Key dimensions are like a shared ledger; consensus, privacy, and a smart contract make it special. The best instance in claim process denials happens when some information is missing in the application but with the help of a smart contract in the blockchain, the field helps to accompany the transaction and provides all the data without interruption and speeds up the process of filing a claim. In the privacy point of view, the transaction gets converted into the hash and then created as the block in connection with the previous block. It creates a chain of blocks that makes healthcare-related blockchain. Medical records created are converted into blockchain and based on the patient's agreement. The intended practitioner can view the data [32,33].

10.3 Personalized Healthcare

A personalized healthcare concept is the milestone in the medical field and is used more in tumor treatment based on the criticality and type. In the earlier days, leukoma treatment began with a number of tests, chemotherapy, and the same medical dosages. But with the birth of personalized healthcare, biomarkers aids to provide efficacy and safety treatment for the sick. Genetic information is also integrated to provide apt treatment with the correct dosage. Blood samples are collected and fed into the decision support system created specifically for the decision-making process. It receives input from practitioners, researchers, and similar datasets to provide specific insights. Personalized medicine is the common word used adjacently with this concept and it is in the very infant stage. In the U.S., the first cell-based therapy was performed in 2018 for a sequence of treatment priced US$475,000. Other than cancer, the concept could be applied for Alzheimer's or diabetes in which the relationship exists in mutation of DNA. A prediction at the earlier stage can secure a patient's life. Genomes structure plays at the backend with information technology methods to provide the best fitting treatment for those who need it [34–36].

10.4 Elderly Healthcare

For older people around 65 years of age, taking care of their health is a major challenge in all nations. Aging causes a slowdown in the activities of day-to-day life and the immune system gets crippled. In the case of comorbidities, the scenario is worse. Aging includes hearing loss, eye problems, neck and knee joint pains, dementia, diabetes, depression, and pulmonary disease. Based on the living environment and living style, conditions may vary. In the case of elderly people, caretaking is the primary component to ensure someone is there for them, and psychologically older people feel more insecure. In the modern world, moral support is lacking and that has an impact on health. As mentioned in the earlier part of the chapter the labor cost is growing high and scarce. Virtual aids help in monitoring and tracking the health of sick people. Elderly people require more medical attention, and recovery also takes time. Many frameworks are proposed and used in different nations to help the effective management of elderly people [37,38].

According to data bridge market research, the cost to take care of elderly people is increasing across the globe, and the count of elderly people is also growing. As illustrated in Figure 10.5, the forthcoming years have greater market growth and its predicted to reach US$1,944,028.05 million by the end of 2027. The major factor that constitutes growth is the awareness of care service to provide a home facility by paying more monetary benefits. The nuclear family and movement of people in pursuit of jobs in more urban locations create an aloof feeling for many elderly people. An alarming increase in chronic disease and continuous medication increases growth.

Types of services provided for the elderly by home health agencies are the following:

- Speech therapy
- Counseling
- Physical therapy
- Occupational therapy
- Nursing services
- Medication management
- Homemaker services
- Nutrition services
- Physical services

Nursing services are the most wanted followed by physical therapy. Based on illness, a combination of services helps in providing a peaceful life for elderly people. Digital assistance helps in the combination with home care services to ensure integrity.

10.5 Patient Management

Patient management is a very crucial part of the healthcare sector. It includes several elements like patient clinic relation, examination, diagnosis, and treatment. To have a desirable treatment output, patient management becomes a very important factor. A well-organized system is required to handle a patient from day one. A patient's medical and social history plays a recognizable role in timely diagnosis and treatment. The traditional way of patient management works by maintaining the health records of patients in physical format. It requires a separate record section every clinical to have a good output. But there are a lot of disadvantages to the traditional method. It's impossible to store records for long years in an organized way. With an increasing population and increasing patients, patient management becomes a cumbersome task for clinics, so there is a requirement for a technology-based patient management system.

Patient management can be completely modernized by technological applications. From appointment fixing to drug intake reminders, it can be incorporated in a single system. Traditional patients go to the clinic only when they have severe issues and it is unplanned. This leads to a mess in the front office and long waiting hours. Rather than this, there can be a regular reminder to the patient to have check-ups so that it can be scheduled in a hassle freeway. In this case, ailments can be diagnosed early. Rather than storing patient records physically, everything can be stored in cloud-based storage which can be accessed across the globe at any time. So a unified system can lead to a uniform healthcare system in every corner of the world. There is also telemedicine and other easy ways to consult physicians. A clinic can keep track of a patient's medications and remind them to take drugs and when to restock. So an enhanced patient management system with technology is required to have a superior healthcare service. Based on the disease patient management is taken into consideration. In this work, Alzheimer's and diabetes-based patient management is discussed [39–41].

10.6 Conclusion

Healthcare is an essential element of every society. In smart cities where every element of life is taken to the next level by the involvement of AI, IoT, and blockchain, a strong and effective healthcare system is a basic part. The influence of AI, IoT, and blockchain has an immediate and intense positive effect on the healthcare sector. This paves the way for quick diagnosis, timely treatment, and maintaining stable medication. The smart system assists all workers in every aspect. It also reduces errors and plays a critical role in taking care of elderly people. Smart patient management is highly efficient. It's very clear that the digital environment with the set of algorithms makes the field of medical vibrant. The direction for future work may be directed to take advantage of healthcare by putting together the features of blockchain and IoT with the goal of supporting network scalability and low-end gadgets.

References

[1] Grimson, J., and W. Grimson. 2002. "Health Care in the Information Society: Evolution or Revolution?" *International Journal of Medical Informatics* 66 (1–3): 25–29.

[2] Hasnat, M., S. A. Mamun, F. Hossain, and S. Hossain. 2019. "IoT Based Smart Healthcare." Master Thesis, United International University Dhaka, Bangladesh.

[3] Lin, S. H., and M. Y. Chen. 2019. "Artificial Intelligence in Smart Health: Investigation of Theory and Practice." *Hu Li Za Zhi* 66 (2): 7–13.

[4] Zhang, P., D. C. Schmidt, J. White, and G. Lenz. 2018. "Blockchain Technology Use Cases in Healthcare." In *Advances in Computers*, Vol. 111, pp. 1–41. Elsevier.

[5] Canlas, R. D. 2009. "Data Mining in Healthcare: Current Applications and Issues." In *School of Information Systems & Management*. Australia: Carnegie Mellon University.

[6] Erstad, T. L. 2003. "Analyzing Computer-Based Patient Records: A Review of Literature." *Journal of Healthcare Information Management* 17 (4): 51–57.

[7] Mehta, V. K., P. S. Deb, and R. A. O. D. Subba. 1994. "Application of Computer Techniques in Medicine." *Medical Journal Armed Forces India* 50 (3): 215–218.

[8] Razzak, M. I., S. Naz, and A. Zaib. 2018. "Deep Learning for Medical Image Processing: Overview, Challenges and the Future." In *Classification in BioApps*, pp. 323–350. Cham: Springer.

[9] Ţăranu, I. 2016. "Data Mining in Healthcare: Decision Making and Precision." *Database Systems Journal* 6 (4): 33–40.

[10] Zhao, Y., L. Liu, Y. Qi, F. Lou, J. Zhang, and W. Ma. 2019. "Evaluation and Design of Public Health Information Management System for Primary Health Care Units Based on Medical and Health Information." *Journal of Infection and Public Health* 13 (4): 491–496.

[11] Mayer, A. H., C. A. da Costa, and R. D. R. Righi. 2019. "Electronic Health Records in a Blockchain: A Systematic Review." *Health Informatics Journal* 26 (2): 1273–1288.

[12] El Zouka, H. A., and M. M. Hosni. 2019. "Secure IoT Communications for Smart Healthcare Monitoring System." *Internet of Things*: 100036.

[13] Sakr, S., and A. Elgammal. 2016. "Towards a Comprehensive Data Analytics Framework for Smart Healthcare Services." *Big Data Research* 4: 44–58.

[14] Syed, L., S. Jabeen, S. Manimala, and A. Alsaeedi. 2019. "Smart Healthcare Framework for Ambient Assisted Living Using IoMT and Big Data Analytics Techniques." *Future Generation Computer Systems* 101: 136–151.

[15] https://www2.deloitte.com/content/dam/Deloitte/global/Documents/Life-Sciences-Health-Care/gx-lshc-digital-transformation-and-interoperability.pdf. Accessed 20 June 2020.

[16] Jararweh, Y., M. Al- Ayyoub, and E. Benkhelifa. 2020. "An Experimental Framework for Future Smart Cities Using Data Fusion and Software Defined Systems: The Case of Environmental Monitoring for Smart Healthcare." *Future Generation Computer Systems* 107: 883–897.

[17] Davenport, T. H., and R. Kalakota. 2019. "The Potential for Artificial Intelligence in Healthcare." *Future Healthcare Journal* 6 (2): 94.

[18] Wiljer, D., and Z. Hakim. 2019. "Developing an Artificial Intelligence–Enabled Health Care Practice: Rewiring Health Care Professions for Better Care." *Journal of Medical Imaging and Radiation Sciences* 50 (4): S8–S14.

[19] Bates, D. W., S. Saria, L. Ohno-Machado, A. Shah, and G. Escobar. 2014. "Big Data in Health Care: Using Analytics to Identify and Manage High-Risk and High-Cost Patients." *Health Affairs* 33 (7): 1123–1131.

[20] Jameson, J. L., and D. L. Longo. 2015. "Precision Medicine—Personalized, Problematic, and Promising." *Obstetrical & Gynecological Survey* 70 (10): 612–614.

[21] Krumholz, H. M. 2014. "Big Data and New Knowledge in Medicine: The Thinking, Training, and Tools Needed for a Learning Health System." *Health Affairs* 33 (7): 1163–1170.

[22] Parikh, R. B., M. Kakad, and D. W. Bates. 2016. "Integrating Predictive Analytics into High-Value Care: The Dawn of Precision Delivery." *JAMA* 315 (7): 651–652.

[23] Parikh, R. B., J. S. Schwartz, and A. S. Navathe. 2017. "Beyond Genes and Molecules-A Precision Delivery Initiative for Precision Medicine." *The New England Journal of Medicine* 376 (17): 1609.

[24] https://builtin.com/artificial-intelligence/artificial-intelligence-healthcare. Accessed June 20 2020.

[25] Davenport, T. H., T. Hongsermeier, and K. A. McCord. 2018. "Using AI to Improve Electronic Health Records." *Harvard Business Review* 12: 1–6.

[26] https://hbr.org/2018/05/10-promising-ai-applications-in-health-care. Accessed June 20 2020.

[27] https://www.databridgemarketresearch.com/request-a-sample/?dbmr=global-elderly-care-market. Accessed June 20 2020.

[28] https://www.philips.com/a-w/about/news/archive/features/20200107-six-areas-in-which-ai-is-changing-the-future-of-healthcare.html. Accessed June 20 2020.

[29] Fischer, G. S., R. da Rosa Righi, G. de Oliveira Ramos, C. A. da Costa, J. J. Rodrigues, et al. 2020. "ElHealth: Using Internet of Things and Data Prediction for Elastic Management of Human Resources in Smart Hospitals." *Engineering Applications of Artificial Intelligence* 87: 103285.

[30] Kanase, P., and S. Gaikwad. 2016. "Smart Hospitals Using Internet of Things (IoT)." *International Research Journal of Engineering and Technology (IRJET)* 3 (3).

[31] Thakare, V., and G. Khire. 2014. "Role of Emerging Technology for Building Smart Hospital Information System." *Procedia Economics and Finance* 11: 583–588.

[32] Tanwar, S., K. Parekh, and R. Evans. 2020. "Blockchain-Based Electronic Healthcare Record System for Healthcare 4.0 Applications." *Journal of Information Security and Applications* 50: 102407.

[33] Tripathi, G., M. A. Ahad, and S. Paiva. 2019. "S2HS-A Blockchain Based Approach for Smart Healthcare System." In *Healthcare*, 100391. Elsevier.

[34] Abbas, A., M. Ali, M. U. S. Khan, and S. U. Khan. 2016. "Personalized Healthcare Cloud Services for Disease Risk Assessment and Wellness Management Using Social Media." *Pervasive and Mobile Computing* 28: 81–99.

[35] Feldman, K., D. Davis, and N. V. Chawla. 2015. "Scaling and Contextualizing Personalized Healthcare: A Case Study of Disease Prediction Algorithm Integration." *Journal of Biomedical Informatics* 57: 377–385.

[36] Fuentes-Garí, M., E. Velliou, R. Misener, E. Pefani, M. Rende, N. Panoskaltsis, E. N. Pistikopoulos, et al. 2015. "A Systematic Framework for the Design, Simulation and Optimization of Personalized Healthcare: Making and Healing Blood." *Computers & Chemical Engineering* 81: 80–93.

[37] Costa, A., A. Yelshyna, T. C. Moreira, F. C. Andrade, V. Julian, and P. Novais. 2017. "A Legal Framework for an Elderly Healthcare Platform: A Privacy and Data Protection Overview." *Computer Law & Security Review* 33 (5): 647–658.

[38] Khawandi, S., B. Daya, and P. Chauvet. 2011. "Implementation of a Monitoring System for Fall Detection in Elderly Healthcare." *Procedia Computer Science* 3: 216–220.

[39] Frisoni, G. B., F. Barkhof, D. Altomare, J. Berkhof, M. Boccardi, E. Canzoneri, and R. Gismondi. 2019. "AMYPAD Diagnostic and Patient Management Study: Rationale and Design." *Alzheimer's & Dementia* 15 (3): 388–399.

[40] Kwon, H. S., J. H. Cho, H. S. Kim, J. H. Lee, B. R. Song, J. A. Oh, H. Y. Son, et al.. 2004. "Development of Web-Based Diabetic Patient Management System Using Short Message Service (SMS)." *Diabetes Research and Clinical Practice* 66: S133–S137.

[41] Lusted, L. B. 1971. "Decision-Making Studies in Patient Management." *New England Journal of Medicine* 284 (8): 416–424.

11

Influence of AI, BC and IoT for Healthcare – II

T. Subha, R. Ranjana, and T. Sheela

Sri Sairam Engineering College, Chennai, Tamil Nadu, India

CONTENTS

11.1 Introduction

Automation has revolutionized all industries and affected all walks of life. One of the major industries that adopted automation and analytics is the healthcare industry. It is one of the industries with a vast domain and consists of clinical trials, pharmaceuticals, and medical equipment. As it is, the healthcare industry strives to improve the lives of people. There are several studies that report about the deaths that occur in hospitals due to human error. The cause for these errors is attributed to the storage and access mechanism of details about the patients. It has also been reported in studies that three out of four errors would have been averted if proper patient information systems were in place. Electronic health records

(EHR) were developed in the early 1960s and they are used to basically store important patient details like medical history, drug information, prescriptions, lab tests reports, medical x-rays and images, and billing information.

11.1.1 What Is EHR?

EHR or the electronic health record is responsible for storing and maintaining the patient's data in a digital form; they are patient records that can be accessed instantly and securely by authorized users [1]. The basic necessity of building an EHR is to save the medical history and treatment plans for a specific user. However, based on the data available in the EHR, broader healthcare can be provided to the patient. Presently EHRs have become a vital part of healthcare infrastructure and they can be used to:

- Store and manipulate a patient's medical history, drugs administered, vaccination dates, images, test results, and more.
- Allow the usage of diagnostic tools that help make decisions about the care the patient needs.
- Consult experts across geographical boundaries by providing access to vital medical data, thereby making state-of-the-art medical care feasible for everyone.
- The basic medical tests taken at one healthcare center need not be repeated at the next facility if all the patients move from one hospital to another [2].

The basic patient healthcare information that an EHR carries could be, but not limited to:

- Basic patient personal profile including age, demographics
- Medical history
- Allergies to drugs if any
- Laboratory tests results
- Prescriptions and drugs administered
- X-ray and scan images
- Patient progress
- Disease diagnosis

EHRs are not a simple replacement of the paper record of the patient. It is converting the patient history along with relevant details into a digital format that can be accessed securely by multiple healthcare workers.

11.1.2 General Structure of the Healthcare System Using EHR

In this section, a brief description of the open EHR [3] architecture is given. Open EHR is an architecture that has been standardized based on research from numerous projects and standards. The architecture has been refined based on many real-time inputs that are obtained after several years of use. The architecture is therefore very generic and can be used for many applications other than the original healthcare EHR. A similar reference model can also be used for animal clinics because the architecture discusses concepts involved in service and administrative events that are relevant to a caring system. In another perspective, though it is defended that open EHR is patient-centric and a shared healthcare record system, it is not limited only to that and can also be used in special situations like radiology record maintenance. The deployed EHR can be categorized based on scope of the applications and the subject whose data is being managed.

Figure 11.1 shows the architecture of an EHR based on open EHR; each requirement is enclosed within a flat oval bubble. The top left bubble represents the generic record for any kind of subject. Subsequent requirements for corresponding living subjects are added at the lower left. The largest bubble, on the left, represents the requirements pertaining to 'local health records for human care,' which includes images, hospital patient records called EPR and other details. The important requirements are ordered until the bottom row of the diagram. One of the main objectives of building an EHR is to implement an integrated record that could be accessed by multiple healthcare authorities which is actually an integrated informational framework. Since the open EHR takes a generic approach, the components and the applications built to satisfy a healthcare management system can also be used for any other record management system. The following section explains the requirements and details that shaped the open EHR into its present form.

11.1.2.1 Generic Care Record Requirements

The generic details that need to be present in the open EHR are anonymization of the patient when records are needed for a research purpose such as auditing and accountability. The details must be collected in a generic format independent of the healthcare provider and the application. The software that manages the records must be flexible and easy to operate and maintain.

11.1.2.2 Healthcare Record (EPR)

Open EHR specifies certain requirements for local health records or EPR. The basic requirement is to support all types of test data and accommodate different types of units for the values observed. The system must support predominantly used natural languages and also allow for translations between these.

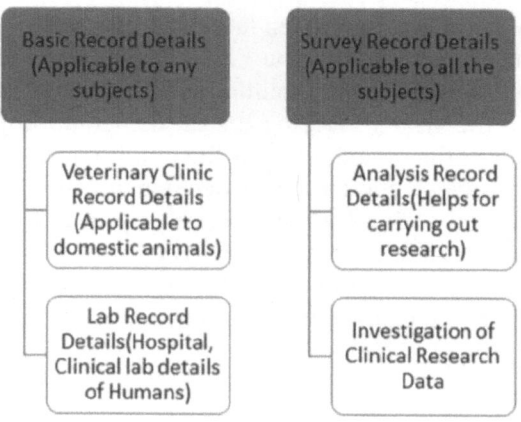

FIGURE 11.1
Sample Architecture of a Healthcare System Using Electronic Healthcare Records [3].

11.1.2.3 Shared Care EHR

For the shared care EHR, the requirements specified in open EHR include the support of preventing patients' identities. The method of sharing the EHRs and the interoperability between healthcare providers is provided by automated workflows.

11.1.3 Advantages of EHR

The major advantages of using an EHR is that it is accessible and available anytime and anywhere irrespective of the patient's location. The patient also has access to their medical history, giving the confidence of knowing their health status. Coordinated healthcare is where specialists from multiple healthcare disciplines can coordinate across geographical boundaries to provide the best treatment for a patient. This coordinated practice will also improve diagnostics and ensure a quality treatment plan. Patients can save costs that occur when basic tests and X-rays are required if a person visits a new healthcare facility. It allows the patient to be independent of a particular healthcare provider, as the details are readily available without the requirement of new laboratory tests.

EHRs have the capacity to increase communication between physicians and their patients [4]. EHRs definitely increase the burden on the clerical staff but bring in easy management and tracking of a patient's treatment plan. If the EHRs are implemented using a web-based application, then it gives patients access to view their medical records, progress, and lab test results. EHRs also avert the problem of files that are lost or misplaced thereby overcoming the major problem faced by the paper-based methods.

11.1.4 How to Implement an EHR?

Having said the advantages of EHR, the next question is how a cost effective EHR can be built. Many studies [5] have categorized the steps in building an EHR into two:

1. The pre-implementation phase
2. The implementation phase

In the pre-implementation phase a thorough study of the process that needs to be followed in creating an EHR that focuses on data integrity is done. The inputs of people involved in the system like the healthcare professionals, lab technicians, and also patients are taken and discussed regarding the building of the EHR. Timely feedback is collected and the workflow must be refined. Education and training need to be given if needed to those who are reluctant to adapt to the system. During the implementation phase, brainstorm sessions are organized to adapt the system to meet the requirements. A change management process must be properly defined and it must be decided how to move the patient data into the system. Security measures with robust authorization protocols must be defined. Levels of accountability in the case of a data breach need to be defined. Data anonymization techniques need to be in place in case data is given for research and analytics purposes. Sufficient training should be given to staff for getting adapted to the system.

11.1.5 Problems with EHR

Although implementation of EHR has many merits, there are a few disadvantages. The first and foremost is the burden on the medical professionals who need to get used to the system. The cost involved could be relatively higher than the merits for independent medical physicians. It is also required that experienced IT staff are required to work with EHRs, so this adds to the cost of EHR. The training and adaptation time requires additional time and effort from medical professionals. During the initial phase of usage, when staff tends to adapt to the system, errors are prone to occur resulting in a risky environment. EHRs are bound to be useful if and only if they have robust interoperability features.

11.2 Applications of Machine Learning Techniques for EHR Maintenance

Artificial intelligence (AI) has given a new dimension to healthcare analytics and diagnosis. Machine learning (ML) algorithms tend to provide accurate insights into diagnosis by accessing training data and allowing physicians to decide on an effective treatment plan [6]. EHRs use artificial techniques to create user-friendly interfaces and automate the process of data collection. Voice recognition systems that use natural language processing tools are suggested to improve the clinical documentation process.

AI may be beneficial in certain processes like to answer routine frequently asked questions (FAQs) from mailboxes, medicine refills, and result notification [6]. It can also be used to prioritize patient visits and treatments for a physician.

11.3 Application of Internet of Things in Healthcare

The healthcare industry has obtained a new dimension with mobility solutions that enables interoperability, machine to machine data communication that enhances healthcare services. Internet of Things (IoT) devices can be used to collect, process, analyze, and report data in a real-time environment thus avoiding storage of data. The data analysis process could very well happen over the cloud and the final report could be generated for further analysis. This analysis helps to make a better decision on an efficient patient treatment plan [7].

Real-time tracking of chronic patients is made feasible using medical IoT devices that gather the vital parameters of a patient and transfers them to a physician for intervention if needed [8]. These applications also provide instructions to caretakers thereby keeping a constant check on patients without the patient necessarily being in the hospital. If an emergency occurs, the patient can contact the doctor through the applications. Nowadays, remote healthcare assistance is gaining prominence. Machines that can deliver drugs to patients based on the prescription are under study.

The major hinge in the usage of IoT in healthcare is the security [9] as patient data is one of the most sensitive data and could be breached. IoT devices lack a proper data protocol or any standard. There is also no confined regulation governing the ownership of data that IoT devices collect. This could be a prospective avenue for cyber attacks that can compromise systems and get access to personal patient data. These attacks may result in filing of fraudulent insurance claims and buying of drugs.

11.4 Application of Blockchain in Healthcare: A Case Study

Block chain technologies have revolutionized the healthcare industry by offering decentralized patient history, tracking drug prescriptions, and providing secured payment options. In this section a case study pertaining to the application of blockchain technology for patient EHR is discussed. The system uses NFC (Near Field Communication) cards and blockchain for secure usage and maintenance of patients' medical records.

An NFC-based healthcare device is designed to reduce the burden of paperwork [10]. NFC tags are integrated circles that store data that can be read by an NFC-enabled device. NFC stands for field communication that is standard low technology, and is the easiest way to communicate between electronic devices that allow for transactions. When a patient visits the hospital, they will be given a unique tag number by a doctor. With previous drugs taken, along with an x-ray or scan reports, the doctor uses this tag to insert a tap into his device using the information displayed on the device. The objective of this work is to summarize the state of the art and to identify key themes in health research. With the help of this technology, the patient does not need to carry all of their medical records and the doctor can access patient information using the NFC marker or share it for research purposes [11].

A framework for managing and sharing EMR patient care information is shown in Figure 11.2 [12]. In collaboration with a hospital, a framework is developed that ensures privacy, security, and availability over EHR information. The proposed work will considerably cut back the turnaround for EMR sharing, improving decision making for treatment.

By providing a tool for achieving consensus among distributed organizations without relying on a single trusted team, blockchain technology will ensure data security, control sensitive data, and enable the management of healthcare data for patients and unique actors in the medical domain. Blockchain technology in healthcare is the unstable and unchanging nature of the fluctuations in the appearance and integrity of sensitive medical data. The program has secured access to patients' medical histories and treatment by updating information and accessing all medical records for patients.

11.5 Detecting Fraudulent Behavior

One of the major areas of applications of AI techniques in the healthcare sector is detecting and averting fraudulent behavior. The fraud detection in healthcare could be categorized into three types based on the person who is involved in the healthcare fraud [13]. They are the healthcare providers, subscribers, and carriers. The providers are the physicians, the healthcare facility, and the testing facilities. The insurance subscribers are the customers who opt for medical insurance like patients or the patient's employer. Insurance carriers are third parties who collect premiums from users called subscribers. The carriers will pay the medical costs for the customers concerned. Based on the activity and the person involved, fraudulent behavior can be classified as the following [14,15]:

1. Fraud by the service provider: Fraudulent in the billing procedure. Billing more than what service was given to the patient. Giving the highest possible treatment for a higher claim. Faulty diagnosis and prescribing tests that are medically not necessary. Representing non-covered treatments as ones that are under coverage.

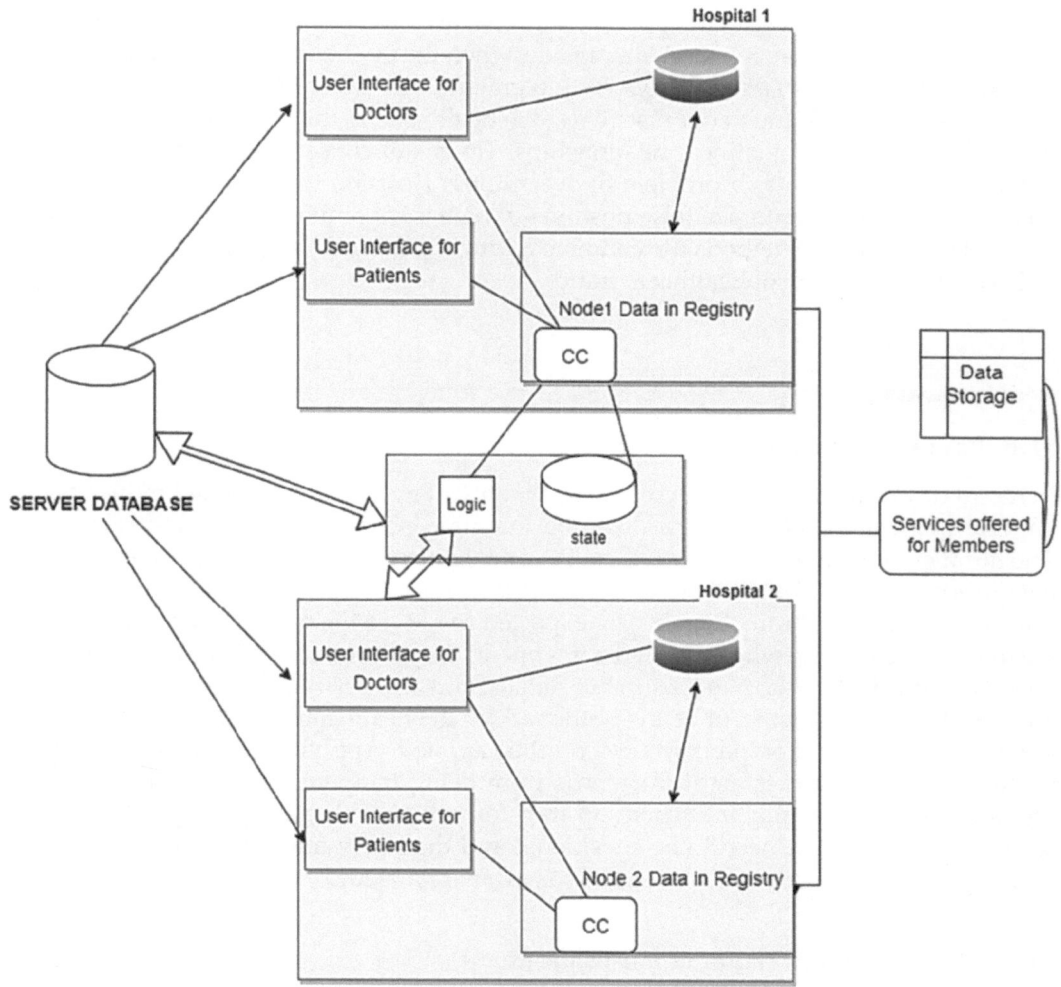

FIGURE 11.2

A Framework for Managing and Sharing EMR Patient Care Data Using Blockchain Technology [12].

2. Fraud by the subscriber: Producing a false employee record enabling eligibility for a low premium. Claiming medical insurance for services not received. Illegal claim wherein the claim is not made for the right person.

3. Fraud by the carrier: Claiming for false reimbursement and producing false service statements.

Machine learning techniques are used to identify fraud detection in healthcare and can be classified based on the learning method used as supervised and unsupervised methods [16]. These two methods are discussed at length in this section.

11.5.1 Application of Supervised and Unsupervised ML Methods

Several researches have been undertaken to study the significance of using supervised methods like neural networks, decision trees, and Bayesian networks for healthcare fraud detection. Most of the research reports neural networks and decision trees to be most accurate. Some literature also reports better results when the methods are combined. For example, in [17] the authors used fuzzy sets integrated with a Bayesian classifier to identify fraudulent claims. Neural networks are used in healthcare as they are found to be effective in handling data with nonlinear relationships. The major concern with a neural network, however, is that it creates a problem of over fitting. Decision trees can also be used with adaptive training to minimize false positives, which is minimizing costs for false identification. Unsupervised methods are combined with supervised learning techniques to provide accurate detection of healthcare fraud.

11.6 Payments through Cryptocurrency

In recent days Bitcoin has been accepted as a payment method for more and more business ventures. Nowadays Bitcoins are also used in online eCommerce stores and restaurants. The number of businesses that are using Bitcoin are increasing as we apply them in different areas of our lives [18]. Bitcoins are entering the pharmaceutical industry and are getting introduced into healthcare. Bitcoins are expected to reduce the number of procedures that we require to complete when we need a treatment. Procedures involve getting the appointment with the doctor, verifying insurance details, and also verifying and updating medical records. This is considered to be the bottleneck of the healthcare service. This is one area where blockchain and cryptocurrency could be applied. Applying blockchain techniques in healthcare can protect medical records from being tampered. Hence blockchain and cryptocurrency can highly influence and transform the healthcare industry [19]. Also, the security issues that are need to be considered and the safety measures needs to be taken while using cryptocurrency had been discussed in detail [20].

11.6.1 Cryptocurrency Platforms in Healthcare

1. **Medicalchain (MTN):** Blockchain technique is used to secure medical records and thereby ensure quality of those records. MTN is a project that uses blockchain for securing healthcare records. This ensures that the medical records are not tampered by any means. The main aim of the project is to provide instant collaboration between patients and healthcare professionals. The project enables them to use healthcare records effectively with multiple levels of access. The medical chain works over Ethereum (ERC-20 tone) [21,22].

2. **Dentacoin (DCN):** Ethereum blockchain is used to build DCN. The project targets the dental care system and makes it efficient by improving the accessibility and payment systems with DCN coin. The platform token is intended to encourage social giving by acknowledging users. Dentacoin is based on Ethereum. It aims to avoid middlemen and restore authority to dental professionals and patients, thereby eliminating any additional expenses.

The following are the key features:

- **DentaCare mobile app:** This is a mobile app that guides users to maintain proper dental hygiene.

- **Blockchain-based diagnostic platform used:** This feature provides a way for patients to accurately rate dentists, providing dentists with market research value. When you make a review, you are provided with a DCN that you may use for future treatment.
- **Healthcare information:** Dentacoin's blockchain allows for secure data exchange and storage of medical records (to be shared between dentists and patients on permissions).
- **Trading platform:** On this platform, users can sell dental products and equipment using DCN.

3. **Patientry (POTY):** It is a medical chain data-based project. It uses blockchain by which physicians, patients, and hospitals can manage their data securely. Data management involves the process of storing, accessing, and transferring data. The main use is to provide a data-secured environment for healthcare industry patients and doctors. This project aims to change the method by which electronic data in the healthcare industry is handled by using blockchain technology. Using PTOY, patients can purchase additional storage facilities from the hospital system. This token has three main purposes:

- Maintain good standards for the healthcare industry
- Effectively manage paid transactions
- Balance the storage space of the network equally between doctors and patient platforms.

4. **Medibloc (MEDX):** MediBloc is described as 'a healthcare information environment built with blockchain technology for patients, healthcare providers, and investigators.'

Since the platform utilizes blockchain technology, one can keep track of everything including healthcare records access, number of doctor visits, and other updates. When all the records are in a single place it is easy for users to manage their record. This also helps in getting better quality of health service and promoting innovation in the healthcare industry. Important features are as follows:

- The data ownership will be redistributed, and personal information will be held by people and they can design who can access their document.
- When blockchain is used, there is more security with a localizer system against unwanted access.
- Data that is stored is accessed only by the trusted healthcare providers so any chance of alteration of data is minimal.

5. **Shivom (OMX):** Shivom is similar to all examples as it focuses on health promotion. The major difference between Shivom and other applications is that Shivom focuses on collaborating between DNA and biotech with the Healthcare Institute and research centres. Having the goal to become the world's largest data centre is huge. The major focus is the preservation analysis of DNA data which will foster the development of drugs. Although the Shivom group pioneers in genetics there are many companies that are working on genetics and digital money.

6. **Solvecare (SOLVE):** Solvecare is considered a reputable platform that can manage healthcare plans and extract benefits. As with a few other examples, authority is given to patients when talking about widespread information. If such authority is given, it is hoped that people will benefit from better healthcare. It is also believed

that medical services will be improved, and payments by the healthcare manager will be on time and precise.

7. **Docademic (MTC):** It is a service platform available globally. Built on Ethereum the platform helps users get access to free and quality healthcare service. The platform also uses a blockchain that enables easy transfer of trustworthy information. Some of the features off the platform are as follows:

- An application that could be installed from the Google Play Store, and provides 'round the clock healthcare service. Here the time taken to connect a doctor to a patient is very minimal and it is used by many people around the world.
- The platform also provides training for registered professionals. This is done by providing real-time metal data when the transactions are being stored on the blockchain.
- The project also uses AI techniques so patients will be able to get treatment and diagnosis based on AI techniques.

Humanscape (HUM): Patients with chronic illness can benefit with assistance provided by HUM. Assistance is avoided by enabling the users to interact with others through social activities thereby providing emotional support. The product claims that many people lack knowledge about diseases and therefore could not be treated properly. One of the key solutions could be to gather data from a large number of patients and provide a single source so that the entire population can be freed from the disease.

8. **Lympo (LYM):** The main aim of Lympo is to monitor health and sports data. It is a distributed application that encourages users to live a free and healthy lifestyle that is suggested by the system and created and controlled by the user. While exercising people will try to monitor their details. The application determines how many calories have been burnt or it monitors body mass. Blockchain technology is used and the users can even monetize their data by giving their personal data to an industry. The industry in turn will use information to create systems that can guide increasing fitness. Before Lympo there was not any platform that could provide this kind of data.

9. **Farmatrust (FFT):** Farmatrust is a novel application that leverages the security technique of blockchain to eliminate the sales of illegal drugs. It provides a robust and complaint framework for pharmaceutical companies. Counterfeit drugs can harm more patients than incorrect diagnosis and treatment. This problem could be averted by using FFT. Pharmaceutical companies can use FFT for their product tracking and management in the supply chain.

Benefits of cryptocurrency platforms in healthcare:

- Comparatively low transaction fees. Typically, a miner who gets cryptocurrency in reward has to pay zero or minimum transaction fees.
- Secured transaction free from hacking and fraud.
- Enables immediate settlement.
- Universal recognition as this is not bound by exchange rates, transaction charges and interest rates of any country. Therefore, there are no hassles in making global transactions.
- Full ownership to the digital currency which cannot be frozen or given limited access by any third-party entity.

- Full identity protection.
- Easy accessibility.

Electronic Health Record (EHR):

- Electronic Health Record (EHR) in a sense can be called a digital copy of a patient's paper-based record. We use it to add, store, and share information of a patient with doctors, nurses, pharmacies, and laboratories. Providers and insurance companies can use digital money for transactions, which may be the best way to pay for out-sourced services.
- Health information exchange (HIE) is one of the major goals of the healthcare industry. Through a secure blockchain network, we can keep the patient's health information private and protected when sharing it across the health network.
- If the U.S. healthcare industry adopts the trend of tokenization, those delivering better care may receive incentives in the form of crypto-tokens as soon as they report on measures. It is an ultra-fast way to pay them for rendering services. The measures correlate with reports that they need to present to the Centers for Medicare & Medicaid Services (CMS). These reports will be evaluated for imbursement programs. It also incentivizes them to do a good job.
- Moreover, cryptocurrency will help the U.S. government cut down on healthcare costs. When the government pays only for the value-based services i.e., the quality rather than the quantity of service, it will automatically reduce expenses.
- Doctors won't cash in patients' visits in value-based care. Only when the patients get better and report healthier outcomes will this convert into payments for the physicians. These outcomes are actually patient-reported outcomes (PROs), which happen through questionnaires and feedback forms.

11.7 Robo-Dentists

Nowadays dental issues have become quite common. A study has revealed that 60–90 percent of children and almost all adults are exposed to dental problems worldwide. It is also observed that dental disease has been found in people at the ages of 35–45. About 30 percent of the elderly people have almost lost all their teeth. It is known that dental problems could be painful and require appropriate treatment at the correct time. There is a global labor shortage for dentists which makes timely treatment impossible. Recently robots are being created that lot in the field of dental care. Robots with precision motion can improve efficiency of dental procedures by avoiding human errors.

Robots should be deployed in dental care under two different categories. They can be used for training purposes for simulating a human reaction during dental treatment. And they can also be used for assisting dental procedures. For an assistant general procedure multiple designs have been proposed of which Yomi, released by Neocis Company, is advanced and unique in its own way. Yomi was examined and found appropriate to perform dental implant surgery by the FDA in 2017. Having a robust navigation system, it is able to provide guidance regarding the precise location of teeth that need to be repaired. The proposed dental robot has performed a clinical dental implant surgery successfully [23].

11.7.1 Application of Robots in Dental Care

1. Dental patient robots: Dental robot design simulates a real-time treatment scenario. Popularly known as Phantom these robots are used for the purpose of training students to enhance their practical skills.

- Showa Hanako is a dental robot that can perform a variety of patient emotion and gestures. It can blink, sneeze, cough, roll its eyes, and also show fatigue of opening its mouth for a long time. Created by Showa University, Japan, it can mimic minute gestures of humans perfectly.

- Geminoid DK was developed by the Advanced Telecommunications Research Institute of Japan. The novelty of the robot is that it uses motion capture technology that can exactly mimic human expressions during a dental procedure, precisely at various neck positions.

- Simroid: The Nippon Dental University Kokoro created the robot named Simroid. It was mainly used to understand the attitude of the student while they perform a dental procedure. The robot has a sensor in its mouth that it can express pain and discomfort. It also has a delicate skin-like structure that will tear if the mount is opened too wide. The robot answers questions so that the students can develop interaction skills. The robot can also assess the treatment based on the parameters of the sensors and camera, which record the entire treatment scenario [24,25].

2. Endo micro robot: This robot aims to provide accurate and quality root canal therapy. It has advanced endodontic technology integrated with online supervision. Using this it can perform perfect root canal treatment with automated drilling, cleaning, and filling. It helps the dentists with its precise treatment as there is little error and less stress for the dentists [26].

3. Dental nanorobots: Nanorobots are the size of nanometres and made up of nanomaterials. These robots are used for treating cavities, tooth hypersensitivity, anesthesia, dental realignment, local drug delivery, and minor tooth repair. These robots are found to be very fast and accurate [27].

4. Surgical robots: Surgical robots are used in dental surgery. The surgeon must preprogram the robots using an interactive procedure. The robot does the preprogrammed task in the operation thereby assisting the dental surgeon in his surgery [28].

5. Dental implantology robot: The recent invention in the field of dental surgery is getting assistance in dental implants. This is done by building a three-dimensional model of the patient's jaw, which can be obtained from a CT scan of the patient. The dental robots are then used to drill the jaw and provide precise surgical guidance [29].

6. Robotic dental drill: This is another recent advancement in dental surgery. In this technique, thin needles are inserted into the patient's gums. These will help to locate the bone of the patient, which is transferred to a computer. The computer has a preloaded CT image. Integrating these two data, a drill guide is provided [30].

7. Tooth-arrangement robot: The dental care of tooth arrangement is called prosthetic dentistry and these robots are deployed in it. These robots help in creating a complete dental prosthesis that has six degrees of freedom. Then virtual tooth arrangement software helps fix dentures precisely [31].

8. Orthodontic archwire bending robot: Robotic technology is used for automatically bending orthodontic archwires to a specific shape. The bending equipment, known

as SureSmile archwire bending robot, is a combination of gripping tools and a resistive heating system combined with the application of CAD/CAM, 3D imaging, and computers for the fabrication of orthodontic appliances [32].

Advantages of robots in dentistry:

- Precision vision of the affected area (3D)
- Usage of digital camera zooming
- Reduced fulcrum effect
- Motion scaling
- There is no hand tremor

Disadvantages of robots in dentistry:

- High costs robotic system
- Sometime difficult to access the patient
- Bulky size and maintenance system
- Absence of surgeon from the treatment field
- Probability of failure

11.8 Conclusion

This chapter briefs about the advent of automation and usage of AI techniques in the healthcare sector. It can be noted that the application of it in healthcare begins with a modest automation of patient records called EHRs. It can also be noted that application of ML algorithms helps for better patient diagnostics and treatment plans. The usage of blockchain technology for secured maintenance of healthcare records was explained with a case study. The advantage of using AI in healthcare transaction processing helps in identifying fraudulent behavior. Finally, the application of cryptocurrency in medical payments and the usage of robots for dental assistance was also discussed.

References

[1] Parkin, E. 2016. "A Paperless NHS: Electronic Health Records." In *Briefing Paper*, House of Commons Library, London, UK.

[2] Liaw, S.-T., G. Powell-Davies, C. Pearce, H. Britt, L. Mc Glynn, and M. F. Harris. 2016. "Optimizing the Use of Observational Electronic Health Record Data: Current Issues, Evolving Opportunities, Strategies and Scope for Collaboration." *Australian Family Physician* 45: 153–156.

[3] https://specifications.openehr.org/releases/BASE/latest/architecture_overview.html.

[4] Lau, F., M. Price, J. Boyd, C. Partridge, H. Bell, and R. Raworth. 2012. "Impact of Electronic Medical Record on Physician Practice in Office Settings: A Systematic Review." *BMC Medical Informatics and Decision Making* 12.

[5] Radhakrishna, K., B. R. Goud, A. Kasthuri, A. Waghmare, and T. Raj. 2014. "Electronic Health Records and Information Portability: A Pilot Study in a Rural Primary Healthcare Center in India." *Perspectives in Health Information Management* 11.

[6] Jiang, Fei, Yong Jiang, Hui Zhi, Yi Dong, Hao Li, Sufeng Ma, Yilong Wang, Qiang Dong, Haipeng Shen, and Yongjun Wang. 2017. "Artificial Intelligence in Healthcare: Past, Present and Future." *BMJ* 2 (4). doi: 10.1136/svn-2017-000101.

[7] Islam, S. M. Riazul, Daehan Kwak, Md Humaun Kabir, M. Hossain, and K. Kwak. 2015. "The Internet of Things for Health Care: A Comprehensive Survey." *IEEE Access*. doi: 10.1109/ACCESS.2015.2437951.

[8] Zhao, C. W., and Y. Nakahira. 2011. "*Medical Application on IoT*." In *International Conference on Computer Theory and Applications*, pp. 660–665.

[9] *Windriver.com*. 2013. "White Paper: Security in the Internet of Things – Lessons from the Past for the Connected Future."

[10] Sethia, Divya Shikha, and Saran Huzur Saran. 2014. "*NFC Based Secure Mobile Healthcare System*." In *IEEE Transactions on Communication Systems and Network (COMSNETS)*. E-ISBN: 978-1-4799-3635-9. doi: 10.1109/COMSNETS.2014.6734919.

[11] Devendran, A., T. Bhuvaneshwari, and A. K. Krishnan. 2012. "Mobile Healthcare System Using NFC Technology." *International Journal of Computer Science Issues (IJCSI)* 9 (3): 10. ISSN:1694-0814.

[12] Thavasi, Sheela, Raja TRS, and Vaidyam Jawahar Harikumar. 2020. "Electronic Health Record Management and Analysis." *Test Engineering and Management* 83.

[13] Li, Jing, Kuei-Ying Huang, Jionghua Jin, and Jianjun Shi. 2007. "A Survey on Statistical Methods for Health Care Fraud Detection." In *Health Care Manage Science*. Springer. doi: 10.1007/s10729-007-9045-4.

[14] Li, J., J. Jin, and J. Shi. 2008. "Causation-Based T2 Decomposition for Multivariate Process Monitoring and Diagnosis." *Journal of Quality Technology*. doi: 10.1080/00224065.2008.11917712.

[15] Yang, W. S. 2003. "A Process Pattern Mining Framework for the Detection of Health Care Fraud and Abuse." Ph.D. thesis, National Sun Yat-Sen University, Taiwan.

[16] Bauder, Richard A., and Taghi M. Khoshgoftaar. 2016. "*A Novel Method for Fraudulent Medicare Claims Detection from Expected Payment Deviations (Application Paper)*." In *Information Reuse and Integration (IRI), 2016 IEEE 17th International Conference*, pp. 11–19. doi: 10.1109/IRI.2016.11

[17] Chan, C. L., and C. H. Lan. 2001. "*A Data Mining Technique Combining Fuzzy Sets Theory and Bayesian Classifier—An Application of Auditing the Health Insurance Fee*." In *Proceedings of the International Conference on Artificial Intelligence*, pp. 402–408.

[18] https://www.businessinsider.in/The-cost-of-bitcoin-payments-is-skyrocketing-because-the-network-is-totally-overloaded/articleshow/62301717.cms.

[19] www.cnbc.com/2017/12/19/big-transactions-fees-are-a-problem-for-bitcoin.html.

[20] Subha, T. 2020. "Assessing Security Features of Blockchain Technology." *Blockchain Technology and Applications, New York:Auerbach Publications*: 1–20. doi: 10.1201/978100308148.

[21] https://www.cryptostache.com/2018/01/01/top-5-cryptocurrency-exchanges-lowest-fees-2018/.

[22] https://wethecryptos.net/healthcare-cryptocurrencies-and-blockchains/.

[23] Rawtiya, M., K. Verma, P. Sethi, and K. Loomba. 2014. "Application of Robotics in Dentistry." *Indian Journal of Dental Advancement* 6: 1700–1706. , doi: 10.5866/2014.641700.

[24] Schulz, M. J., V. N. Shao, and Y. Yun. 2009. "*Nanomedicine Design of Particles, Sensors, Motors, Implants, Robots, and Devices*." *Artech House*, p. 10. ISBN: 9781596932791.

[25] Bansal, A., V. Bansal, G. Popli, N. Keshri, G. Khare, and S. Goel. 2016. "Robots in Head and Neck Surgery." *Journal of Applied Dental Medical Science* 2: 168–175. doi: 10.1155/2012/286563.

[26] Dong, J., S. Hong, and G. Hesselgren. 2006. "*A Study on the Development of Endodontic Micro Robot*." In *Proceedings of the 2006 IJME-INTERTECH Conference*, 8 (1).

[27] Lumbini, P., P. Agarwal, M. Kalra, and K. M. Krishna. 2014. "Nanorobotics in Dentistry." *Annals of Dental Speciality* 2: 95–96.

[28] Lueth, T. C., A. Hein, J. Albrecht, M. Demirtas, S. Zachow, E. Heissler, M. Klein, H. Menneking, G. Hommel, and J. Bier. 1998. "*A Surgical Robot System for Maxillofacial Surgery*." In *IEEE International Conference on Industrial Electronics, Control, and Instrumentation (IECON)*, Aachen, Germany, pp. 2470–2475. doi: 10.1109/IECON.1998.724114.

[29] Xiaojun, W. C., and L. A. Yanping. 2005. *"Computer-Aided Oral Implantology System."* In *IEEE Engineering in Medicine and Biology 27th Annual Conference*, Shanghai, China, pp. 3312–3315.

[30] Palep, J. H. 2009. "Robotic Assisted Minimally Invasive Surgery." *Journal of Minimally Invasive Surgery* 5 (1): 1–7.

[31] Zhang, Y., J. Ma, Y. Zhao, L. Peijun, and Y. Wang. 2008. *"Kinematic Analysis of Tooth-Arrangement Robot with Serial-Parallel Joints."* In *IEEE*, Hunan, China, pp. 624–628.

[32] Rigelsford, J. 2004. "Robotic Bending of Orthodontic Archwires." *Industrial Robot* 31 (6): 321–335.

12

How IoT, AI, and Blockchain Will Revolutionize Business

Yogesh Sharma

Maharaja Agrasen Institute of Technology, G.G.S.I.P. University, New Delhi, India

B. Balamurugan

SCSE, Galgotias University, Greater Noida, Delhi-NCR, India

Nidhi Snegar

Maharaja Agrasen Institute of Technology, G.G.S.I.P. University, New Delhi, India

A. Ilavendhan

SCSE, Galgotias University, Greater Noida, Delhi-NCR, India

CONTENTS

12.1 Introduction

Many companies are working on business models and strategies. The business strategies are changing day by day in order to give the best to the customers, as well as on-time delivery at lower cost while maintaining the profit. There have been many revolutionary innovations done so far that have changed the way of doing business. It is important to note companies, policy makers, and the government are using the benefits of emerging technologies in order to open up a path for business competition while keeping in mind customer satisfaction.

These emerging technologies have revolutionized business procedures that lead to higher efficiency and ease process management with an overall increase in the productivity [1]. Business companies are now getting ready to work on business models, coming with the arrival of emerging technologies like Internet of Things (IoT), artificial intelligence (AI) and blockchain. IoT can resolves issues related with connectivity for data, meaning the IoT-enabled devices could be useful in collecting data from various resources and can also exchange data [2] whereas AI would be very useful in improving machine automation and performing tasks more efficiently as compared to humans [3]. Blockchain can make the integration of IoT and AI a more powerful technology [4–6] which gives a more solid security and privacy to the data collected and used while maintaining transparency between the stake holders. For example, if we consider a use case of a smart solar charge station, which is a dApps. This charging station would be a blockchain-powered solar charging station combined with IoT. The blockchain network is created on a smart contract for solar charging stations and is deployed. Once any vehicle arrives for charging, the IoT device monitors the status of the smart contract associated with the charging station and only those vehicle get permission to charge, which is part of the blockchain network. Thus we have seen how IoT in combination with blockchain could be useful. Now, if all three emerging technologies work together for any organization or industry it could provide a huge benefit and profit for that industry. We can say that convergence of these three technologies will benefit industries in a more powerful manner while reducing risks. The IoT devices will help in gathering real-time data [7]. AI will protect data from hackers and malwares [8] whereas the blockchain technology will provide transparency and security with all the stakeholders [9]. IoT, AI, and blockchain when combined together will show that the transformation is going to be unique and this would be the first time in modern history that three transformational technologies have appeared within the same generation, and the complete conclusion is yet to arrive. Now before looking into more of convergence of these three technologies let's look into the brief information about the three technologies.

12.2 Internet of Things

Internet of Things (IoT) is the concept of connecting any device to the internet and to other devices. All devices in the network interact with each to collect and share information and it is depicted in Figure 12.1. IoT is the collection of some hardware devices combined with sensing devices [10]. The hardware for IoT is divided into two categories: the general devices and sensing devices [11]. The general devices do the embedded processing and connectivity for the platforms they are connected with either by a wired network or a wireless interface [12]. They are the main component for data collections and information processing. Current home appliances are a classic example of such devices that are controlled by a sensor. These sensors help solve common problems. Apart from sensors, actuators are another important device in IoT that perform similar functions with different capabilities [13]. They work as an interface between sensors and machines and collect various information like humidity and light intensity. This information is computed using the edge layer which typically assists between the cloud and the sensor [14]. They are the layers that store the intermittent transfer of

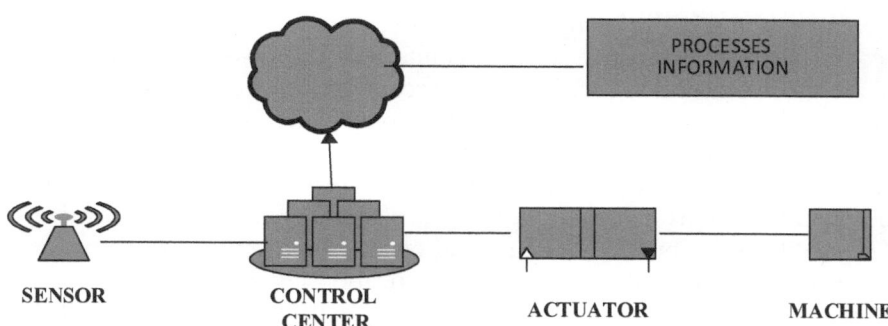

FIGURE 12.1
Internet of Things.

information. Finally the backend server of the cloud processes this information. both the sensors and actuators are the chief component of IoT. The sensor measure temperature, humidity, light intensity, and other key parameters of a home environment.

12.2.1 IoT Architecture

There are four layers in the IoT architecture. The base layer consists of IoT devices; this includes all components like sensor with the ability to sense and compute and connect to another device. The second layer is the IOT gateway or aggregation layer [15]. This layer significantly aggregates data from various sensors. These two layers form the definition engine and set the rules for data aggregation. The next layer is based on the cloud and is called the processing engine or event processing layer. It has numerous algorithms and data processing elements that are ultimately displayed on a dashboard. This layer basically processes the data obtained from the sensor layer. The last layer is called the application layer or API management layer. It acts as an interface between a third party application and infrastructure. The entire landscape is supported by the device managers and identity and access managers, which are useful for security of the architecture shown in Figure 12.2.

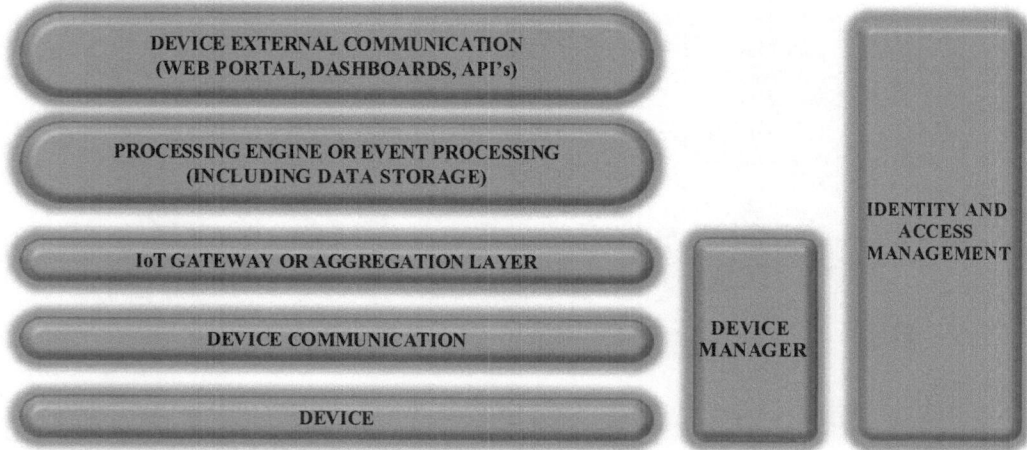

FIGURE 12.2
IoT Device Architecture.

The first layer in the IoT reference architecture is the device layer. At the device layer various components like sensors are interconnected like how Bluetooth is connected through a mobile phone and ZigBee [16] is connected through a ZigBee gateway. The other different devices include raspberry-Pi [17] that is connected to the Ethernet through Wi-Fi. This is directly connected to the communication layers, which is part of the second layer. The communications layer or gateway layer has rest protocols and other applications of the protocols [18]. Both the layers are tightly coupled and generate an enormous amount of data. The next layer is a bus layer or aggregation layer, and acts as a message broker [19]. It forms a bridge between data and the communication layer for the sensors. This is an important layer for three reasons. This layer supports an HTTP server and or MQTT Broker [20]. It aggregates and combines communications through gateway and bridges and transforms data between different protocols.

The next layer is the event processing and analytics layer. This layer drives data and provides transformation to generated data [19]. It provides the ability to do data processing, and the data is stored in the database. The client layer is used to create a web-based engine to interact with external APIs. This information can be fed into the API management systems. This layer helps create a dashboard and provides a view of analytics event processing. This layer helps communicate the system's outside the network using machine to machine communication [21].

IoT reference architecture is depicted in Figure 12.3. According to a survey by Cisco [22], it is expected that by 2020, there would around 50 billion connected devices, all of which will be generating a huge volume of data and of course nothing will remain static, especially when it comes to IoT. So as time passes, this number will keep on growing and the volume of data being generated will also grow. The biggest challenge with IoT is that there is no standard reference model for IoT [23] that guides how IoT

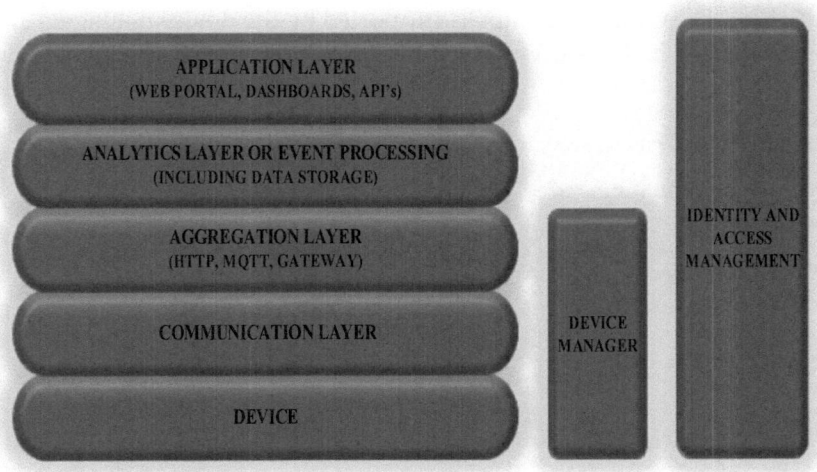

FIGURE 12.3
IoT Reference Architecture.

from different vendors can be used together to solve any problems, and also they all uses their own different protocols like Wi-Fi, ZigBee, or Bluetooth and there is no standardization on how all of these devices can be used together to create a single solution [24]. Hence if there is a drop in the standardization, the barrier drops, and more and more devices would be able to communicate with each other, which in turn increase the data collection and transferred to the data centres on a massive scale.

12.3 Artificial Intelligence

AI, as the name says it all, is artificially formed intelligence fed into computers and shows human-like thought processes like learning, reasoning, and even works like humans. Thus, make the computers work more effectively by using enhanced programming methods.

AI is present in every single organization from the banking and finance industry to medical sciences and also in the aerospace industry. It has been seen that many banks have numerous activities on a day-to-day basis that need to be done accurately and most of these activities take up a lot of time and effort from bank employees. At times there is also chance of human error in these activities. Banks and financial institution are investing lot of money in their operations in order to manage their operations. The use of AI-enabled operations helps these institutions to work efficiently with better results.

In basic terms, AI is a broad area of computer science that makes machines seems like they have human intelligence [3]. So it is not only programming a computer to drive a car by obeying traffic signals, but it is when that program also learns to exhibit science of human like road rage. The term artificial intelligence was first coined back in 1956 by Dartmouth Professor John McCarthy. He called together a group of computer scientists and mathematicians to see if machines could learn like a young child does, using trial and error to develop formal reasoning. The project proposal says that they would figure out how to make machines 'use language, form abstractions and concepts, solve kinds of problems now reserved for humans, and improve themselves'. Since then AI remains for the most part in university, classrooms, and even scientific labs. Now AI has become a big thing. The amount of data generated in last decade has pushed industries to follow technology and infuse it into the markets with cash and new applications. This makes industries more profitable and reliable for their customers.

Machine learning (ML) and deep learning are considered to be the subsets of AI. That means ML and deep learning are the ways in which we can achieve AI and the relationship is shown in Figure 12.4. ML is a technology that targets to teach computers the way to perform tasks with a set of data but without any specific programming but only numerical and statistical approaches could be used along with artificial neural networks to encode learning directly into the predesigned model [25].

Deep learning is another subset of AI, which is also a subfield of ML. Deep learning comes in where ML fails. Deep learning can be applied using an artificial neural network.

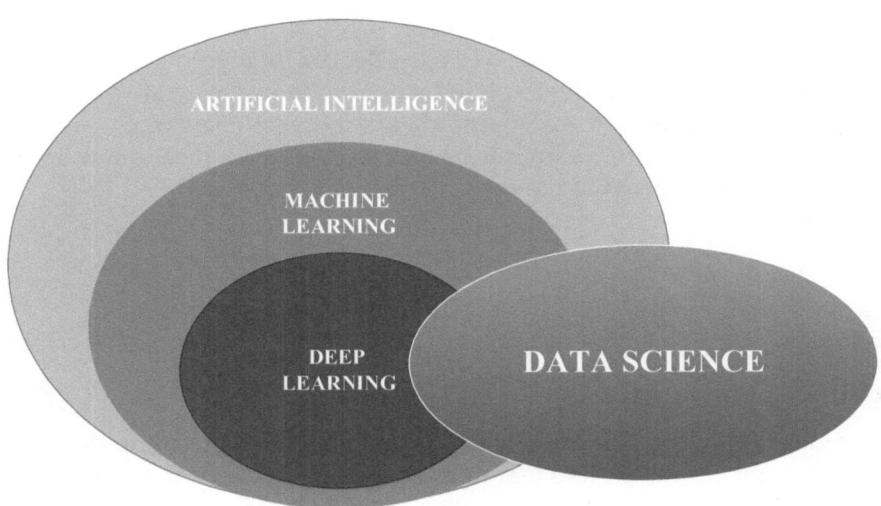

FIGURE 12.4
Artificial Intelligence.

12.4 Blockchain

A blockchain network is a decentralized and distributed network where all transaction and events are validated and secured by all parties involved in the network [26]. A blockchain network can provide a high level of security for the network users as all the transactions occurring in the network are completely anonymous to the users. This technology came into light in the year 2008 when Satoshi Nakamoto in his white paper gave the concept of cryptocurrencies and the idea was implemented in 2009. Cryptocurrency has become a very well-known currency in various industries and organizations.

The blockchain technology was earlier developed for cryptocurrencies but technology has evolved many folds and it is being used in many sectors. The organizations using blockchain technology are benefitting from the technology as the technology provides security and privacy to the industry, which involve many parties. A blockchain network is a tamper-less distributed ledger implemented in a decentralized network, that is a network where there is no single/central controller of the network [27]. This kind of network can be used in a community of users where the transaction can be recorded on a distributed ledger.

The blockchain network provides a feature of immutability hence any transaction, once recorded on the network, cannot be changed by any of the party involved in the network. Blockchain technology is a basic technology for the creation of cryptocurrencies as a cryptographic function places a major role in the creation [28]. The public and private key can be utilized for digitally signing transactions and provides a high level of security for the system.

The core idea behind blockchain technology is to eliminate the middleman or the third party from the network. Let's take for an example if a person buys something from a store, when the payment is done it involve the bank as the third party between

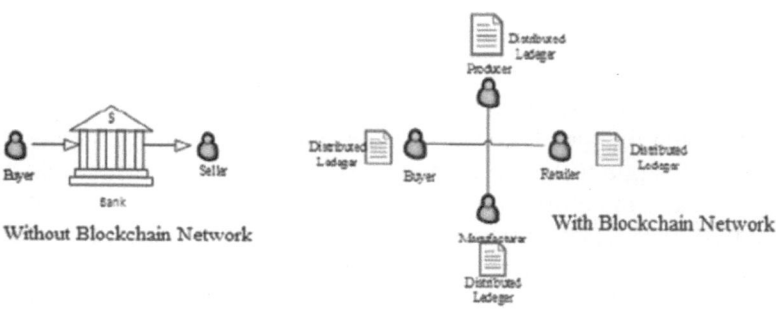

FIGURE 12.5
Centralized and Decentralized Network.

the buyer and the seller as shown in Figure 12.5. With the blockchain technology coming into picture the buyer and seller would be on the same decentralized network and both the parties can now make the transaction on a distributed ledger, thus removing the bank from the network. All payments done in a blockchain network can be done using a component of blockchain technology called a smart contract (a self-executing code).

Different organizations thinking of implementing blockchain technology should first understand the concept of the technology. When an organization goes to make a change in the database, in a simple network the changes could be done in the database any time when required, but once the database or the information is maintained in a blockchain network, changing of the information would be difficult.

There are two types of blockchain, a permissioned blockchain and a permissionless blockchain [29]. Permissioned blockchain are mostly used in business with blockchain. A permissioned blockchain is not accessible to any anonymous user. Users who are part of the blockchain or enabled by the users of blockchain can only access the blockchain such as viewing of ledger or chain. These blockchains are used between stakeholders who trust each other and are secured with a username/password. Various organizations are now moving their business to a blockchain network giving them a decentralized network combined with a level of security.

12.5 Business Models on the Intersection of IoT, AI, and Blockchain

Entrepreneurs always looks for great business ideas, and new technologies are an important source of new ideas [30]. AI has been a source of great ideas for a decade. Many business companies have built or have used many business models built upon AI which have given not only a profit to the companies, but benefits to the companies and the customers too. The business models produced using emerging technologies like AI, IoT, or blockchain have already increased the efficiencies of both industries and organisations [31]. What if all the three technologies are combined together in a product? The combination of the three or two technologies could not only increase the efficiency but will go on to benefit both the customer and the company. Let's see the merger of these emerging technologies in different fields and related business models.

12.5.1 Supply Chain

While many vendors analysts and system integrators look at new disruptive technologies such as IoT, AI, and blockchain individually. The real value to the supply chain comes from combining these technologies together. Organizations conducting paper-based business transactions will struggle to leverage disruptive technologies without extensive manual rework [32]. To obtain maximum business value companies need to ensure they have an end-to-end digital foundation or backbone in place. The digital foundation allows companies to seamlessly exchange electronic business transactions with 100 percent of trading partners in a digital business ecosystem [33]. Once the digital foundation is in place, then at that point companies can leverage the combination of IoT, AI, and blockchain to enable an autonomous supply chain. Leveraging an IoT platform allows companies to create a digital twin of a piece of equipment or serviceable asset. For example, IoT sensor information can be leveraged to not only identify a shipment location while moving through the supply chain but also monitor shipment condition. We can also remotely monitor the condition of the truck systems. AI derives insights from information moving across the digital business ecosystem whether analysing transaction information to determine the best performing trading partner or analysing IoT centre information to determine if a piece of equipment is about to break down. AI helps companies optimize and drive efficiencies across an end-to-end supply chain [34]. Blockchain is a relatively new disruptive technology, and in its simplest form is a trustless, immutable distributed ledger that can archive information from across the extended enterprise. From a supply chain point of view the most popular use case relates to track and trace and knowing the provenance of goods or raw materials. An autonomous supply chain will help companies establish a highly intelligent connected self-aware and trusted environment [35]. An environment that leverages deeper insights to supply chain performance to refine business processes improves traceability of goods and records and secures an archive of all digital interactions between a company and its trading partner community.

A company called Road Lunch is a smart digital freight that brings intelligent and simple yet sophisticated solutions to freight and digital logistics. They work with trucking and shipping companies so the truck goes to pick up the freight, and then they go through the process of delivery and transit multiple pickups and drop-offs. IoT allows them to take that smart contract and automate a position of goods and where it is going. IoT and AI allow them to not only provide real-time updates to the smart contract, further automating, but then with AI they can give a customer a good experience. These technologies allow people and associated parties in logistics to be able to do freight and fleet management, and lower all of their payables or receivables overhead. Blockchain means that they can get a company called Road Lunch free because they have an already digitized capacity at the trailer level. The sensors and the mobile apps allow them to do all that intelligent load matching in real time and they also have a whole transaction history. Blockchain and IoT-Based Supply Chain is shown in Figure 12.6.

12.5.2 Agriculture

Global demand for agriculture and food produce is going to increase by 69 percent to feed 9.6 billion people by 2050 [36]. Emerging technologies can help growers and consumers to collaborate and build a quality-centric and trustworthy system to meet this demand in a sustainable way [37]. This industry has undergone major developments over the last century. Two factors were most responsible for this surge in productivity: engines and the widespread availability of electricity and inclusion of technologies like IOT, AI, and blockchain.

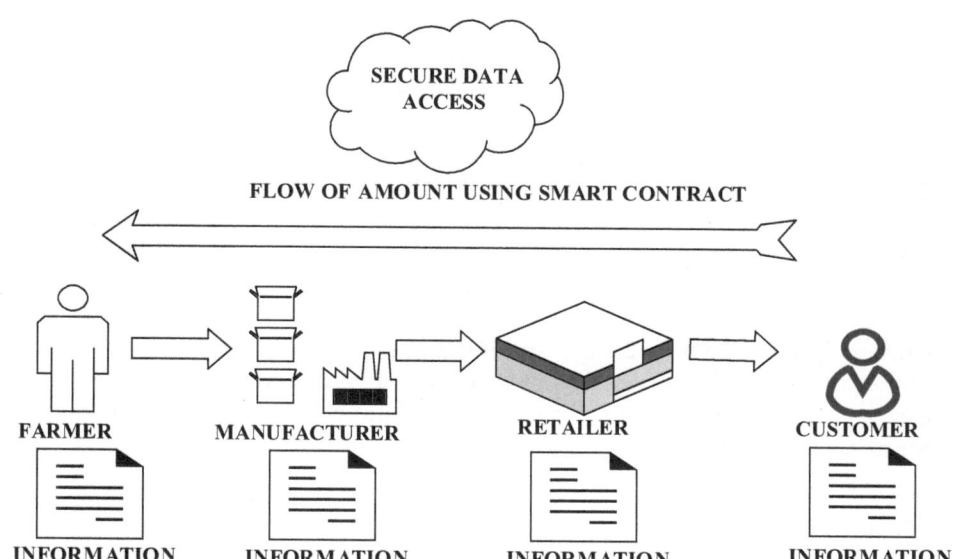

FIGURE 12.6
Blockchain and IoT-Based Supply Chain.

The combination of IoT, AI, and blockchain technologies can enable a circular economy in agriculture. IOT helps farmers to manage the vital parameters of soil and water at a micronutrient level [38]. On the other side consumers are benefitting in the form of reliable quality assurance. AI can help farmers provide knowledge in the form of crop advisory and protection of market dynamics to retailer. Finally, blockchain enables trust and revenue assurance between growers and consumers [39]. The growers and the consumers will empower through a unified agriculture supply chain with the use of blockchain and smart contracts that will ensures food information, accessibility, and transparency via trust, knowledge, and reliability.

Today the innovations on our immediate horizon include autonomous pickers; researchers have already created one that gathers strawberries twice as fast as humans. Robots or drones that can precisely remove weeds or shoot them with a targeted spritz of pesticide [40], using 90 percent less chemicals than a conventional blanket sprayer [5]. IoT sensors and cameras will record and check crop growth and alert farmers on their smartphones for problems, or when it is the best time to harvest. Farmers can take a soil sample, liquidize it, then analyse its pH and phosphorous levels – all in real time.

Companies like Agribotix have already commercialized software that analyses drone-captured infrared images to spot unhealthy vegetation. ML will regularly improve the system's ability to differentiate between varieties of crops and weeds that threaten them [41]. A company called Mavrx contracts 100 pilots to fly light aircrafts that are outfitted with multispectral cameras on data-gathering missions over large farms throughout the country. There are some other companies that are creating analytics software to act as farm-management systems [42], allowing growers of all sizes to deal with this big data. In indoor farming, the upside is that artificial lights and climate-controlled buildings allow crops to grow day and night, year-round, producing a significantly higher yield per square foot than an outdoor farm. One of the advancements in indoor farming is the Open Agriculture Initiative [43,44], which aims to create a 'catalogue of climates' so temperature and humidity can be set to recreate the perfect conditions for crops to grow indoors.

Millions of people entering the middle class every year in developing countries are demanding tens of millions of pounds of additional meat. Who would haveve thought Fitbits could be for livestock too? Cows are being fitted with smart collars that monitor if they're sick or if they are moving around more, which is a sign of fertility. Researchers at Scotland's Rural College are analysing cow breath. Exhaled ketones and sulphides reveal potential problems with an animal's diet [45]. Thermal imaging cameras spot inflamed udders to provide earlier treatment to combat a bacterial infection known as mastitis [46], one of the costliest setbacks in the dairy industry. There are 3D cameras that quickly measure the weight and muscle mass of cattle so they're sold at their beefiest. Companies have even begun positioning microphones above pig pens to detect coughs, giving sick animals the treatment they need a full 12 days earlier than before. Less antibiotics are used if fewer animals become ill for shorter lengths of time. A system of just three cameras, developed by researchers in Belgium, tracks the movements of thousands of chickens to analyse their behavior and spots over 90 percent of possible problems [47].

12.5.3 Healthcare

Emerging technologies like AI, IoT, and blockchain technologies may work wonders in the healthcare industry and for healthcare customers. IoT devices, by their very nature, create lots of data like gigabytes and terabytes of data that people need to be able to look at. IoT use cases have been difficult to monetize or at least have a long scale in terms of return on investment. Integration of AI and IoT can solve the problem of data. AI will actually take this data and be able to construct meaningful insights and meaningful interpretations. To meet the growing demands of businesses, we need to develop both skill sets and capabilities in order to meet the needs of healthcare industry. With these technologies, healthcare providers can build their own model and also reuse models that being tried, tested, and there are templates that will allow customers to consume the services only. ML is able to actually build, train, and deploy the models. Adding the capabilities of blockchain trust can be included in the business model. The combination of the three technologies really takes business into that stepwise incremental investment profile.

Blockchain is about immutability. That means some information is stored in the ledger or that one construct of a blockchain is very difficult to change. The model in a healthcare supply chain business blockchain can be used for traceability and visibility in the supply chain. Blockchain is more like all partners joining an ecosystem with a common cryptographic storage point and shared data [48,49]. In this way the supply chain domain can share information in the ledger that can be instantly seen with other participants. We can think a blockchain is a network of business participants. Centrally there lies a ledger and that ledger is nothing but a data store that you cannot change, and through this immutability you can store something that will be existing there eventually. We can share the data through this process with those who are joining in the network, so these are the three four characteristics immutability, finality, track and trains, and provenance [50].

The concept of IoT that means smart things grow rapidly in healthcare alone producing billions of data. Healthcare has many definite use cases where this application of IoT is being used. A smart watch that monitors trackable devices and our smartphone are examples of all those things that can capture a lot of data and those data can accumulate. Another famous use case of IoT is home health monitoring and elderly monitoring. IoT can really help some preventive actions in healthcare like if you are having a symptom of a stroke or you are have a symptom of a mental illness and are treated post facto or very late, but now by sending vital parameters to the doctor even as a form of warning is a very helpful thing to care provider. The other area that is taking up healthcare is the operational management part in the hospital; a person can track assets and there are concepts of like a smart bed that will tell if patient has consumed medicine or not [51]. IoT means connecting lots of devices and generating a huge lot of data that can be taken care of by AI.

A computer vision technology is a sub-branch of an AI and is used to predict anemia by training all the red blood cells (RBC), white blood cells (WBC), blast cells, and memory cells by putting all the data into the AI program and giving a new slide to it [52]. AI can identify the cells there and give prediction results in percentage. But this technology is not making a decision, it will be given to the doctors and they will decide. If it is 92 percent it is anemic, or not. A pathological opinion can be easily provided with the help of AI.

If unstructured data can be changed into a standard homogeneous format using interoperability techniques then predictive analytic processing of data will become meaningful, and control of disparity standard can be made.

Blockchain may be the heart of many of the solutions but when we are going for a digital transformation, blockchain does not work in isolation because we always start with a customer problem and that customer problem may be a combination of blockchain, IoT, and AI.

One of the best examples of integration of these technologies in healthcare is the concept of Cold Chain, which is that during the supply chain process many of the perishable medicines come in a freezer container [53]. There are some anomalies or problems in the freezer container. Today when we receive a problem with medicine, we know that some problem happened with the freezer and the medicine. We cannot use that medicine that is a critical for some patient. If we could think of a blockchain solution with IoT such that the freezer container has some IoT device that senses the temperature and whenever there is a fluctuation of temperature it sends and notifies all the participants. Like this, AI is making a clinical decision support system, not a clinical decision-making system, for the healthcare industry and all these technologies are going to help clinicians become more efficient in the decision-making process.

12.5.4 Finance and Banking

AI, IoT, and blockchain have the potential to revolutionize the finance industry. FinTech is the first to implement blockchain into financial technology and was set by institutional organizations since they make large bets on digital assets. ING is working on the integration of blockchain into its system. HSBC is already planning to transfer twenty billion U.S. dollars in assets to a new blockchain custodian platform in the year 2019 [54]. Other valuable examples come from large international banks, Commerce Bank, credit Zeus, and UBS already made their first transactions onto the Dutch voice HQLA X platform. Additionally, according to a report in Coindesk.com, Northern Trust is testing blockchain for selling bonds and parts to mass investors. BNY Mellon, MUFG, Credit Agricole, and JP Morgan have joined the blockchain race, too, which brings us to the conclusion that blockchain is not just universal but also a cosmopolitan technology. Blockchain can be transformed into a tool of international diplomacy also. The Democratic Party of Thailand had chosen its leader with the help of blockchain [55]. In 2019, Indonesia adopted this technology to greatly facilitate the counting process and reduce the risk of fraud. In the U.S. blockchain voting is already being tested in several states. It is predicted that 51 percent of enterprises will switch to AI with blockchain integration and that blockchain and AI make a nice couple [56]. The benefits are mutual because blockchain increases the usefulness of AI, and AI can be used to increase the security of the blockchain. In addition, tools based on blockchain are easier to use and more convenient for the average user. Blockchain will help make AI more coherent and understandable because it records all data and variables adopted as part of ML. According to a website (www.ec.europa.eu), in 2019, the European Investment Fund announced a launch of a program aiming to allocate up to 100 million euros to support companies that specialize in AI and blockchain. The total funding taking private investment into account should be 400 million euros. By combining all the available advanced technologies, companies will be able to make their forecasts faster and with more accuracy, as well as optimize the supply chains and quickly compare their products and services to new markets. Another trending area of fitting for blockchain implementation is IoT since the number of devices connected to IoT has recently exceeded the 26 billion mark, which is hard to bypass in such a huge industry. Smart Homes equipped with various devices and special sensors should function like clockwork reliably and safely because IoT implies a permanent connection to the network, which is why blockchain is one of the best solutions to satisfy this industry. Blockchain will allow smart home devices to make automatic micro transactions; for example, a customer will be able to set up an automatic payment system for their Uber trips or for product workers due to its nature. The blockchain will carry out these transactions many times faster. According to a Gartner study, 75 percent of organizations that have implemented IoT technology have either integrated blockchain or plan to do so. In the industries where blockchain, IoT, and AI could be applied, they will simplify optimize and automate workflows.

Blockchain is taking over the world by storm. It will disrupt and revolutionize many financial banking institutions. Blockchain technology can be considered a mathematical model for processing, securing, and finalizing transactions in the form of cryptocurrencies like traditional banks for trillions of transactions are being processed every day, which is the reason why it will disrupt the financial industry. To understand the process of blockchain in banking is quite simple by following one example. Let's say you make a bet with someone whether Real Madrid or Spain will win the World Cup. Each of you picks a team, and blockchain allows a smart contract to initiate running. Each party would send $100 U.S. dollars and the smart contract would hold the 200 states until a winner has been determined. The smart contract would automatically transfer the entire amount to the winner. Once the smart contract is running on blockchain it cannot be changed or stopped. Now over 60 percent of financial institutions are planning to use blockchain technology for international money transfers, 23 percent for security clearing and settlement, and over 20 percent for KYC know your customer regulations and services for anti-money laundering [57]. The three key factors of blockchain technology that are responsible for disrupting the banking industry irrespective of providing a decentralized platform for secure transaction are that transactions are extremely fast, processing fees are minimal, and recorded data is immutable. Whether in financal or personal banking the wait times depend on deposits and are often long and frustrating; it typically takes two to six business days to process whereas using blockchain the same can be achieved in about 10 seconds. Now every year 150–300 trillion dollars of transactions are made across the globe with an average transaction fee of 10 percent [58]. It takes about two to six business days to transfer the money, and financial institutions are spending up to 500 million dollars just to keep up with KYC regulation. With blockchain technology transactions can be made within seconds and for pennies on the dollar and it will remove the customer identification cost as on a distributed ledger. Confirmations are done effectively by everyone on the network thus it will eliminate intermediaries and toll extraction in the process. Blockchain technology will not only have the potential to drastically speed up transactions locally and across the globe, but it will also save banks millions of dollars. These are some of the reasons why banks around the world are currently investing in blockchain, and various banks are seeking patents for blockchain. A number of banks like Citibank Bank of New York Mellon, Goldman Sachs, and JP Morgan Chase have introduced their own cryptocurrency.

Another technology that is going to change the way we do banking is AI. In banking, AI is estimated to be a 450 billion U.S. dollars opportunity for banks. It is going to make the experience of working in the banking industry very competitive, more effective, and efficient. AI could be the big opportunity not only for banks but also for companies; the industry is trying to explore better ways of bringing banking to the customers. Now there has been research from business insider intelligence, which has come out with certain findings that say the potential cost saving for banks from artificial application intelligence application is estimated to be around 450 billion U.S. dollars. It is going to replace front and middle office accounting for 460 billion U.S. dollars. The future of the banking industry is going to be less manual and more based on AI. Banks will not be hiring people but they would be using AI to serve customers to get better work done. Now 80 percent of banks are extremely aware of the possible benefits of AI [59]. Many banks are planning to deploy AI services. Banks with over 100 billion U.S. dollars in assets say they are already implementing AI strategies and finding ways to bring AI to service. Certain AI cases have already gained prominence across bank operations with chatbots in the front office and anti-payment fraud in the middle office. Companies that already started using AI applications are Capital One, City Bank HSBC, JPMorgan, and U.S. Bank. AI applications offer the greatest cost saving for the bank because most time and resources are used by a bank in dealing with customers at the front and middle office, and if this can be automated through AI, they can send time elsewhere. Now banks are leveraging AI on the front end to smooth customer identification. When we talk about a front-end banking service, we mean improving the customer identification, or the customer authentication. AI will also interact with the customers by mimicking live employees through chatbots and voice assistants. It will deepen customer relationships and provide personalized insights and recommendations. AI is also being implemented by banks with middle office functions to detect and prevent payment fraud. This is very important as integration of AI and blockchain could detect fraudulent happenings in the bank. Anti-money laundering and "know your customer" regulatory checks can be performed by these technologies. These strategies, when employed by banks, are going to make banks efficient and effective. There will be less scope of any fraud and customer dissatisfaction. This is going to be the future of banking and this is certainly going to cause a huge disruption in the market in the years to come.

12.5.5 Social Network

The integration of blockchain and artificial networks into social networks is another trend. This alliance could make life simpler for everyone involved and it would be easier to contain scandal outbreaks, prevent privacy violations, control data storage, and maintain relevance of content. This technology can guarantee that all data published on social networks will not be tracked or duplicated even after being deleted. This would allow the user to feel more protected as they could control all essential aspects of their data. However, there is just one small issue: convincing social network platforms to implement blockchain (Maybe then we will finally stop putting stickers on our webcams).

12.6 Emerging Trends in Financial Services

Financial technology has always been an important part of the banking sector. According to blog.mokotechnology.com "The year 2020 marks the beginning of a significant phase in the financial services industry." Founded on a slew of disruptive innovations of the previous decade, a majority of industry partners are continuing to digitize and automate their processes, leverage data and analytics to steer strategic business decisions, and develop new service delivery cultures to tune up their customer experiences. Thus, the financial services industry is opening up the idea of ecosystems and partnerships between challengers and traditional banks. For financial service providers and other connected businesses, staying primed about the road ahead will be instrumental in pivoting their next moves. These are some of the most popular emerging trends on the financial service horizon:

> Emergence of the sharing economy peer-to-peer (P2P) payment platforms such as PayPal, Venmo, and Zelle have inspired consumers to route for money without approaching traditional institutions.

The popularity of these platforms has encouraged not only big legacy banks to develop their own versions of a similar offering, but also drive non-traditional players such as Google, Amazon, and Facebook to improve their e-wallet offering [60]. FinTech start-ups often need resources geared with specific domain and technical skill sets. We should be thankful to the sharing economy that they can access on-demand professionals that match their eligibility criteria, ready to take up ad-hoc engagements at a relatively affordable budget. Thus, the sharing economy has made procuring ideal resources cheaper and the most efficient as compared to the hiring of a part-time or a permanent employee.

The role of blockchain technology is in innovation and cross-border payments [61]; the blockchain is undoing outdated business models. According to Accenture and Spain's Santander, this technology is expected to save as much as $20 billion U.S. dollars in annual operating costs for the banking and financial services industry. The adoption of blockchain technologies will become critical for small and mid-size enterprise (SMEs) globally [62], as it will enable improved liquidity and reduced operational costs, freeing up valuable resources for reinvestment. Apart from this, blockchain has been instrumental in driving and effortless cross-border payments.

In 2017 SWIFT GPI was launched by the Society of Worldwide Interbank Financial Telecommunication. It strives to develop existing messaging and processing systems that connect over 10,000 banks. Recently JPMorgan Chase had announced its individual cryptocurrency, JPMCoin, meant to undertake issues faced by the banking firms in the cross-border payment arena.

The role of cognitive intelligence in the financial services industry, AI, and robotics is going beyond customer services and is expected to broaden industry prospects. Risk assessment and analytics, logistics, investments, and supply chain management can all be automated using these technologies to provide a steadier and more dynamic process.

These technologies will help realize the benefits of optimizing costs while enhancing operations. For instance, Canada-based TD Bank set up an Innovation Centre of Excellence. This centre provides a platform for bank-wide experimentation to diminish operational complexity and enrich the consumer experience.

Other latest innovations in the industry built on AI are robot advisors like Pepper, Nao, and Lakshmi; Biometric-based authentications; and voice commerce. A rise in the dependencies on cloud providers in order to reduce the cost related to IT causes financial institutions to use software-as-a-service cloud services for data storage, consumer relationship management (CRM) platforms, and human resources [63]. But 2020 will see an evolution in its usage to cover billings, loan management, cross border exchanges offering a smoother end-use experience. However financial services providers should also stay geared to deal with its impending cybersecurity threats.

Banks are adopting new standards of cloud solutions like 10X and Thought Machines [64], and more players will follow in 2020 and ahead. For instance, Deutsche Bank Luxembourg, adopted the Avaloq Banking Suite, which enabled them to provide their customers their entire suite of services through a single cash ledger while reducing complexity, risk, and expenses especially related to wealth management.

Demand for the in-app real-time micropayments and digital wallet payment, previously micropayments, had been restricted to messaging applications like telegram, but big technology firms are introducing payment services of their own. The year 2020 might witness a rush of developers crowding blockchain and digital assets to develop solutions to satisfy the high volume demands for in-app real-time micropayments. There is a rising consumer demographic trend towards a preference for digital wallets. This has led some top banks to offer comprehensive mobile banking applications. While a handful of banks have grown into the digital wallet domain, this vertical is seeing a slow but steady adoption growth.

Cybersecurity will be the topmost priority. The convenience of digital banking is advantageous, however, it is also can be misused by hackers leading to cyberthefts. In fact 70 percent of the breaches take advantage of the end user as opposed to the bank's gap in cybersecurity [65]. The centralization of data collected through cloud computing and decentralizing its access may help to build more data security layers. Localizing the data and avoiding third party intermediaries will ensure that the financial organizations have a greater authority over how the data is reported and distributed.

The year 2017 saw cryptocurrencies like Bitcoin, and Ethereum converted from a margin interest to a mainstream investment. It was noted that in a year, the rate of a single Bitcoin went from under $1,000 U.S. dollars to approximately $18,000 U.S. dollars; however, some analysts believed that cryptocurrency could substitute the global financial system soon. However, in 2019, the biggest cryptohack occurred when the Japanese cryptoexchange, Coincheck, got exhausted of NEM coins worth about $534 million U.S. dollars.

Software providers will attempt to offer flexible cybersecurity solutions by integrating advanced technologies, such as AI and cloud computing, to facilitate swift and reliable threat exposure and alleviation. The financial services realm is witnessing a sustained and aggressive focus on digitization and the adoption of new and emerging technologies to bring in operational effectiveness, improve speed to market, and achieve superior customer experiences. As the growth in the FinTech companies continues to stimulate, it will be very exciting to see how the competition shapes the industry in the coming years.

12.7 Catalyst of Digital Transformation

Digital does not refer to the creation of new websites or the addition of e-commerce stores getting likes, retweets or social posts, or even the use of technology. So what does digital mean? In this context the word digital is a synonym for the piece of change that is occurring in today's world driven by the rapid adoption of technology [66]. This is putting existing organizations under tremendous pressure, and in many cases driving them into the land of irrelevance. Some are seeing signs that others just don't because of rapid technology adoption. How our customers engage with us is changing the operating system of how we create a new and sustainable competitive advantage and how we must change if we are to keep peers. We are seeing two types of organizations; some are those that are just doing digital, and a new kind of digital innovator. The digital innovators are winning. They are disrupting every conceivable marketplace enabled by this new technology adoption. The biggest mistake we are seeing is that many organizations are simply digitizing existing services and calling it a digital transformation. It's really not. What is Digital Transformation?

"Digital transformation is a journey of strategic plan and organizational change". It starts by empowering the teams with new methods to create highly responsive strategy and a fearless culture of innovation" [67]. It is the right leadership that creates the high-performing innovative organizations that are delivered by marketers and technologists, principally this is digital transformation.

12.7.1 Mobile Phone as Catalyst of Digital Transformation

Mobile devices have played an important role in providing a kind of freedom in the interaction between customers and companies from anywhere. Nobile applications used by customers for products and purchases have benefited the companies. The apps allow the companies to reach a wider number of customers and offer new or multiple ways for marketing, making the employees more productive.

According to an article in *TIME*, around 41 percent of all Americans have only one mobile phone, and 66 percent people do not have landlines at their home. This means the marketers will expect huge difficulties in communication with customers in the traditional manners as they used to do.

Customers are rapidly becoming perpetually connected. Customers are more connected than ever and as a marketer it is their role to meet customers where they are but we know that it is not just the mobile device. As part of this bigger shift occurring, digital behaviors are splintering across multiple devices and at the same point, they're becoming more flexible and more frequently available to customers. The customers are more connected to people, to information, and to companies than ever before. So just to put some numbers around how much the customer is connected, according to a report, about 70 percent of online adults today have a smartphone, 72 percent also have a laptop, and 46 percent have a tablet. So the mobile phone is not replacing other digital connections, rather it is an additive creating this perpetually connected customer. It is important to know that this is more than just a shift in devices and time spent. It is actually leading to a fundamental shift in the way that the customers are expecting to interact and what is leading the step is a large amount of time that they are spending. According to a report, it has been seen that 70 percent of the customers use smartphone and they're spending almost 67 hours per month on their mobile device. This is based on some in phone tracking that Forrester has done. Marketers can see that customers are spending more than two hours a day on their mobile device and they are accessing multiple apps and sites. This mean the average per month is that the customer in the U.S. accesses 26 different apps and 52 different websites. There are similar numbers for tablets as well and this are only going to continue to grow and continue to advance. This is happening globally and we can see the projection is increasing and it will continue to do.

Most of the companies know that in order to increase the efficiency and benefit they have to become a digital enterprise. The new era of digital technologies and the amount of data available from them could make companies use data that may improve efficiency and also give complete information about interests of customers [68].

The mobile phone has increased the pace and the amount of interactions placed between customers and companies. The reports are based on interactions and allowed the companies to access the data in a timely manner. The data thus collected can be monitored continuously from multiple social media accounts, the purchase done, and customer interactions. The location sharing of the customer could give marketers a target specific locations during the most opportunistic time.

In the coming five years the mobile solutions will become more important. In a recent survey done by CIO, it was shown that around 71 percent of mobile solutions were more important that of IoT and cloud computing, but there are some senior officials that chose cloud computing as more important as compared to mobile solutions. However starting with the mobile solution could be the first step for organizations to digitally transform their business and industries.

12.8 Conclusion

The technologies, namely IoT, AI, and blockchain, are the emerging technologies. Many organizations have used them according to their type of work and products, and these technologies have indeed given profits and benefits to the organizations. These technologies have increased the efficiency of the industries. In this chapter we have discussed how these three emerging technologies could revolutionize business. Until now the technologies have shown improvement when either of them is used. Now if the three technologies are combined together and work together the technologies could create wonders in the coming future. This means the business models would be able to gather data using IoT, it would be able to analyse the gathered data efficiently using AI, and the security and privacy would be provided by blockchain technology. We have discussed the use case business models based on the emerging technologies and thus the type of use cases would definitely provide much better throughput.

References

[1] Wheelwright, S. C., and K. B. Clark. 1992. *Revolutionizing Product Development: Quantum Leaps in Speed, Efficiency, and Quality*. https://books.google.com/books?

[2] Medagliani, P., J. Leguay, A. Duda, and F. Rousseau. 2014. "Internet of Things Applications: From Research and Innovation to Market Deployment." https://archivesic.ccsd.cnrs.fr/UNIV-PMF_GRENOBLE/hal-01073761v1.

[3] Shabbir, Jahanzaib, and Tarique Anwer. 2018. "Artificial Intelligence and Its Role in Near Future." http://arxiv.org/abs/1804.01396.

[4] Daniels, Jeff, Saman Sargolzaei, Arman Sargolzaei, Tareq Ahram, Phillip A. Laplante, and Ben Amaba. 2018. "The Internet of Things, Artificial Intelligence, Blockchain, and Professionalism." *IT Professional* 20 (6): 15–19. doi: 10.1109/MITP.2018.2875770.

[5] Kshetri, Nir. 2017. "Can Blockchain Strengthen the Internet of Things?" *IT Professional* 19 (4): 68–72. doi: 10.1109/MITP.2017.3051335.

[6] Liu, Jin. 2018. "Business Models Based on IoT, AI and Blockchain." *Uppsala Universitet*, p. 33. http://www.teknik.uu.se/student-en/.

[7] Yasumoto, Keiichi, Hirozumi Yamaguchi, and Hiroshi Shigeno. 2016. "Survey of Real-Time Processing Technologies of IoT Data Streams." *Journal of Information Processing* 24: 195–202. doi: 10.2197/ipsjjip.24.195.

[8] Laurence Aimee. 2019. "The Impact of Artificial Intelligence on Cyber Security." https://www.cpomagazine.com/cyber-security/the-impact-of-artificial-intelligence-on-cyber-security/.

[9] Sharma, Yogesh. 2020. "A Survey on Privacy Preserving Methods of Electronic Medical Record Using Blockchain." *Journal of Mechanics of Continua and Mathematical Sciences* 15 (2): 32–47. doi: 10.26782/jmcms.2020.02.00004.

[10] Marques, Gonçalo, Rui Pitarma, Nuno M. Garcia, and Nuno Pombo. 2019. "Internet of Things Architectures, Technologies, Applications, Challenges, and Future Directions for Enhanced Living Environments and Healthcare Systems: A Review." *Electronics (Switzerland)*. doi: 10.3390/electronics8101081.

[11] Li, Shancang, Theo Tryfonas, and Honglei Li. 2016. "The Internet of Things: A Security Point of View." *Internet Research* 26 (2): 337–359. doi: 10.1108/IntR-07-2014-0173.

[12] Serpanos, Dimitrios, and Marilyn Wolf. 2018. *Internet-of-Things (IoT) Systems: Architectures, Algorithms, Methodologies.* Springer International Publishing.

[13] Misra Joydeep. 2017. "IoT System | Sensors and Actuators." https://bridgera.com/iot-system-sensors-actuators/.

[14] Carvalho, Otávio, Manuel Garcia, Eduardo Roloff, Emmanuell Diaz Carreño, and Philippe O. A. Navaux. 2018. "IoT Workload Distribution Impact Between Edge and Cloud Computing in a Smart Grid Application." *Communications in Computer and Information Science* 796: 203–217. Springer Verlag. doi: 10.1007/978-3-319-73353-1_14.

[15] Sethi, P., S. R. Sarangi, and Journal of Electrical and Computer Engineering, and undefined. 2017. "Internet of Things: Architectures, Protocols, and Applications." *Hindawi.Com.* Accessed June 9 2020. https://www.hindawi.com/journals/jece/2017/9324035/abs/.

[16] Drew, Gislason. 2010. "ZigBee Applications – Part 1: Sending and Receiving Data." https://www.eetimes.com/zigbee-applications-part-1-sending-and-receiving-data/#.

[17] Eben, Upton. 2012. "What Is a Raspberry Pi?" https://www.raspberrypi.org/.

[18] Wagner, Wagner Luís, Tarcísio Da Rocha, and Edward David Moreno. 2015. *"GoThings: An Application-Layer Gateway Architecture for the Internet of Things."* In *WEBIST 2015 – 11th International Conference on Web Information Systems and Technologies, Proceedings*, pp. 135–140. doi: 10.5220/0005493701350140.

[19] Fremantle, Paul. 2014. "A Reference Architecture for the Internet of Things." *WSO2 White Paper*, October 2015: 21. doi: 10.13140/RG.2.2.20158.89922.

[20] Hou, Lu, Shaohang Zhao, Xiong Xiong, Kan Zheng, Periklis Chatzimisios, M. Shamim Hossain, and Wei Chen. 2016. "Internet of Things Cloud: Architecture and Implementation." *IEEE Communications Magazine* 54 (11): 32–39. doi: 10.1109/MCOM.2016.1600398CM.

[21] Zhu, Julie Yixuan, Bo Tang, and Victor O. K. Li. 2019. "A Five-Layer Architecture for Big Data Processing and Analytics." *International Journal of Big Data Intelligence* 6 (1): 38. doi: 10.1504/ijbdi.2019.097399.

[22] Evans, Dave. 2011. "The Internet of Things How the Next Evolution of the Internet Is Changing Everything." *CISCO White Paper* 1: 1–11.

[23] Lee, Suk Kyu, Mungyu Bae, and Hwangnam Kim. 2017. "Future of IoT Networks: A Survey." *Applied Sciences (Switzerland)* 7 (10). doi: 10.3390/app7101072.

[24] Nikhade, Sudhir G. 2015. *"Wireless Sensor Network System Using Raspberry Pi and Zigbee for Environmental Monitoring Applications."* In *2015 International Conference on Smart Technologies and Management for Computing, Communication, Controls, Energy and Materials, ICSTM 2015 – Proceedings*, pp. 376–381. doi: 10.1109/ICSTM.2015.7225445.

[25] Mariette Awad, Rahul Khanna. 2015. "Machine Learning." In *Efficient Learning Machines*, pp. 1–18. Berkeley, CA: Apress.

[26] Javed, Muhammad Umar, Mubariz Rehman, Nadeem Javaid, Abdulaziz Aldegheishem, Nabil Alrajeh, and Muhammad Tahir. 2020. "Blockchain-Based Secure Data Storage for Distributed Vehicular Networks." *Applied Sciences (Switzerland)* 10 (6): 2011. doi: 10.3390/app10062011.

[27] Gamage, H. T. M., H. D. Weerasinghe, and N. G. J. Dias. 2020. "A Survey on Blockchain Technology Concepts, Applications, and Issues." *SN Computer Science* 1 (2). doi: 10.1007/s42979-020-00123-0.

[28] Yaga, Dylan, Peter Mell, Nik Roby, and Karen Scarfone. 2018. "Blockchain Technology Overview – National Institute of Standards and Technology Internal Report 8202." *NIST Interagency/Internal Report*: 1–57. doi: 10.6028/NIST.IR.8202.

[29] Wust, Karl, and Arthur Gervais. 2018. *"Do You Need a Blockchain?"* In *Proceedings – 2018 Crypto Valley Conference on Blockchain Technology, CVCBT 2018*, pp. 45–54. doi: 10.1109/CVCBT.2018.00011.

[30] Gabrielsson, Jonas, and Diamanto Politis. 2012. "Work Experience and the Generation of New Business Ideas among Entrepreneurs: An Integrated Learning Framework." *International Journal of Entrepreneurial Behaviour and Research* 18 (1): 48–74. doi: 10.1108/13552551211201376.

[31] Makridakis, Spyros, Antonis Polemitis, George Giaglis, and Soula Louca. 2018. "Blockchain: The Next Breakthrough in the Rapid Progress of AI." *Artificial Intelligence – Emerging Trends and Applications*. doi: 10.5772/intechopen.75668.

[32] Volberda, Henk, Frans van den Bosch, and Kevin Heij. 2017. "Reinventing Business Models: How Firms Cope with Disruption." In *Reinventing Business Models: How Firms Cope with Disruption*. doi: 10.1093/oso/9780198792048.001.0001.

[33] Moore, James Frederick, and James F. Moore. 2006. "Business Ecosystems and the View of the Firm Business Ecosystems and the View from the Firm." *The Antitrust Bulletin* 51 (1): 31–75. doi: 10.1177/0003603X0605100103.

[34] Patil, Sandeep Omprakash, and S. Ramachandaran. 2019. "Artificial Intelligence and High-Tech Supply Chains –Infosys."

[35] Wentworth, Craig. 2018. "The Supply Chain Gets Smarter." pp. 1–12.

[36] Alexandratos, N., J. Bruinsma, and G. Bödeker. 2006. "World Agriculture: Towards 2030/2050." Food and Agriculture Interim Report. Organization of the United Nations, FAO, Rome.

[37] Forum, World Economic. 2018. "Our Shared Digital Future Building an Inclusive, Trustworthy and Sustainable Digital Society."

[38] Na, Abdullah, William Isaac, Shashank Varshney, and Ekram Khan. 2017. *"An IoT Based System for Remote Monitoring of Soil Characteristics."* In *2016 International Conference on Information Technology, InCITe 2016 – The Next Generation IT Summit on the Theme – Internet of Things: Connect Your Worlds*, pp. 316–320. doi: 10.1109/INCITE.2016.7857638.

[39] Sylvester, Gerard. 2019. "E-Agriculture in Action: Blockchain for Agriculture." In *E-Agriculture in Action: Blockchain for Agriculture*. Food and Agriculture Organization of the United Nations.

[40] Pilz, Karl Heinz, and Simon Feichter. 2017. "How Robots Will Revolutionize Agriculture." *Webspace.Pria.At* 4. http://webspace.pria.at/ecer2017/papers/Paper_17-0597.pdf.

[41] Aitkenhead, M. J., I. A. Dalgetty, C. E. Mullins, A. J. S. McDonald, and N. J. C. Strachan. 2003. "Weed and Crop Discrimination Using Image Analysis and Artificial Intelligence Methods." *Computers and Electronics in Agriculture* 39 (3): 157–171. doi: 10.1016/S0168-1699(03)00076-0.

[42] Saiz-Rubio, Verónica, and Francisco Rovira-Más. 2020. "From Smart Farming Towards Agriculture 5.0: A Review on Crop Data Management." *Agronomy*. doi: 10.3390/agronomy10020207.

[43] Castelló Ferrer, Eduardo, Jake Rye, Gordon Brander, Tim Savas, Douglas Chambers, Hildreth England, and Caleb Harper. 2019. "Personal Food Computer: A New Device for Controlled-Environment Agriculture." *Advances in Intelligent Systems and Computing* 881: 1077–1096. Springer Verlag. doi: 10.1007/978-3-030-02683-7_79.

[44] Rowley, Trevor. 2019. *The Origins of Open Field Agriculture: The Origins of Open Field Agriculture*. doi: 10.4324/9780429059230.

[45] Dobbelaar, P., T. Mottram, C. Nyabadza, P. Hobbs, R. J. Elliott-Martin, and Y. H. Schukken. 1996. "Detection of Ketosis in Dairy Cows by Analysis of Exhaled Breath." *Veterinary Quarterly* 18 (4): 151–152. doi: 10.1080/01652176.1996.9694638.

[46] Fagiolo, A., and O. Lai. 2007. "Mastitis in Buffalo." *Italian Journal of Animal Science* 6 (2): 200–206.

[47] Rowe, Elizabeth, Marian Stamp Dawkins, and Sabine G. Gebhardt-Henrich. 2019. "A Systematic Review of Precision Livestock Farming in the Poultry Sector: Is Technology Focussed on Improving Bird Welfare?" *Animals*. doi: 10.3390/ani9090614.

[48] Raj, Pethuru, and Ganesh Chandra Deka. 2018. "Blockchain Technology : Platforms, Tools and Use Cases." *Advances in Computers* 111: 1–41. https://books.google.com/books.

[49] Schumacher, A. 2017. *Blockchain & Healthcare – 2017 Strategy Guide*. Munich: Axel Schumacher.

[50] Carter, C., and L. Koh. 2018. "Blockchain Disruption in Transport," p. 48. http://ts.catapult.org.uk/Blockchain/.

[51] Ghersi, Ignacio, Mario Mariño, and Mónica Teresita Miralles. 2018. "Smart Medical Beds in Patient-Care Environments of the Twenty-First Century: A State-of-Art Survey." *BMC Medical Informatics and Decision Making* 18 (1): 1–12. doi: 10.1186/s12911-018-0643-5.

[52] Alam, Mohammad Mahmudul, and Mohammad Tariqul Islam. 2019. "Machine Learning Approach of Automatic Identification and Counting of Blood Cells." *Healthcare Technology Letters* 6 (4): 103–108. doi: 10.1049/htl.2018.5098.

[53] Bishara, Rafik H. 2006. "Cold Chain Management – An Essential Component of the Global Pharmaceutical Supply Chain." *American Pharmaceutical Review*: 1–4.

[54] Wilson, Tom, and Lawrence White. 2019. "HSBC Swaps Paper Records for Blockchain to Track $20 Billion Worth of Assets." https://www.reuters.com/article/us-hsbc-hldg-blockchain/hsbc-swaps-paper-records-for-blockchain-to-track-20-billion-worth-of-assets-idUSK-BN1Y11X2.

[55] Tan, Aaron. 2018. "Thailand's Democrat Party Holds Election with Blockchain." https://www.computerweekly.com/news/252452435/Thailands-Democrat-Party-holds-election-with-blockchain.

[56] Meijer, Carlo R. W. De. 2019. "What May We Expect for Blockchain and the Crypto Markets in 2020?" https://www.finextra.com/blogposting/18285/what-may-we-expect-for-blockchain-and-the-crypto-markets-in-2020#:~:text=According to them more than, to AI with blockchain integration.

[57] Mori, Taketoshi. 2016. "Financial Technology: Blockchain and Securities Settlement." *Journal of Securities Operations & Custody* 8 (3): 208–227.

[58] Bansal, Sukriti, Philip Bruno, Olivier Denecker, Madhav Goparaju, and Marc Niederkorn. 2018. "Global Payments 2018: A Dynamic Industry Continues to Break New Ground." *Global Banking McKinsey*.

[59] Fethi, Meryem Duygun, and Fotios Pasiouras. 2010. "Assessing Bank Efficiency and Performance with Operational Research and Artificial Intelligence Techniques: A Survey." *European Journal of Operational Research*. North-Holland. doi: 10.1016/j.ejor.2009.08.003.

[60] Rohan, Sounak. 2019. "Future of Digital Payments." https://www.infosys.com/services/digital-interaction/documents/future-digital-payments.pdf.

[61] Guo, Ye, and Chen Liang. 2016. "Blockchain Application and Outlook in the Banking Industry." *Financial Innovation*. SpringerOpen. doi: 10.1186/s40854-016-0034-9.

[62] Ilbiz, Ethem, and Susanne Durst. 2019. "The Appropriation of Blockchain for Small and Medium-Sized Enterprises." *Journal of Innovation Management* 7 (1): 26–45. doi: 10.24840/2183-0606_007.001_0004.

[63] Oracle Cloud for Industries. 2015. "Cloud Computing in Financial Services: A Banker's Guide." *White Paper, Oracle Industries*

[64] Xavier, Lhuer, Phil Tuddenham, Sandhosh Kumar, and Brian Ledbetter. 2019. "Next-Generation Core Banking Platforms: A Golden Ticket?" https://www.mckinsey.com/industries/financial-services/our-insights/banking-matters/next-generation-core-banking-platforms-a-golden-ticket.

[65] Yan, Ye, Yi Qian, Hamid Sharif, and David Tipper. 2012. "A Survey on Cyber Security for Smart Grid Communications." *IEEE Communications Surveys and Tutorials*. Institute of Electrical and Electronics Engineers Inc. doi: 10.1109/SURV.2012.010912.00035.

[66] Gebayew, Chernet, Inkreswari Retno Hardini, Goklas Henry Agus Panjaitan, Novianto Budi Kurniawan, and Suhardi. 2018. *"A Systematic Literature Review on Digital Transformation."* In *2018 International Conference on Information Technology Systems and Innovation, ICITSI 2018 – Proceedings*, pp. 260–265. Institute of Electrical and Electronics Engineers Inc. doi: 10.1109/ICITSI.2018.8695912.

[67] Ziyadin, S., S. Suieubayeva, and A. Utegenova. 2020. "Digital Transformation in Business." *Lecture Notes in Networks and Systems* 84: 408–415. Springer. doi: 10.1007/978-3-030-27015-5_49.

[68] Carolan, Lisa. n.d. "How Mobile Acts as a Catalyst in Digital Transformation." https://www.exsquared.com/blog/how-mobile-acts-as-a-catalyst-in-digital-transformation/.

13

Storage, System Security and Access Control for Big Data IoT

T. Lucia Agnes Beena and T. Kokilavani
St. Josephs College, Tiruchirappalli, Tamil Nadu, India

D. I. George Amalarethinam
Jamal Mohamed College, Tiruchirappalli, Tamil Nadu, India

CONTENTS

13.1 Introduction

The world's top five companies, IBM, Google, Intel, Microsoft, and Cisco have started spending money on Internet of Things (IoT) to empower their business and to more easily handle the global competitive pressure. A separate division in IBM with 1,400 employees is operating in IoT. Google has officially started adding IoT products. Intel's sales division is driven by IoT. According to Global System for Mobile Communications (GSMA) Intelligence, numerous IoT devices are forecasted to scale up to 25 billion globally by 2025. The embracing of IoT amplifies the total volume of data generated, converting the industrial data into industrial big data. This digital renovation of industry enhanced by IoT adoption opens up new ways for companies to adopt new strategies to handle a huge volume of data, thereby optimizing their performance by gathering, filtering, processing, and analyzing data through the whole product lifecycle.

Big data is an appropriate platform to extract information from IoT data. IoT devices produce continuous streams of data in a flexible mode. It is essential to handle the high volume of stream data in an uninterrupted flow. Real-time analytics are required in an IoT environment. Realizing real-time analytics in an IoT environment is demanding because:

- A huge number of IoT devices generate volumes of data
- Processing and analyzing data must be done at low latencies
- Dedicated visualization and reporting techniques are needed
- The requirement of common protocols.

Recent developments in big data analytics used for IoT-generated data can crack the real-time analytics of IoT. A cloud-based solution that affords scalability, flexibility, compliance, and a refined architecture can be implemented to accumulate all IoT data, as opposed to built-in data systems that require being persistently updated whenever the data load is increased. If a company has sensitive data requiring fine-tuned security, a private cloud is highly preferable for data storage. Some organizations utilize NoSQL databases like Apache CouchDB to store IoT data as they offer minimum latency and great throughput. The NoSQL databases are schema-less and can handle elasticity, providing users the choice to update with more innovative events.

IoT devices are of different kinds and generate heterogeneous data that carry different types of data security risks. IoT-based security is still in a genesis stage. Attacks of any kind not only steal data, but also produce risks like damage to the sensors connected to the network. Therefore an identifier is needed for every device for the purpose of authentication.

The enterprises can be confident in getting the right details from the correct source. Proper configuration is indispensable in an IoT system with a fine-tuned access control policy to ensure the connectivity of various IoT devices in the network. At the same time, the IoT environment should focus on naming and identity management, information privacy, objects safety and security, interoperability and standardization, data confidentiality and encryption, spectrum, and green IoT. This chapter converses about various storage strategies, big data security techniques, and access control mechanisms for big data IoT.

13.2 Big Data Integration with IoT

The evolution of IoT opened up the new service providers who need to build innovative, novel, connected products. Analysts' prediction conveys numerous fresh IoT services may connect billions of new IoT devices over the next few years [1]. The IoT refers the creation of incredible volume of data and a compilation of bulk data. The organizations willing to adopt the IoT technology need to apply big data analytics to overcome the significant obstacles while deploying IoT [2]. In order to collect needed data and avoid unnecessary data, interfaces have to be designed for existing applications; required data can be collected from partners or third parties and physical environment around the business. Figure 13.1 illustrates the flow of big data in IoT applications [3].

13.2.1 Key Requirements of Big Data Analytics in IoT

Big data analysis has a great capability in mining significant details from the data generated by IoT sensors. In fact, the basic requirements for big data and IoT determine the

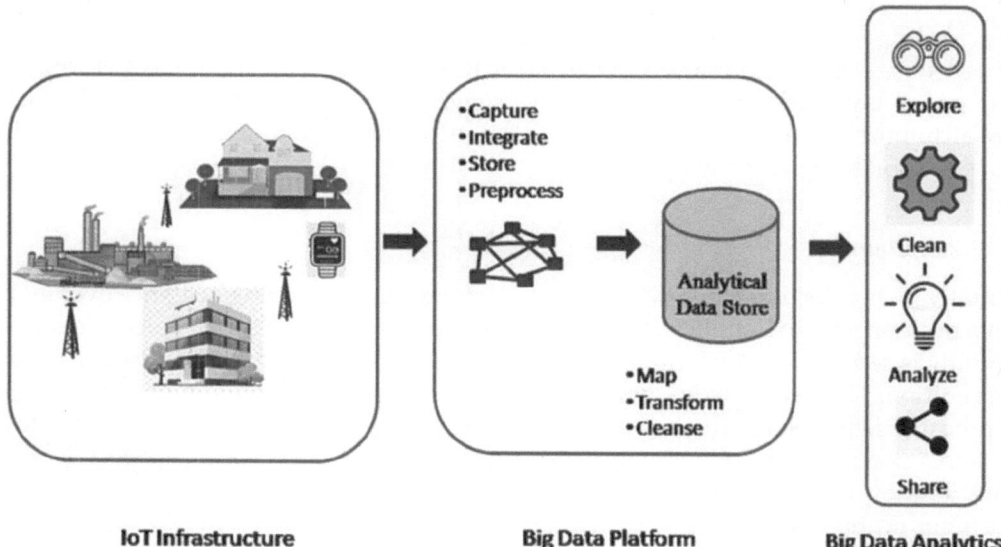

FIGURE 13.1
The Flow Big Data in IoT Applications.

practical and purposeful specifications for data analytics. Big data analysis in the IoT environment plays an important role in enriching IoT services are connectivity, storage, quality of services, and benchmarks [3].

13.2.2 Big Data Analytical Solutions

IoT enables a sequence of connections among various objects, people, data, and applications with the use of internet for interactive services and remote control. Therefore, the IoT network needs an operational control platform that can gather, analyze, and process the raw data retrieved from various sensor nodes. Some types of applications generate a very huge volume of data due to the availability of more sensors in the IoT environment. The innovative technologies or novel architectural patterns are necessary for data collection, data storage, data processing, and data retrieval. The following sections discuss various solutions.

13.2.2.1 Apache Hadoop

Apache Hadoop [4] is a non-proprietary IoT big data processing software framework that forms the basis of many a large IoT network. This open-source software (OSS) enables support and organization across industries and even between business competitors. Additionally, the initiation of open-source IoT throughout industrial applications makes even smaller businesses able to acquire the benefit of IoT technologies without the requirement of any high-cost dedicated development team. The most important components of Hadoop architecture are the Hadoop Distributed File System (HDFS) and the MapReduce programming model. HDFS is preferably utilized for storing data and MapReduce is exploited for processing data in a distributed way.

13.2.2.2 ThingSpeak

ThingSpeak [5] is an open-source IoT platform to build IoT prototype rapidly. ThingSpeak provide http and mqtt APIs for transferring IoT device data to the ThingSpeak cloud. The Matlab widget is used to analyze and visualize IOT device data on ThingSpeak. ThingSpeak can trigger any action based on data input. ThingSpeak supports many types of IoT device like arduino, nodemcu, and Raspberry Pi. The IoT device data from the ThingSpeak cloud can be exported for deep analysis.

13.2.2.3 Countly

Countly [6] is an open-source platform for enterprise and marketing analytics over mobile, web, desktop, and IoT applications with the perception of data independence and security. Countly is accessible in two different editions. There is a self-hosted edition and a community edition. The self-hosted or private cloud enterprise edition includes premium plugins and customizable SLA. The community edition contains basic plugins and a free-to-use non-commercial license. Countly has more than 15 open-source SDKs to collect data from web, desktop, mobile, smart watch, smart TV, IoT, and set-top box applications. It supports push notifications, crash analytics of Android apps, crash analytics of iOS apps, and Web analytics. As Countly application server logic is stateless, it can be deployed as a cluster of any size. Countly enterprise is scalable to any size of customer base and data volume when combined with MongoDB sharding.

13.2.2.4 AT&T IoT Platform

The AT&T IoT platform [7] provides fully customizable control of user data. Hence the data can be aggregated and run through custom applications. There is also a provision for integrating user data with external services to send data anywhere in the globe. The AT&T global SIM connectivity feature offers greater reliability, security, and mobility. The customizable IoT dashboards aggregate, compare, and track user devices' data to gain real-time insight on devices. Industry vertical solution templates are used in orchestration with rule engines to identify events and trigger actions and notifications.

13.2.2.5 Axonize

Axonize [8] has developed a disruptive, multi-app architecture that enables the deployment of fully customized smart solutions across all applications, verticals, and device types, within a short period of six to eight days. Customers can create complete smart facilities with very few resources in a short duration. This facilitates no upfront investment and scales fast to show ROI. Some of the benefits of adopting Axonize are ROI, managed services, security, a no-code platform, integrability, and scalability.

13.3 Storage Technologies

IoT is a rapidly growing technology and its applications are used in various business areas from manufacturing to utility. The IoT sensors generate data from various areas like agriculture, healthcare, finance, and transportation. Storing huge data collected from these sensors is a big challenge. The cloud may be used to store data, but it is not appropriate for storing real-time data and executing real-time analytics.

13.3.1 Key-Value Database

The simple form of storing data is based on key-value pairs. Key is uniquely used to identify the data and value implies the actual data stored. The value can be of any data type whereas key will be of a string type. This is the simple method of NoSQL for storing data in a relational database. This way of a storing key-value pair is easy to understand and easily scalable. It also provides opportunities for bulk storage, high concurrency, and fast lookups [9,10]. Redis and Riak are some of the examples of key-value pair storage representation. Figure 13.2 shows the mechanism of key value data store.

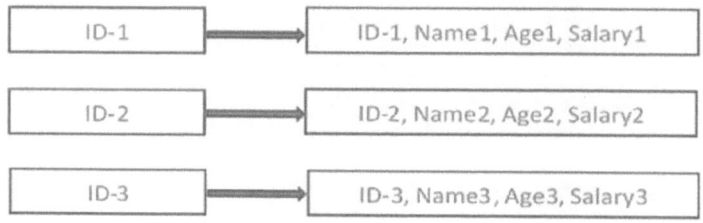

FIGURE 13.2
The Mechanism Key Value Data Store.

13.3.2 Column-Oriented Database

The column-oriented data store method of storage representation is suitable for data mining and data analysis applications [11]. This method stores data in a tabular form of columns; the performance of the columnar approach is better than the conventional row-wise database systems. The column-oriented data store efficiently reads and writes data. The map reduce model is compatible with the column-oriented data store. HBase and Cassandra are examples for column-oriented database mechanism. Figure 13.3 shows the column-oriented data store-based data representation.

13.3.3 Document-Oriented Database

A single document is employed for storing each data and its associated records. Horizontal scaling is supported in document-oriented data store. To analyse a large amount of data, this type of storage representation is used. The advantage of this data store is that different fields can be included in different documents and there is no need of adding the same empty fields that waste space in some documents in a collection. The documents can be referred to using object-id. Complex data structures like dictionaries, trees, and collections can be handled by document-oriented data store. But it is not suitable for handling data with relationships and data with duplication. Thus the IoT applications that generate variable fields are appropriate for using this type of data store. Figure 13.4 shows the

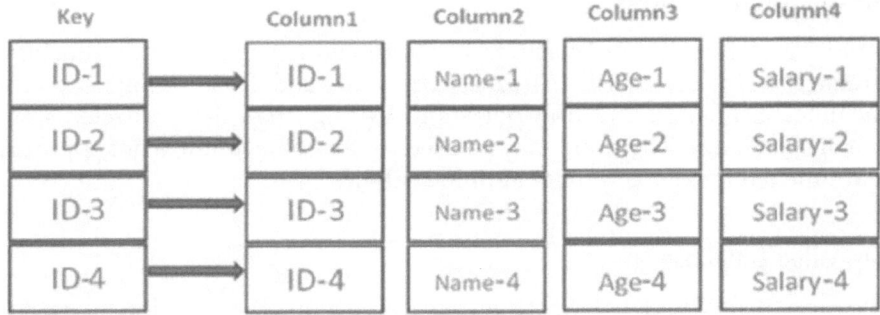

FIGURE 13.3
Column-Oriented Data Store.

```
Document 1
{
Id: 1;
Name: "John Smith";
Dob: "1970-24-07"
}
```

```
Document 2
{
Id: 2;
fullName:
{
First:"Sarah"
Last:"Jones"
}
Dob: "1980-04-11"
}
```

FIGURE 13.4
Document-Oriented Data Store.

document-oriented data store representation. MongoDB and CouchDB are examples for open source NoSQL document-oriented data stores.

13.3.4 Graph Database

The graph data store is best suited for representing and querying interconnected data. These types of databases store information that is similar to the human way of thinking about data. The nodes that are related to each other are located across different machines, then the graph data store cannot be used since it is not suitable for horizontal scaling. A massive amount of interconnected data is easily stored and handled by a graph database, but cannot handle scaling. A graph database will maintain data using graphs consist of nodes and edges in which the objects are represented as nodes and the relationships between objects are represented as edges. Every node can point to its adjacent node using a direct pointer called index free adjacency [12]. Connections are the key concept in a graph data store and it can handle both semi-structured and unstructured data as well. But it is very hard to obtain sharding from the graph data store. For instance, Neo4J is also an open-source software based on graph data store. Figure 13.5 shows the graph data store model for storage representation.

13.3.5 Cloud-Based IoT Platforms

IoT platforms are important components of the IoT environment. To make it cheaper and easier for users, developers and businesses, the IoT environment provides built-in tools and abilities. The cloud providers who wish to extend their business into the IoT environment offer cloud-based IoT platforms. Infrastructure as-a-service (IaaS) provides solutions in the form of a backend like processing power and hosting space for various applications. The backend infrastructure designed for applications may be restructured or enhanced and incorporated into IoT platforms [13]. Some of the cloud-based IoT platforms are discussed next.

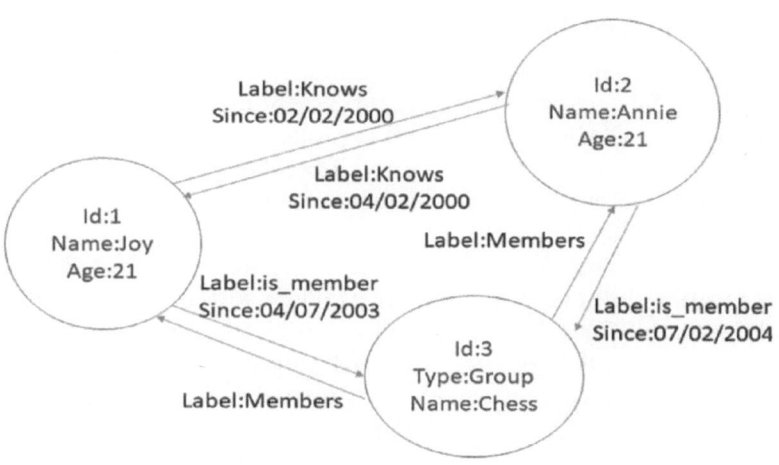

FIGURE 13.5
Graph Data Store.

13.3.5.1 AWS IoT Suite

People are using many devices in hospitals, factories, cars, homes, offices, and in many other places. With the widespread use of these devices, solutions are needed to connect them and collect data for storing and analyzing. To figure out IoT solutions for different use cases virtually across a variety of devices, Amazon Web Services (AWS) IoT provides wide functionality and covers the edge to the cloud. The devices can be made even smarter without using any connectivity because AWS IoT incorporates artificial intelligence (AI) services. When the business requirements grow, AWS IoT will easily scale, hence it is used by leading industrialists around the world. Complete security is provided by AWS IoT. With the help of these security features, preventive security policies and immediate response to potential security issues can be easily created [14].

The AWS IoT suite is used by industries for prediction, home for automation and security, commercial applications for health monitoring, traffic monitoring, and more. It provides analytics services, connectivity and control services, and device software.

13.3.5.2 Google Cloud IoT

Google Cloud IoT is a platform that has a set of tools to establish the connection, storage, processing, and analyzing of data at the edge and inside the cloud as well. This platform has integrated software along with machine learning (ML) skills to satisfy the requirement by maintaining scalable and completely managed cloud services. Google Cloud IoT provides complete services for real-time business needs by collecting data from geographically dispersed devices in the cloud and at the edge. For ad hoc analysis Google BigQuery is used and for advanced analytics cloud machine learning engine is used by applying ML. The reports and results can be visualized in Google Data Studio. It improves operational efficiency by providing external support to the devices from various top manufacturers like Intel. With the use of Google Maps, Google Cloud IoT enhances the business solution with location intelligence. It is used to identify the location of assets in real time during their movement. Their position can be tracked with high precision even when IoT data are available in isolated regions, indoors, or spread across several cities [15].

13.3.5.3 Microsoft Azure IoT Suite

The Azure IoT is used to develop businesses and industries in order to face the future with the IoT. Using AI and ML, Azure IoT quickly processes huge amounts of data from different IoT devices. Azure IoT [16] is used by all organizations to improve productivity and reduce waste. It is globally available and it provides cost effective and scalable services for IoT applications and devices. It helps customers in their decision making by advanced analytics [16].

Microsoft Azure IoT Suite provides cloud services to IoT assets for connecting, monitoring, and controlling billions of devices. It provides a two-way communication services among billions of IoT devices and the cloud environment. It provides reliable and highly secure communication between the IoT applications and its devices. Using Azure IoT Hub, a solution can be provided for virtually connected devices. Microsoft Azure ensures a secured communication channel between IoT devices while sending or receiving data. It authenticates all connected devices to maintain security by providing identities. It provides scalability and built-in device management for connecting and managing devices [17,18].

13.3.5.4 Salesforce IoT

Salesforce is one of the pioneers in cloud computing. In recent days, they are excelling in IoT applications. For storing and processing data obtained from IoT devices, Salesforce launched an IoT cloud platform. Scalable and secure solutions are offered by Salesforce IoT, which helps industries transform their business models. From the developer's point of view, setting up is not simpler, but the ease-of-use technology has it more accessible. Two main services called IoT scale and IoT explorer are offered by Salesforce IoT.

Salesforce has predefined rules and logic to trigger an action, and it gets data as input from the devices connected in an IoT system. It operates as a nonlinear workflow engine. It is a state machine that manages the transformation of people and objects from one state to another state. A state is a rule or set of rules for a set of people or objects to which they can be applies. Workflows created by developers react to different events and related data, which are beneficial to users. In this method, complete set of sequences to cover all possible combinations of events and related data need not be predefined [13].

13.3.5.5 IBM Watson Internet of Things

IBM Watson IoT is a cloud service platform that is introduced to derive data and values from IoT devices. This platform connects, collects, and processes IoT data securely, easily and quickly. It uses the IBM cloud, which makes the organizations able to scale and adapt to the changing business needs quickly without losing the privacy and security. The IBM Watson IoT platform visualizes the AI-driven data from the cloud for better analytics. It offers various analytical functions to enhance, supplement, and gain insights from IoT data in a simple and instinctive way. It uses the blockchain service to enable secured information to share among the organization's business network [19,20].

13.4 Security Attacks at the IoT Layered Architectures and Protection with Security Mechanisms

IoT is a promising concept involving an extensive environment of connected devices and services, such as sensors, cars, smart objects, consumer products, and industrial components. Hence, IoT focuses on very essential challenges in terms of safety and security. The arising security issues in IoT are not newly introduced, rather they occur by exploiting networking technologies [21]. Yet, IoT implementation poses various security risks, threats, and challenges that have to be handled efficiently in order to avoid serious causes. Safeguarding IoT systems is significant since it includes a huge number of devices, applications, diagnostic tools, backend services, and applications. Moreover, these challenges have to be addressed to guarantee the security in IoT services and products.

There are different IoT architectures that have been introduced by the number of researchers in the literature. According to some researchers, there are three layers in IoT architecture (application, network, and perception layer) [22]. Due to the continuous development of IoT, four layers of IoT architecture include application, support, network, and perception, and are also proposed [23]. Considering the security and privacy, An IoT architecture with five layers comprised of business, application, processing, transport, and perception is also proposed in the literature. Figure 13.6 depicts the IoT layered architecture. This section explores layer-wise security attacks and the protection mechanisms [24,25].

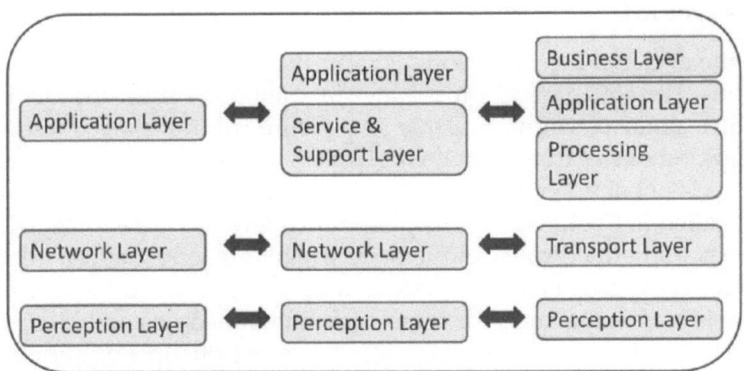

FIGURE 13.6
IoT-Layered Architecture.

13.4.1 Perception Layer

Based on the applications, sensors are chosen and deployed in this layer called the perception layer. Many types of sensors like RFID and 2-D barcode are connected with objects to collect data about location, environment, movement, and vibration. These sensors may become the target for attackers. Some of the popular security threats that arises in the perception layer are tabulated in Table 13.1.

13.4.1.1 Protection Mechanism

The perception layer threats are solved by the following mechanisms:

- Hash-base encryption and lightweight cyptography can be utilized.
- Public key infrastructure (PKI) like protocol can be embraced.
- Secure authorization system may be implemented.
- Embedded security framework can be adopted.

13.4.2 Transport Layer

The transport layer is referred as the transmission/network layer in certain circumstances. It connects the perception layer and processing layer. The physical objects collect from the sensors are transmitted to the smart things present in the network through a wired/wireless medium. Hence it is extremely vulnerable to attacks. It has major security issues in the network concerning reliability and validation of information. General security warnings and troubles that occur in the transmission level are given in Table 13.2.

13.4.2.1 Protection Mechanism

Network layer threats are solved by the following mechanisms:

- Identity management Framework can be used to overcome denial of service.
- Framework has to be designed to recognize the attackers and remove them from the network.

TABLE 13.1

Security Threats at Perception Layer

Threats	Description
Eavesdropping	It is an unofficial instantaneous attack where personal interactions, such as call history, fax, tweets, or video conferences are captured by an intruder. It steals the detail that is broadcasted through the network. This attack is made through unlocked transmission to acquire the details being transmitted and received.
Node Capture	It is a dangerous attack to be handled in this layer. An intruder acquires the control of a gateway node. By acquiring control, the intruder gets knowledge of the details being shared between communicating parties and the key utilized for secure communication by accessing memory.
Fake Node and Malicious	Fake information is fed to the system by an intruder by including a node to the system and to hit the environment. The intent is to stop transmitting real information. A fake node burns up expensive power of actual nodes and probably takes the organization of the network in a motive to demolish it.
Sinkhole Attack	In a sinkhole attack, the unattended nodes in the network for long periods become a target for attackers. Through these nodes the information is extracted from all the nodes around it.
Selective Forwarding Attack	Wicked nodes are selectively filtering certain packets and they crash them out in the network. The crashed packets may hold essential insightful data for future dealings.
Witch Attack	A wicked IoT node takes advantage of a breakdown of a legal node. When the legal node fails, the actual link is used by the wicked node for future communication, resulting in data thrashing.
Replay Attack	An intruder listens to the discussion between the sender and recipient and steals the reliable details from the sender. Using the reliable details of the sender, the intruder communicates to the victim. On seeing the encrypted message the receiver considers it as a proper demand and does the necessary act as expected by the attacker.
Timing Attack	This attack is targeted on devices that are weak in computing resources. The intruder discovers susceptibly vulnerable details preserved in the system. They scrutinize the time limit for the structure to answer to different inquires and cryptographic algorithms.

TABLE 13.2

Security Threats at Transport Layer

Threats	Description
Denial of Service (DoS) Attack	In a DoS attack the real users are denied utilizing devices or other network resources. This attack is performed by streaming the access to the targeted devices or network resources with unnecessary needs by making the real users not use them.
Man-in-The-Middle (MiTM) Attack	In MiTM attack, the intruders quietly acquire and alter the message between sender and receiver who think that they are directly conversing with each other. Since an attacker controls the communication, it produces a severe risk to online safety since the intruders has the facility to acquire and alter information in actual time.
Storage Attack	The users' details are stored on storage devices or the cloud. There is a possibility that the attacker may assail both storage devices and the cloud to amend the user detail. This causes replication of information associated with access of the users, which may provide a chance for different types of people to make an attack on the storage.
Exploit Attack	An exploit attack is any illegitimate operation that comes as software or a chain of instructions or blocks of data. It takes benefit of safety susceptibilities in the software or hardware. This attack is made for achieving the power of the system and to steal details gathered in a network.

- SDN with IoT acts as the controller and protects the IoT agent also.
- Reputation system based mechanisms are capable of detecting and preventing intruders, for which there must be cooperation among the nodes in the communication protocol.
- To detect and prevent the intruders at the cluster level, a strong infringement discovery system is required.

13.4.3 Processing Layer

In between the application layer and transport layer exists the processing layer. This layer is to remove additional data that is insignificant, and it extorts valuable data, which is an important step in handling big data IoT. This process improves performance of IoT. The frequent attacks in the processing layer intended to reduce the routine of IoT are listed in Table 13.3.

13.4.3.1 Protection Mechanism

- Using anti-virus/malware software and keeping it up to date will protect the system [26].
- Utilizing multiple strong passwords will protect devices.
- Updating the operating systems, browsers, and plugins often to scrap any security vulnerabilities.
- Avoiding IoT devices exposed directly to the internet, and by running port scans on all machines, protection is ensured.
- High-interaction honeypots are used to divert hackers from the real data center. It also allows gaining knowledge of their behavior in greater detail, without any disruption to data center or clouding performance [27].

13.4.4 Application Layer

Extending solutions to IoT applications is laid on the part of application layer. It is important to realize strong security to all the devices used in IoT applications due to their low processing capability and less storage space. The common security warnings of the application layer are displayed in Table 13.4.

TABLE 13.3

Security Threats at Processing Layer

Threats	Description
Exhaustion	The intruder uses weak spots to agitate the working of the IoT arrangement. It happens after attacks, such as a DoS hit that makes the network busy for clients, or as a result of other hits that target to exhaust the system facilities, such as memory and battery.
Malwares	This attack uses spyware, adware, viruses, worms, and Trojans horses to connect with the network. The attack's aim is to take action against the requirements of the network to acquire the confidentially of information.

TABLE 13.4

Security Threats at Application Layer

Threats	Description
Cross Site Scripting	It is an insertion attack that permits an intruder to add a client-side script in a reliable site. As a result, an intruder can fully modify the contents of the software as their wish and illegally use the original information.
Malicious Code Attack	It is a set of instructions that can be added in any part of programs planned to introduce horrible effects to injure the system. It is difficult to prevent these instructions with the use of anti-virus tools. It is activated by itself or a program requiring a client's interest to execute an action.
The Ability of Dealing with Mass Data	The great dealing of devices and an enormous amount of data communication between consumers cause its inability to serve the requirements. Therefore, it causes network disturbance and loss of data.

13.4.4.1 Protection Mechanism

- Preference-based privacy protection can be provided by an arbitrator that behaves as a bridge and ensures security connecting the provider and client.
- An access control mechanism is a simple mechanism that provides security to users.
- OpenHab provides security by simple enrollment but fails to tolerate device disparity.
- IoTOne can give solutions to problems that arise in the OpenHab technology. The clients verify identity by sending the request to server and providing the service by themselves.
- In an identity-based security framework, the framework has the guidelines to control and supervise resources and the users with reference to the rules described by the admin.

13.4.5 Business Layer

The business layer manages and governs applications, earning prototypes of IoT. It must also take care of users' privacy. It has the facility to find out how information can be twisted, stocked up, and altered. The attackers can misuse an application by taking control of the business layer by modifying business activities. Weaknesses in an application may lead to liberation in security. Familiar troubles concerning business-layer safety are listed in Table 13.5.

TABLE 13.5

Security Threats at Business Layer

Threats	Description
Business Logic Attack	There are quite a lot of regular errors in this layer, namely, wrong programs fed by a programmer, input validation, password recovery, validation, and encryption techniques. The attackers take lead of an error in coding to alter and modify the exchange of details between a user and the appropriate database used by the application.
Zero-Day Attack	The attacker identifies a security hole in the software, which is unknown to a seller. The attacker takes the authority of the application beyond the user's approval and awareness.

13.4.5.1 Protection Mechanism

A virtual identity (VID) is used to secure client information from unauthorized users. The user requests the service provider for a VID. The service provider, upon receiving the client's demographic details, creates the client's VID. The user information is revealed to anyone, but only with the permission of the user. The attackers cannot access the information without the knowledge of the user. So, the VID structure guarantees privacy of clients' data and protects them from invaders and unapproved entry.

13.5 Access Control Mechanisms

Data generated by the internet is increasing every day which leads to an increase in IoT-based cloud services. IoT devices are used in various fields like healthcare, marketing, weather forecasting, and security management [28]. The main focuses in addressing security issues for data produced from IoT devices are authentication and access control. The security clearance mechanism has to be easily controllable and adaptable. As IoT devices are included in human lifestyle (in tools like fridges, watches, etc.) people with diverse proficiency need to be implicated in security clearance activities. An effective access control mechanism should satisfy three constraints, namely [29]:

- Confidentiality: To prevent unauthorized access of resources
- Availability: To ensure access to the authorized users whenever resources are needed
- Integrity: To prevent modification of resources without authorization

When users try to access data, a strong protection mechanism needs to be applied to check the user's permission to obtain the data and the allowable conditions to access the data. To ensure security in data storage, different types of access control mechanisms are followed in IoT applications. The enormous amount of heterogeneous data generated by IoT devices needs to be analyzed using big data analytics. The data produced by IoT devices are semi-structured like bank/credit card transactions, current location of a device, and measurements from human body. Traditional database management systems cannot support huge volumes of data in terms of performance, efficacy, flexibility, and scalability. Databases like NoSQL and MapReduce mechanisms are used for the systematic analysis of semi-structured data. A fine-grained access control mechanism is an effective approach for protecting personal and sensitive data. For customized access control, context management is used through which constrained access can be given to data based on exact time periods or geographical locations [30].

The access control mechanism can be categorized as the folowing:

- Platform-specific approach: The mechanisms derived for a single system are classified under this category.
- Platform-independent approach: These approaches do not target specific systems, but they are more general and they can be applied to any platform.
- Domain-specific approaches: These approaches target domain-specific IoT data, which may include both platform-specific and platform-independent mechanisms.

13.5.1 Platform-Specific Approaches

The storage systems used for IoT-integrated big data must be capable of dealing with large volumes, large velocity, and large varieties of data. The databases that outperform traditional database systems used are the in-memory databases and columnar databases. But there is no standard for big data storage systems [31].

Platform-specific approaches can be used only with the platform for which they have been defined. The MapReduce and NoSQL databases are used for these types of systems.

13.5.1.1 MapReduce

The core component of Hadoop framework is the MapReduce, which is used for data processing. It has an implementation for processing huge data sets in parallel [32]. The MapReduce module checks access rights [33], then proceeds in reducing and delivering the response to the user query. A MapReduce model comprises of a map task and a reduce task. The map task processes a set of key/value pairs, and a set of intermediate key/value pairs are produced using filtering and sorting functions. In map task a set of data is broken down into tuples of small set. The reduce task does the reverse process of map task by combining smaller data set to original larger tuples. The reduce task is executed only after the map task. Figure 13.7 depicts the MapReduce framework working model.

The data received from IoT devices are separated into multiple chunks of data in the MapReduce systems. The partitioned data is distributed among a group of nodes. The analysis of data is done in parallel by MapReduce tasks, which are customized by the users. The users define the functions of MapReduce tasks. These tasks first extract the key/value pairs and then manipulate the flow of the key-value pairs. The MapReduce framework can handle both unstructured data and semi-structured data from various IoT devices.

The Hadoop distributed file system is a structure for warehousing and handling big data across computing clusters using a simple programming model [34]. Applying fine-grained access control to the arbitrary MapReduce jobs is a significant challenge in Hadoop. Research is being carried out to instil entry check rules into surrender works especially for IoT-integrated big data that needs fine-grained encryption for access control policies [35]. To access personal and sensitive data, Hadoop-distributed file systems implement user access limitations. To have control over the user and to accept only permissible data, an access restrictions check is necessary during the MapReduce task. Kerberos protocol is used in Hadoop for protecting the file system from intruder access [36]. In Kerberos protocol, a security voucher is created to authorize a client when they are from one organization

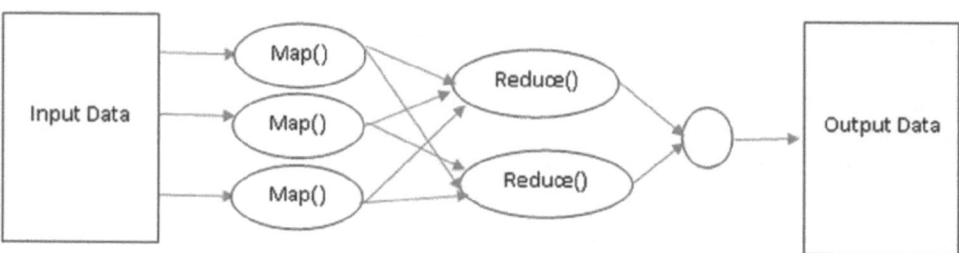

FIGURE 13.7
MapReduce Working Model.

and want the service of another organization. But the voucher is valid for only one day. If the token is needed again, it can be renewed once per week.

Yenumula [34] has proposed a model in which each an authenticated user will get a service voucher for a facility to have a control over the user for accessing the facility within their domain. For each facility and for each service request, the system has to produce a set of entry approval. The user should not exceed their limits for each service request. The system will deny the request when the requested facility is outside the user's permission. If the user does not have any role in the system, then they are considered a hacker. Users trying to access unauthorized information are also considered hackers.

Even though the MapReduce model has different security/privacy policies and issues for different types of records like organized, unorganized, and semi-organized, the fine-grained access control (FGAC) mechanism can be used for all types of data. The FGAC model is supported by many IoT applications like healthcare and finance. Thus some industry experts suggest that FGAC must be added to MapReduce systems since the model is widely accepted [36].

13.5.1.2 NoSQL Data Stores

The database mechanism that is used for data generated by IoT applications must be able to handle data that are small and non-mutable, high volume streams, time-oriented, and have low consistency requirements. Some relational databases are able to store the majority of web application data. Storage representations may be different like a column-oriented table, document-oriented, key-value pair, and graph. But these storage representations may be used to supplement and not completely replace a relational database mechanism.

NoSQL stands for 'Not Only SQL' and it has a different performance characteristic from the existing relational database. The NoSQL database provides better results on some read/writes NoSQL systems that have the property of horizontal scaling that is duplicating and segregating data over different data stores. This feature allows them to support a feature called online transaction processing (OLTP), an enormous amount of simple transactions per second. The transactional property ACID is not provided by NoSQL systems. Instead of that NoSQL uses a new feature called BASE, representing availability and consistency. Higher efficiency and elasticity is achieved when NoSQL includes ACID constraints [37]. NoSQL is a freely accessible, non-relational, and distributed data store. It can be used for real-time web application data as well as IoT integrated big data. NoSQL data store also supports different storage representations like column-oriented data store, document-oriented data store, key-value data store, and graph-based data store.

While storing IoT-integrated big data in the cloud, multi-tiered entry control rules have to be imposed even on a distinct portion of data or a facility [38]. This makes access control a naturally complex problem area. For providing multi-tiered entry control Apache Accumulo software is used at the unit level in a key-value data store.

For the IoT data, it is needed to extend real-time and on-demand access control policies that should consider the parameters like user identity, multi-tenancy, data sanitation, data integration, connection between permitted user, service, and the target facility that is to be obtained. Mansura et al. [38] have proposed an access control management as a service (ACMaaS) for NoSQL datastore, which is used for big data IoT.

Big data is usually stored in the cloud and cloud computing uses a variety of access control policies. Some of them are discretionary access control (DAC), mandatory access control (MAC), role-based access control (RBAC), attribute-based access control (ABAC), and policy-based access control (PBAC). Identity-based methodology is followed in most of these access control mechanisms. In DAC, MAC, and RBAC, the client is permitted to access the requested data when the identity of the client matches a set of ACL rules. In ABAC, not only identity is verified, but other attributes are also verified to allow access. In policy-based access control, a set of entry control rules are defined and instead of the user identity, each access request is confirmed against those defined rules.

RBAC model roles are allocated to clients based on the organizational structure [39]. The client's role determines the authorization to perform operations in the database. Depending on the need, the assigned client roles can be newly created or withdrawn. In RBAC the privileges that are needed to perform operations are only granted. Each client role has multiple permissions associated with it. These roles-related permissions describe the right to perform an operation in a structure on a specific resource. For instance, a read permission can be granted to all workers, but only administrator role can be assigned an approve permission. When organizational functions evolve and change, and when new permissions are introduced, then role associations can be easily updated. Motahera [40] proposed a context-based model which expands the role-based access control used in NoSQL datastores. This prototype estimates and performs protection rules that contain flexible approaches against the vibrant characteristics of IoT data.

13.5.2 Platform-Independent Approaches

Most research in the arena of access control mechanisms for IoT-integrated big data recommend solutions based on a specific platform. Most NoSQL datastores operate with a platform-specific approach. For example, MongoDB can be used only with the specific platform. The heterogenic character of the IoTdata needs solutions to work with multiple different platforms. As a result, the access control mechanisms lead to an ambitious task. Academicians and industrialists started doing a collaborative project to develop unified query language for NoSQL data stores.

SQL++ is a query language developed exclusively for handling semi-structured data generating from IoT applications. SQL++ can also handle the traditional DBMS data. SQL++ includes the data design and query language features of NoSQL, SQL-on-Hadoop, and NewSQL data stores. SQL++ was developed as a development of SQL and it is compatible backwards with SQL. It was developed from SQL in motive to help the SQL users for two reasons. First, since the features of SQL do not support many surveyed databases, it is extended and a new unifying query language SQL++ was proposed. Second, query languages are needed to support semi-structured data generated from IoT.

FORWARD query processor is used to help software retrieve data from new databases. To access data, initially users issue a SQL++ query over the database and the FORWARD processor translates the query into the native query language for the corresponding database. In the middleware, FORWARD will compensate for any semantic discrepancies between SQL++ and the corresponding original database. Figure 13.8 depicts the SQL++ based FORWARD query processor.

JSONiq is another platform-independent language based on Xquery and supported by MongoDB databases. It follows a JSON-based data model. Fine-grained attribute-based access control (ABAC) mechanism is provided through JSONiq and SQL++ unifying query languages for NoSQL data stores [41]. For the data to be analyzed, the ABAC approach derives an in-memory authorized view. The original queries to be executed on the data to

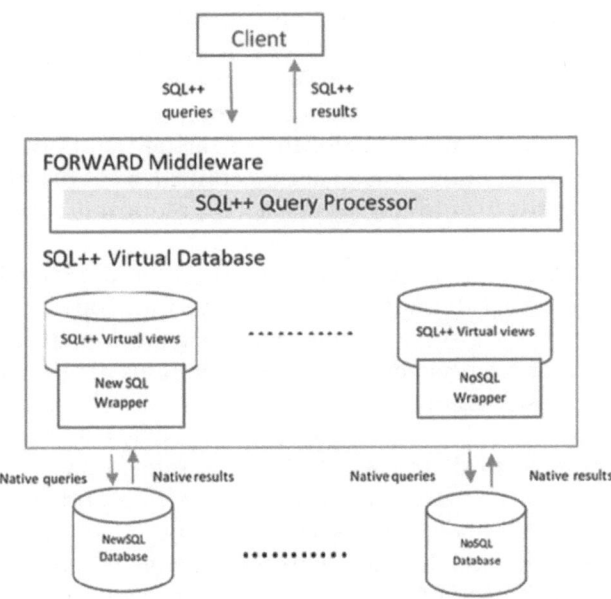

FIGURE 13.8
SQL++-Based FORWARD Query Processor.

be analyzed are executed on the derived in-memory views for enforcing the context-aware access control policies [37].

13.5.3 Domain-Specific Approaches

In this section, the access control mechanism for data stream analytics for IoT is discussed. IoT implementation is increasing hastily and is used in a various fields for the concrete developments of human beings. Recently a wide variety of wearable devices like tracking the movements of patients, monitoring sports activities, and health conditions are used by many people. IoT applications like home automation services are available to manage the safety of living areas. The services may be like smart lights that are switched off automatically when people leave the home, smart locks that give alarm when unauthorized people try to open the door or locker, and so on [36]. Since the information generated from these devices is sensitive and personal, providing safety and confidentiality of these applications forces a major challenge. Several new research projects are being carried out for the confidentiality and safety of IoT applications.

The capability-based access control (CapBAC) model for IoT ecosystems has been proposed by Gusmeroli et al. and Hernandez et al. [41,42]. Access authorization management can be externalized and distributed in this model. The major drawback of the CapBAC model is that it does not consider context awareness for access control.

Ferraiolo et al. [43] proposed a role-based access control (RBAC) which defines the user-role assignment and permission-role assignment. Hu et al. [44] proposed a model called attribute-based access control (ABAC) that stretches RBAC and assigns roles to users in a dynamic way. RBAC and ABAC have been proposed for IoT ecosystems to regulate access control technologies.

13.6 Key Challenges and Future Directions

The hot topic for most research for the past few years is IoT. In the beginning only a limited number of devices could be linked with the internet/intranet. But the advancements in sensors, communication protocols, and RFID have paved the way for connecting billions of heterogeneous devices. From smart phones to smart grids, data are being generated in large amounts every minute [45]. Even though the latest technologies work efficiently for a specific set of problems, due to the distributive nature of IoT, these technologies become a big challenge. Some of the key challenges are:

- Identification of device, interaction between devices, deriving standardization mechanisms, and inter-devices collaboration.
- Since IoT devices and sensors generate data at a very high speed, it needs a large amount of storage space to store the data.
- The IoT devices are heterogeneous pertaining to the types of devices, different manufacturers, communication mechanisms, and different applications. So the data generated from these devices are different in types and semantics. Managing and processing these heterogeneous data is another key challenge.
- An important challenge faced by the data management process is the access control mechanism to be followed for the heterogeneous data. In addition to access control, data management also involves challenges in areas like, data integrity, data heterogeneity, interoperability, data analytics, data aggregation, mobility, and confidentiality and privacy.
- Since different devices use different encryption and decryption mechanisms, it is a big challenge for the data management task to handle these cryptographic methods.
- Archiving large volumes of IoT data for future use is a big challenge to address primarily.
- Availability and attainability of data are key challenges for security management [46].

Design of uniform communication protocols, common global standards for all industries, middleware complications, and highly improved security mechanisms are considered for future research. Identity management can be included within the devices used for IoT with a quick encryption mechanism. Cyber sensors can be implemented to capture data from physical objects that perform actions based on real-time event response. Privacy and protection mechanisms based on requirements can be developed to protect IoT from privacy attacks. Threat models can be identified for reducing eavesdropping attacks. Security methods can be established in a transport layer to provide complete security over the edge. Ensuring over the edge protection at the application level can also be proposed to simplify deployment complexity and reduce the cost for data processing [46].

Blockchain integration with IoT increases the security of extremely susceptible IoT devices. The decentralized framework and use of cryptographic algorithms in blockchain enhance the security of the IoT platform. Hence, it is difficult for a hacker to alter each and every block of blockchain, and it aids in detecting frauds. Processing time is reduced as the most important information is stored in the blockchain network. Combining blockchain with IoT enables the application of ML algorithms on real-time data collected from the IoT devices for analysis. Thus, blockchain-facilitated IoT increases trust, security, performance, data sharing, and the computing power of the IoT network.

13.7 Conclusion

The tremendous growth in IoT devices produces enormous amount of data that needs to be handled, transformed, and evaluated in high frequency. This raises the need for big data analytics in extracting meaningful insights from data spawned by IoT devices. So storage mechanisms, security, and access control mechanisms for big data IoT was discussed in this chapter. The traditional database technologies are not suitable for storing and analyzing real-time data, which may be structured or unstructured. Thus various storage technologies like column-oriented database, key-value database graph database, and cloud-based IoT platforms were explored in this chapter. The unique behavior of IoT sensors presents new threats, risks, and security challenges. The protection mechanism at different IoT layers (business, application, processing, transport, and perception layers) has been thrashed out in this chapter. Effective access control mechanisms that satisfy confidentiality, availability, and integrity and that are adaptable and easily controllable need to be identified for IoT data. Hence platform-specific approaches, platform-independent approaches, and domain-specific approaches were presented in this chapter. Finally, the key challenges and future directions for IoT-integrated big data have been discussed. It was found that the concerns related to the big data IoT paradigm are in their infancy and blockchain technology can be adopted to overcome the challenges.

References

[1] IoT Security Guidelines. https://www.gsma.com/iot/wp-content/uploads/2019/10/CLP.11-v2.1.pdf. Accessed January 20 2020.
[2] Alansari, Zainab, Nor Badrul Anuar, Amirrudin Kamsin, Safeeullah Soomro, Mohammad Riyaz Belgaum, Mahdi H. Miraz, and Jawdat Alshaer. 2018. "Challenges of Internet of Things and Big Data Integration." In *International Conference for Emerging Technologies in Computing*. Springer, Cham, pp. 47–55.
[3] Ahmed, Ejaz, Ibrar Yaqoob, Ibrahim Abaker Targio Hashem, Imran Khan, Abdelmuttlib Ibrahim Abdalla Ahmed, Muhammad Imran, and Athanasios V. Vasilakos. 2017. "The Role of Big Data Analytics in Internet of Things." *Computer Networks* 129: 459–471.
[4] Hadoop, Apache. https://bestcellular.com/apache-hadoop/. Accessed January 21 2020.
[5] ThingSpeak. https://thingspeak.com/. Accessed January 21 2020.
[6] Countly. https://count.ly/product. Accessed January 21 2020.
[7] AT&T IoT Platform. https://iotplatform.att.com/. Accessed January 21 2020.
[8] Axonize. https://www.axonize.com/platform/. Accessed January 21 2020.
[9] Venkatraman, S. et al. 2016. "SQL Versus No SQL Movement with Big Data Analytics." *International Journal of Information Technology and Computer Science*.
[10] Kaur, K., and R. Rani. 2013. "Modeling and Querying Data in NoSQL Databases." *Big Data* (IEEE International Congress).

[11] Karande, N. D., et al. 2018. "A Survey Paper on NoSQL Databases: Key-Value Data Stores and Document Stores." *International Journal of Research in Advent Technology.*

[12] Bagga, Simmi, and Anil Sharma. 2019. "A Review of NoSQL Data Stores." *International Journal of Scientific & Technology Research* 8 (11): 661–666.

[13] "Azure IoT Hub, Managed Service to Enable Bi-directional Communication Between IoT Devices and Azure." https://azure.microsoft.com/en-in/services/iot-hub/?&ef_id=EAIaIQob ChMIo6Slzc_W5wIVRKWWCh1krgw5EAAYASAAEgJ1q_D_BwE:G:s&OCID=AID2000081_ S E M _ X d m A w 1 J 6 & M a r i n I D = X d m A w 1 J 6 _ 3 4 0 8 0 7 3 8 7 9 1 7 _ % 2 B a z u r e % 2 0 %2Biot_b_c__63148366013_kwd-303430196768&lnkd=Google_Azure_Brand&dclid=CJ7rldPP 1ucCFVeRjwodDDAEvg. Accessed February 17 2020.

[14] Sarhan, Amany. 2019. "Cloud-Based IoT Platform: Challenges and Applied Solutions." In *Harnessing the Internet of Everything (IoE) for Accelerated Innovation Opportunities.* IGI Global.

[15] AWS. "IoT, IoT Services for Industrial, Consumer, and Commercial Solutions." https://aws. amazon.com/iot/. Accessed February 16 2020.

[16] "Google Cloud IoT, Unlock Business Insights from Your Global Device Network with an Intelligent IoT Platform." https://cloud.google.com/solutions/iot. Accessed February 16 2020.

[17] "Microsoft Azure IoT Suite – Connecting Your Things to the Cloud." https://azure.microsoft. com/en-in/blog/microsoft-azure-iot-suite-connecting-your-things-to-the-cloud/. Accessed February 17 2020.

[18] "What Is Azure Internet of Things (IoT)?" https://docs.microsoft.com/en-us/azure/iot-fundamentals/iot-introduction. Accessed February 17 2020.

[19] "itransition." https://www.itransition.com/blog/salesforce-iot. Accessed February 20 2020.

[20] "Securely Connect, Manage and Analyze IoT Data with Watson IoT Platform." https://www.ibm. com/internet-of-things/solutions/iot-platform/watson-iot-platform. Accessed February 22 2020.

[21] "Watson IoT Platform – A Fully Managed, Cloud-Hosted Service with Capabilities for Device Registration, Connectivity, Control, Rapid Visualization and Data Storage." https://www.ibm. com/cloud/watson-iot-platform. Accessed February 22 2020.

[22] European Union Agency for Network and Information Security (ENISA). 2017. "Baseline Security Recommendations for IoT in the Context of Critical Information Infrastructures." doi:10.2824/03228.

[23] Said, Omar, and Mehedi Masud. 2013. "Towards Internet of Things: Survey and Future Vision." *International Journal of Computer Networks* 5 (1): 1–17.

[24] Darwish, Dina. 2015. "Improved Layered Architecture for Internet of Things." *International Journal of Computing Academic Research (IJCAR)* 4: 214–223.

[25] Burhan, Muhammad, Rana Asif Rehman, Bilal Khan, and Byung-Seo Kim. 2018. "IoT Elements, Layered Architectures and Security Issues: A Comprehensive Survey." *Sensors* 18 (9): 2796.

[26] Gloukhovtsev, Mikhail. 2018. "IoT Security: Challenges, Solutions & Future Prospects." *DELL Knowledge Sharing Article.*

[27] "IoT: A Malware Story." https://securelist.com/iot-a-malware-story/94451/. Accessed October 15 2019.

[28] "Trending: IoT Malware Attack." 2019. https://www.biz4intellia.com/blog/iot-malware-attack/. Accessed February 22 2020.

[29] Jeong, Yoon-Su, Yong-Tae Kim, and Gil-Cheol Park. 2018. "Efficient Big Data-Based Access Control Mechanism for IoT Cloud Environments." *International Journal of Engineering & Technology* 7 (4.39): 539–544.

[30] Ouaddah, Aafaf, Hajar Mousannif, Anas Abou Elkalam, and Abdellah Ait Ouahman. 2017. "Access Control in the Internet of Things: Big Challenges and New Opportunities." *Computer Networks* 112: 237–262.

[31] Colombo, Pietro, and Elena Ferrari. 2019. "Access Control Technologies for Big Data Management Systems: Literature Review and Future Trends." *Cyber Security* 2 (3).

[32] Strohbach, M., J. Daubert, H. Ravkin, and M. Lischka. 2016. "Big Data Storage." In *New Horizons for a Data-Driven Economy.* Springer, Cham, pp. 119–141.

[33] Ghemawat, Jeffrey Dean Sanjay. 2004. *"MapReduce: Simplified Data Processing on Large Clusters."* In *OSDI'04: Sixth Symposium on Operating System Design and Implementation*, San Francisco, CA, pp. 137–150.

[34] Reddy, Yenumula B., 2013. *"Access Control for Sensitive Data in Hadoop Distributed File Systems."* In *Third International Conference on Advanced Communications and Computation*, INFOCOMP [Conference Paper], pp. 17–22.

[35] "HadoopMapReduceTutorial."https://www.dezyre.com/hadoop-tutorial/hadoop-mapreduce-tutorial. Accessed February 22 2020.

[36] Bertino, Elisa and Elena Ferrari. 2018. "Big Data Security and Privacy." In *A Comprehensive Guide Through the Italian Database Research Over the Last 25 Years*. Springer International Publishing AG, pp. 425–439.

[37] Huseyin, Ulusoy, Murat Kantarcioglu, Erman Pattuk, and Kevin Hamlen. 2014. "Vigiles: Fine-Grained Access Control for MapReduce Systems." *IEEE BigData*.

[38] Habiba, Mansura, and Md. Rafiqul Islam 2015. *"Access Control Management as a Service for NoSQL Big Data as a Service."* In *Second Asian Pacific World Congress on Computer Science and Engineering* [Conference Paper].

[39] Cattell, Rick. 2011. "Scalable SQL and NoSQL Data Stores." *Acm Sigmod Record* 39 (4): 12–27.

[40] Shermin, Motahera. 2013. "An Access Control Model for NoSQL Databases." Electronic Thesis and Dissertation Repository.

[41] Hernández-Ramos, J. L., A. J. Jara, L. Marin, and A. F. Skarmeta. 2013. "Distributed Capability-Based Access Control for the Internet of Things." *Journal of Internet Services and Information Security (JISIS)* 3 (3/4): 1–16.

[42] Gusmeroli, S., S. Piccione, and D. Rotondi. 2013. "A Capability-Based Security Approach to Manage Access Control in the Internet of Things." *Math Comput Model* 58 (5): 1189–1205.

[43] Ferraiolo, David F., Ravi Sandhu, Serban Gavrila, D. Richard Kuhn, and Ramaswamy Chandramouli. 2001. "Proposed NIST Standard for Role-Based Access Control." *ACM Transactions on Information and System Security (TISSEC)* 4 (3): 224–274.

[44] Hu, V. C., M. M. Cogdell. 2014. "Guide to Attribute Based Access Control (ABAC) Definition and Considerations." *National Institute of Standards and Technology*.

[45] Abbasi, Mohammad Asad, Zulfiqar A. Memon, Jamshed Memon, Tahir Q. Syed, and Rabah Alshboul. 2017. "Addressing the Future Data Management Challenges in IoT: A Proposed Framework." *International Journal of Advanced Computer Science and Applications* 8 (5): 197–207.

[46] Kumar, Sathish Alampalayam, Tyler Vealey, and Harshit Srivastava. 2016. *"Security in Internet of Things: Challenges, Solutions and Future Directions."* In *49th Hawaii International Conference on System Sciences. IEEE Computer Society*, pp. 5772–5781.

14

Security Challenges and Mitigation Approaches for Smart Cities

S. Ponmaniraj, Tapas Kumar, V. Gokul Rajan, and Sanjay Sharma
Galgotias University, Greater Noida, Delhi-NCR, India

CONTENTS

14.1 Introduction

According to a statement by an Indian government, every year the city population is growing in terms of millions. An Indian economic survey says that by 2040, India will require better infrastructure worth US$4.5 trillion. The Indian ministry for Housing and Urban Affairs says that India entailed 700–900 million square meters of built-up and profitable space for a new 600 million urbanites for the next ten years. In that circumstance, India's smart cities mission has colossal implication for its future. There is no exact universal definition for the term called smart city. As per the government mission statement, urban areas have to get all features that give first-rate quality of life for its citizens in all the aspects such as ceaseless electricity and water supplies, well-organized cleanliness, waste management, ample conveniences of healthcare and educational services, transportation. Based on a Smart City Plans (SCP) for urban renewals, nearly 100 cities were chosen to implement their projects with more funds. Cities received funds for SCP, and have to complete their projects flanked by 2019 to 2023.

SCP is not a project for integrated urban area development. Still SCP functions with two broad divisions [1]. Firstly, SCP works on area-based development in which SCP undertakes redevelopment for existing business, and upgradation for infrastructures like streamline of water and sewage lines, development of new commercial hubs and public places and more. The second division is Pan-city development, which employs 'smart solutions' by means of technologies. Interleaving of electronic-based devices and equipment across the entire city is used to control and monitor city activities. Based on SCP projects, selected cities have a proposal for 5,151 projects at the cost estimated around $33.8 billian. Few important SCP-based projects implemented by smart cities are as follows [2]:

1. Parking management
2. Emergency response and city incident management
3. Smart bus services: intelligent traffic management system
4. Information and Communication Technology (ICT) and e-governance
5. Mobile app and website for pan city water
6. Air quality monitoring
7. Mobile apps for citizens to report their problems
8. Social, mobile, analytics and cloud (SMAC) center
9. Cleanliness metering
10. Solid waste management

14.1.1 Smart City Development

The objective of smart city development is to furnish a sustainable and disciplined life to its citizens by smart solutions via technology. The mission started in India on June 25, 2015. The Smart City Plan component supports the enhancement, renewal, and extension of cities, Retrofit, Redevelop and Greenfields. Pan city wraps 'smart' solutions for all parts of cities through technology. SCP projects improve lifestyle for humans by changing impoverished neighborhoods into well planned and organized cities for citizens who connect and accommodate solutions through technology, improving Greenfields for expanding populations to get greater livability. Area-based transformations are retrofitting; remodeling of waste and impoverished neighborhoods allowing humans to live in a better place. Well-developed areas are serviced like new and encouraged around cities to make themselves livable for increasing populations.

Comprehensive development provides quality of life, employability, and enhances income for low-income people. A few of Smart City Plans and Projects are listed in Table 14.1 [3].

Table 14.1 contains applications functioning by digital equipment. Those applications produce numerous digital data for transactions. The handling of data being transferred is very sensible and shrewd. Storage and retrieval of data should be more confident and secure to avoid exploitation. Since data is communicated with a centralized controlling system from various parts of cities, any hacker can attack and mistreat this information. Through an accessing path a hacker can intrude any systems and devices, then take all control over the equipment. An intrusion detection system is applied to detect any kind of intrusion approaches found on the communication link and the intrusion preventive system is used to avoid attacks based on the outcome of detected vulnerabilities to bring the best performance over network communication on data transaction. For a smart device network model framework, an edge device has to communicate with the centralized server for data transaction. Formal pictorial representation of smart city development activities and services are shown in Figure 14.1.

TABLE 14.1

List of SCP Projects on Pan City Development

S. No	Component	S. No	Component
1	City navigation system	18	Toll collection
2	Property survey	19	E-challan
3	GIS	20	Smart parking
4	Aadhaar seeding	21	Vehicle tracking: Buses/cars/e-rickshaws
5	Emergency services/disaster management	22	Passenger information system
6	CCTV surveillance	23	Common smart card/fare collection system
7	Command control centre	24	Road signage
8	Data centre	25	Air pollution monitoring
9	E-governance	26	Solid waste management
10	City asset management system	27	Sewerage
11	Document management system	28	SCADA–power
12	Wi-fi	29	Solar farm/solar city
13	Optical fibre	30	Street lighting
14	NMT	31	Solar panels
15	Traffic management system	32	Digital employment exchanges
16	Buses/fleet management	33	Incubation centres
17	E-bus	34	Telemedicine & kiosks

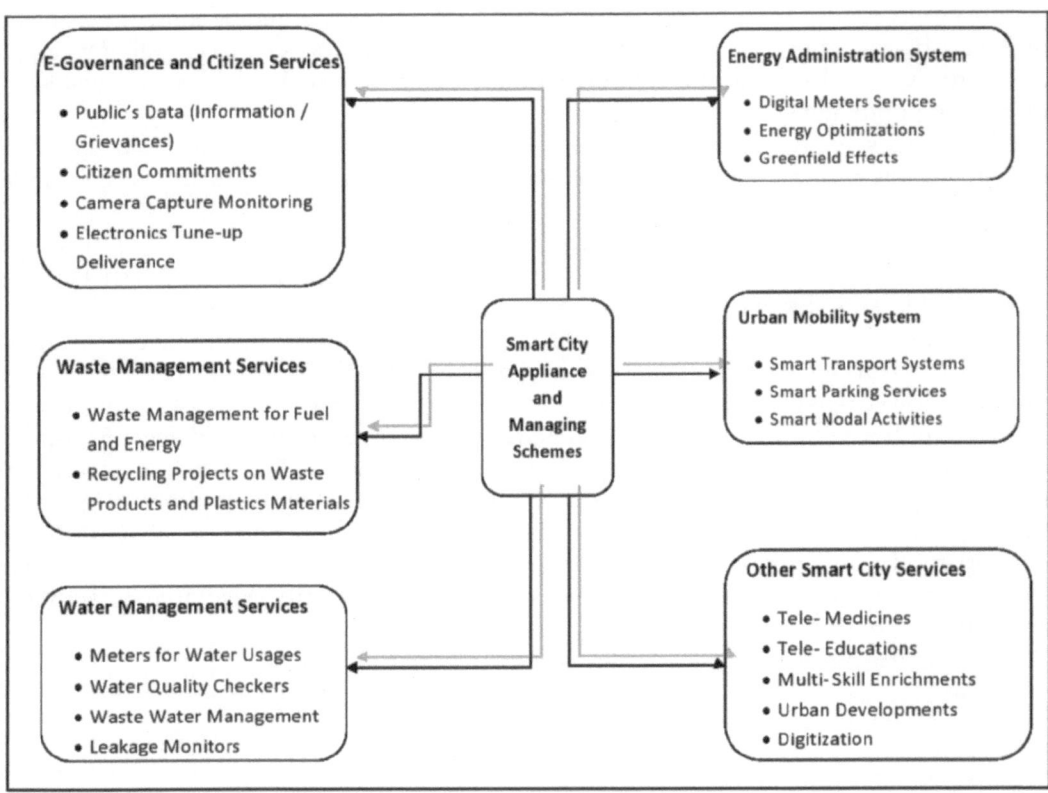

FIGURE 14.1
Pictorial Representation of Smart City Activities (Image Source: http://smartcities.gov.in/upload/uploadfiles/files/Appbased%20Projects_60%20cities.pdf).

14.1.2 IoT Devices and Configuration

Nowadays smart applications are increasing rapidly to satisfy human's call. Many application programs are running on devices that connected through internet and transmitting data to the centralized server. Any devices with sensors, actuators, software, and computer access via internet connection to transmit data are recognized as IoT devices. Without human interventions IoT devices are capable of data transfer from one device to another device or by a human enabling internet connectivity. According to a populations and technologies survey, every individual wanted to access an average of ten IoT devices in one whole lifespan. A modern device collects data from the physical world via its sensors then processes through its complex programs embedded in itself. Smart phones, laptops, gadgets, smoke detectors, fire sensors, smart lighting, and digital televisions are a few examples for smart devices that are activated by multimedia IP addresses. Bridging between the physical and digital worlds makes human livability simpler, easier, and better. Smart devices proceed on two bases as follows:

1. Devices that are capable of connecting to the internet
2. Devices that are integrated with sensors, actuators, functional programs, and built in network connection modules.

14.1.2.1 Life Cycle of IoT Devices

The deployment of IoT devices has to go through the following subsequential life cycle process including to monitor, update, manage, and service the end with decommission [4]. The IoT life cycle is depicted in the Figure 14.2.

In the life cycle diagram in Figure 14.2, firstly IoT devices are configured with functional software programs and built-in hardware modules. Then it will be installed in a proper consign with executable mode. All devices are installed with a configuration and factory configuration reset mode for future access. After installation, data transmission takes place in the cloud-based platform in which all devices and its transactions are controlled and monitored by a centralized server who connects all the devices from various parts of the city via different communication mediums such as sensors and networks. Installation means the startup of a device's services and applications. Decommission is the process of component detachment. It is the last process in the IoT lifecycle and it happens for two main reasons as follows, mission accomplishment and malfunctioning of devices [5]. Once the device is corrupted or malfunctions, then it will decommission for rectifying its problem. Rectifications are done by either factory reset or proper troubleshooting of the tarnished device. Before decommissioning, all service access by its components is stopped then it stops running applications. Then it will detach application components subsequent to terminating its program files. Detached components will be undergo reconfiguration of application parameters settings and testing. If the refinement progress is over, then devices will be deployed back to their positions whenever they were in good condition.

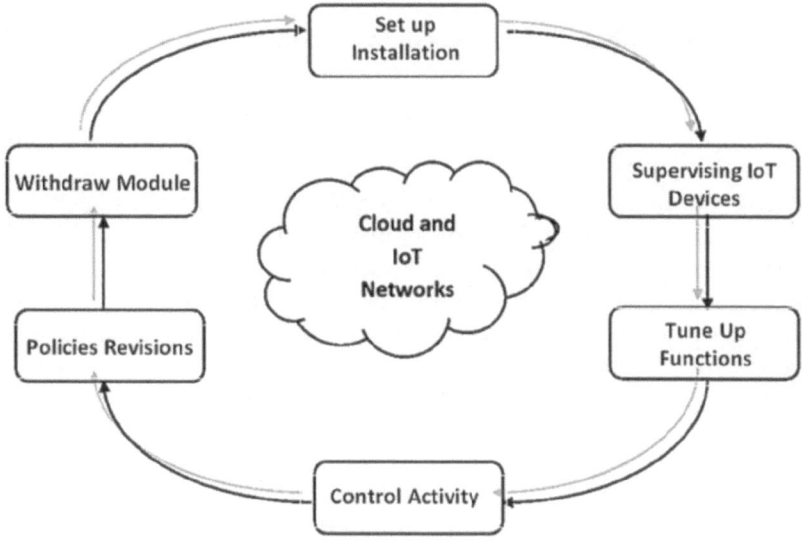

FIGURE 14.2

Life Cycle of IoT Devices (Image Source: https://www.learnbigdatatools.com/life-cycle-management-of-iot-and-iiot-with-device-used-in-iot-and-iiot/).

14.1.2.2 Layered Configuration of IoT Devices

IoT devices are configured based on a three-layer approach. The physical, network, and application layers are the three layers of IoT configurations. Functional details of these layers are briefly explained as follows [6].

Physical Layer: In this layer, a 'smart' device collects data with its sensors from the environment or from any object under surveillance, then it converts this raw data into some useful data. Collection of raw data size is increasing rapidly due to robotic camera systems, voice recognizers, water level detectors, fire sensing alarms, air quality checkers and monitors, and more. These devices collect user data from various level and places based on conditions. However, all collected information will not be useful in every user's expectation, so the collected data will be segregated based on its applications and priorities. For an example, some data's course of action must be handled with higher priority like threat detection and analysis data, time sensitive data, abrupt crash detection and shutdowns, and more. Processing of all data of an organization will lead to delays in progress when higher priority data is requireed in some circumstances. So that, data storage, analysis, and easy retrieval from databases is the vital role in cloud-based activities.

Network Layer: Once data is collected from sensors or actuators, aggregated data must be converted to digital formats from raw states. The environment can give only analog data to 'smart' devices. With analog information a computer cannot make any decision or operate specific applications from devices. A data acquisition system (DAC or DAQ) process carries data from the environment as a sample measurement then it converts sample values into numerical data for manipulation. A DAC is used for analog data to convert digital value. DAC connects with sensor networks to aggregate input values from any object or environment, then converts it into a digital format to route towards any network-oriented systems such as Wi-Fi, Wired/Wireless LAN, or internet for further activities. This information feeds into application layers to perform concern activity.

Application Layer: Once this layer gets data from the network layer, this will be shared with the server for expected operations to be carried out. The user will get their required task done through aggregated, cleaned data served from stage2. This information will be applied to new products and services for the intention of making all 'smart' devices trained.

Each device is pointed out by its 'unique identifier'. The centralized server looks for the device's request and respondes through its unique identifier values. Whatever data are coming from devices are processed in the aforementioned layers and respond, in turn, as the same as how the request came from the devices. Since the back and forth of data happens on multiple devices from various parts of cities, data handling, analysis, and processing is very obsessive. Big data technologies are applied over 'smart' city-based projects to give a remedy for such kinds of fascinations.

14.1.3 Big Data in IoT

Data analytics is vital thing to processing the right data on a smart city project. From a large volume of data sets, finding exact data, cleaning them from noises, and feeding them the right input to many automata learning places are important tasks. Big data perception takes responsibility to heed data processing hurdles in smart city applications. Big data faces three different challenges while handling data sets. The four Vs challenges are called volume, velocity, variety, and veracity [7].

14.1.3.1 Challenges of Big Data

Volume Challenge: Handling large amounts of data sets in the course of storing, processing, and easy access is a big issue in big data. SCP cannot set a quantity border to the data in real-time applications. Every day a minimum of hundreds of terabytes of data are managed in a conventional database management system associated with a relational database concept (RDBMS). Traditional database systems keep optimized data for specific usage (i.e., preprocessed information), meaning to load relevant data only. The big data approach is to store all the data, or whichever is not for the current task. Beyond the present use big data stores all information in a raw format for future use. This data set will be acting like a master data set for the scientist who does research on the relationship between the relevant information and requirements needed to analyze later. It reduces the human error while extracting erroneous data on any transactions.

Velocity Challenge: Velocity in big data means the streaming of data into an enterprise's infrastructure at a high-rate processing speed and minimal latency. Many new technologies are working with these strata based on the complexity of the analysis of the given problem statement. Still all applications are unable to bring finicky solutions to make a data stream. Newly incoming data are much larger than previous records and this leads to aggregate compared with existing data. Complex event processing (CEP) engines are required to process larger amount of incoming data into the stream. NoSQL databases write performance procedure to achieve data stream operations as required in real-time applications based on a variety of complexity analyses [8].

Variety Challenge: Data collection from various sources such social media and other objects or sensors are raw in manner. These data are unstructured in nature and cannot process in any of smart application functions. Rapid growth of unknown data has to be validated and integrated before passing any nebulous data to an application; many scripts and written performances are made to stream by map reduced function [9].

Veracity Challenge: Since big data collects data from various sources, it maintains different formats of data. Data that are not used in a current task are considered noise or abnormal. Those data do not relate with all the application programs. Simply those data are used in restoring and reference for some devices when they have problems or misbehave. Meaningful data must be analyzed from stored information to carry out comparison of original data and its functions. Comparison of big data information is a big challengeable task because a system can deliver good outcomes when it reads faultless or cleaned data. So once the data is collected by the server, the system tries to remove the abnormality of all the collected data.

14.1.3.2 Database versus Big Data

Databases are used to handle a certain volume of data. Traditional databases and RDBMS cannot handle when the data quantity becomes more in terms of terabytes and petabytes. Whereas the Hadoop kind of big data concept works better in handling a of greater volume of data than other traditional database systems. The comparison of traditional databases and big data are given in the following Table 14.2.

TABLE 14.2

Comparison of Traditional Databases and Big Data

S. No	Functional Challenges	Traditional Databases	Big Data
1	Volume	Smaller amount of data quantity (in gigabytes)	Larger amount of data quantity (in terabytes and petabytes)
2	Components	ACID property	Hadoop distributed file system (HDFS), mapreduce, and Hadoop yarn
3	Variety of data	Structured and semi-structured data	Structured, semi-structured, and unstructured data
4	Latency	Very fast accessing of data set retrieval	Low latency
5	Throughput	Low	Very fast
6	Scalability	Vertical scalability or scaling up	Horizontal scalability or scaling out
7	Resources	Ability to add more resources such as hardware, memory, and CPU, etc.	Ability to add more machines to the existing resources
8	Data processing	Online Transaction Processing (OLTP)	Online Analytical Processing (OLAP)
9	License cost	RDBMS is the licensed software and need to pay in order to get access	Hadoop is an open-source software that is available free of cost
10	Example	IBM DB2, MySQL, SAP, Sybase, Teradata, MS-Access, etc.	New York stock exchange (NYSE), social media, jet engine, etc.

14.2 Database Authorization Workflow and Management Structure

In this model authorization verification is performed based on identity validation and the access control rights verification process. The user must have rights to access tables and data sets of any databases. Before getting logged into a security system it has to verify the user's login credentials, and later it should verify access rights of users to continue the admittance of data sets, columns values, and database tables. Figure 14.3 shows authorization and workflow management for the database.

14.2.1 Transaction Failures

There are looms to hack and a system failure in real applications. Every database administrator is looking into some solutions in all the ways towards securing databases from any malfunctions or data loss. Once an administrator gets the call of database failure then they have to stop whatever work they are undergoing and they need to rush to do database recovery. Since all organizations maintaining sensible data, an administrator has more responsibility on failure recovery.

Some Causes for Database Failures

1. System crashes
2. Media failure

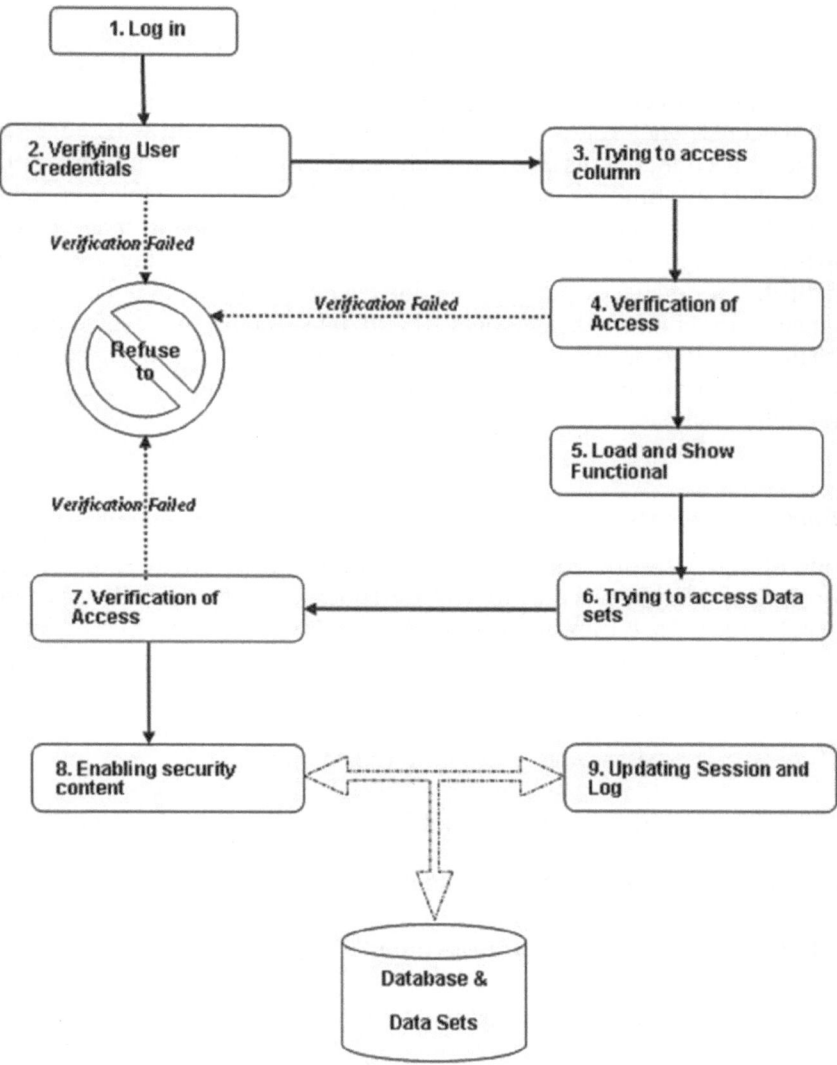

FIGURE 14.3
Authorizations and Workflow Management for Database.

3. Application software errors
4. Security breaches
5. Unexpected conditions
6. Control enforcement
7. Power failure
8. Network link failure
9. Read/write malfunctions
10. Data corruption

14.2.2 Database Control Management and Protocols

Database (DB) transaction prefer atomicity, consistency, isolation, and durability (ACID) properties to ensure perfection. In which atomicity means the completeness of any transaction execution. Consistency refers data maintainability in the right way. For example the consistent state of account details must be maintained properly in both source and destination databases after any funds transaction happens between them. Isolation is the process of serialization in transaction. Updating information on a source and target place must be in serial manner and it should not happen concurrently or parallel with both places. Durability checks for database update perfection even when systems fail during any transactions. It ensures that both databases persist with data and not the reverse process taking place [10]. There are two different modes of functions used to perform errorless transactions such as transaction and concurrency control. For each mode, specific protocols are used to streamline database activities.

14.2.2.1 Transaction Management

A transaction of databases includes complete programs, or an instruction, or a fractional code. Those scripts carry a piece of logical functions on any databases involving more numbers of sub processes. A transaction operation focuses on two main functions, Read(X) and Write(X). Let's go for an example, adding $100 to account 'Y' from account 'X'.

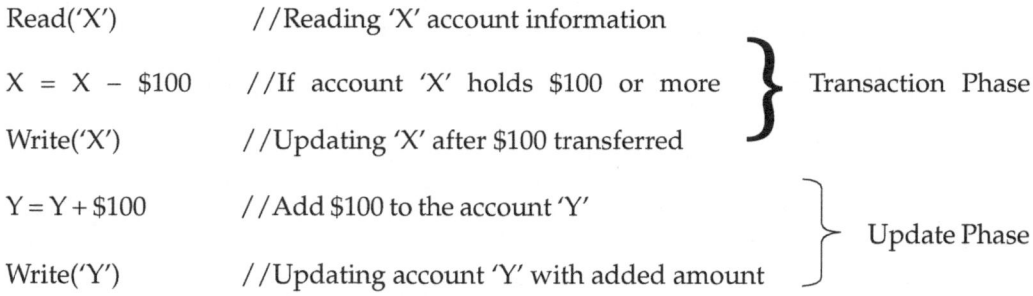

Read('X')	//Reading 'X' account information
X = X − $100	//If account 'X' holds $100 or more
Write('X')	//Updating 'X' after $100 transferred
Y = Y + $100	//Add $100 to the account 'Y'
Write('Y')	//Updating account 'Y' with added amount

Transaction Phase

Update Phase

The transaction is running on five different states based on states of data flow such as active, partially committed, committed, failed, and aborted.

14.2.2.2 Concurrency Protocols

During any transaction, data might get conflicted. Concurrency protocols are taking care of these conflicts on concurrent access. In real-time applications multiuser systems do parallel transactions, meanwhile all users are trying to continue their operations without interfering in others' data and process. Practical inconsistencies are there on accessing concurrent operations over data from sources. This concurrency control is used to avoid conflicts between many processes running simultaneously. Some of the concurrency protocols are used in practice to avoid clashes of any transactions and such protocols are listed next [11].

Two Phase Protocols: This protocol is used to avoid deadlocks from happening on a transaction. It uses shrinking and growing phases to lock and unlock operations. This protocol acquires a new lock after releasing one of its locks on any transactional progress.

Time Stamp Protocols: It ensures the read and write operations of a transaction based on system time or logical count as a time stamp to verify conflicts on every operations. It also guarantees that the serialization of transactions progresses.

Example:

Let takes transactions T1, T2, and T3. T1 enters the process at time 10.00. T2 enters the process at 20.00.

T3 enters the process at 30.00.

Then the priorities for those transactions are in the order of T1, T2, then T3.

Lock Based Protocols: A lock protocol is associated with data items of any process. A transaction takes place when lock is granted by its protocol. Locking control uses a few different modes such as simplistic lock, shared lock, and an exclusive lock based on requirements.

14.3 Database Security System

A traditional database system is compromised and allows any hackers to breach the sensitive or legitimate data of an organization. Internally or externally a database system could be compromised for data breaches. An organization need not worry about external types of attacks rather they have to concentrate on their internal part. Because an employee works inside with legal rights, they can misuse their rights and steal sensitive data for profit from an opposite organization [12].

14.3.1 Internal Attacks

Unbounded Access Rights Beyond Work Nature: An employee having all the database access rights further than their specific task. For example, a staff is in a client contact update progression and has rights to access clients' income sources.

Insufficient Key Management: Storing all sets of keys used in database encryption at one's hard disk is not safe enough and it leads to misuse of an employee's system for exploiting keys.

Database Indiscretion Performance: Database administrators must be aware of attacking methods to give protection. Hackers peruse new thoughts of finding dodge in a database to exploit its sensitive data, so databases must be secure in all ways and administrators have to keep on updating security functions for DB systems from cybercriminals attacks.

Database Backups Pilfer: It does not matter what security tools an organization is using for database protection until 100 percent of its employees are loyal and understand what responsibility they have. Due to betrayal and misuse of the authentic rights of an employee in an organization, all the security protection tools can fail to safeguard legitimate data from a database.

Database Infrastructure: All security software systems cannot perform well in all situations. A hacker uses string manipulations to reach a database backend. This kind of manipulation cannot be predefined or analyzed in software tools. Once the method of attack has been identified then that tool will be updated with a new security function to protect from the same kind of attack since administrators or employees cannot identify the weakness of a database easily when they construct more complex functional databases.

14.3.2 External Attacks

SQL Injection: Hackers using malicious codes and variable to exploit online-facing databases. Applying security functions over web pages and testing them along with firewall protection gives better a solution to preventing online databases from any kind of injection attacks. If a database fails to monitor query-based attacks online, then an organization fails to save its genuine data from SQL injection types of attacks.

Weak Database Structure: Weakness of the database structures allows anyone to intrude in it easily, for example weakness in key management, encryption functions, scrawny firewall setup and more.

Web Security Flaws: Once the system fails to monitor its web access from unauthorized or an illegal user then that system is able to give a chance for hackers to exploit its valuable data. An updated system firewall and network security along with an intrusion detection system (IDS) protect databases from online attacks through malicious websites.

Denial of Service (DoS): DoS is the process of getting continuous requests from a particular system or from multiple sources in order to crash the communication functions over the internet. By the way of passing continuous requests, hackers can do overload any resources on a targeted system to get database access.

14.3.3 Preventive Measures for Database Attacks

Preventive measures for database attacks focus on role-based access control (RBAC). In a layer-based database structural design, data security falls on validation of identification, code verification, access management, Prevention of illegal user access by using oracle parameters and storage techniques [13].

These are some of the preventive methods of a database

1. Validating employee identity
2. Verifying access codes
3. Access control management and interfaces
4. Oracle parameter for database security
5. Data storage process
6. Detecting and avoiding illegal user logging

14.4 Smart City Infrastructure

A city integrated with ubiquitous technologies and services is also referred as a U-City. In which, stability and reliability on centric connected devices plays a vital role. In some significant circumstances any devices on the network flow may be at fault and due to this reason, the entire networks may not pass their valuable instructions or data to other devices. This kind of critical infrastructure must be handled smoothly with some advanced technologies and protective algorithms. All the devices on the smart network must be capable of holding some auto-recover option when the systems fail conditions or malfunction since malfunctions in a real-time application spoil the reliability and stability of the system and make the entire connectivity paralyzed. Overcoming those problems is severely chaotic and hectic to protect sensitive data flows towards centric server. According to four Cs rule connections of smart cities are playing the role along with the help of ICT [14].

The four Cs are listed as follows:

1. Connections between objects and people
2. Collection of data towards context awareness
3. Cloud computations and
4. Communication through wireless links.

Under the aforementioned four Cs, ICT-based smart cities bring the ideas of requirements and challenges, opportunities and benefits, and framework models on the basis of a given application. Those challenges and models are together working on the basic principle of data driven approaches. This model is depicted in the following Figure 14.4.

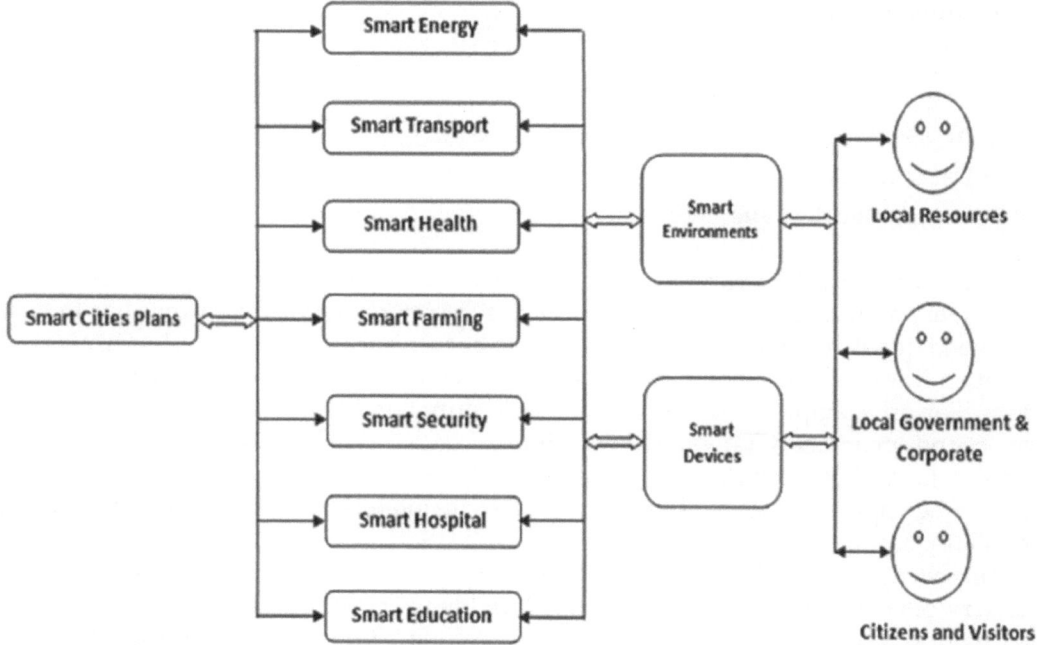

FIGURE 14.4
Sample Data Driven Approaches in Smart City Application..

Since smart cities are functioning and collaborating at all ICT devices for data abstraction towards a specific task, those applications have to report required data from control units. This abstraction would be done through aggregating information from a received request, then it will accumulate the required data from the databases and then fetch the data element that went through edge computing for analysis and confirmation to continue the processes. Once the computing progress is over then the completed data passes through a network connectivity medium to the physical devices, sensors, machines, or to the nodes wherever entailed [15].

14.5 Challenges and Security Issues

Smart cities carry a wide range of an IoT device's communications on sensors networks, mobile ad-hoc networks, ubiquitous computing, and grid computing over wired and wireless network transmissions; there will be some inherent vulnerabilities of security and privacy of sensible data or of the networks. Research is undergoing to identify many security issues in IoT-based activities on networks. Before deploying any IoT devices on these network areas two key security features must be taken into account. Security access control and authentication access control are the two focal areas of any hackers to breach or exploit their target activities [16]. Some of the challenges and security issues of IoT-based smart applications over internet or any network model are identified and listed here:

1. Data excellency management
2. Authentication
3. Integrity
4. Confidentiality
5. Access control
6. Data privacy
7. Geographic data augmentation

Data Excellency Management: In big data transactions from various resources towards corporate and local government bodies, access to better services for citizens has to maintain good quality data. Prerequisite data has to be furnished before delivering any services to urban cities. No data should be missing or be wrongly accessed on any specific applications. For an example collection of vehicle information is done by many manufacturer devices on smart city ICT devices. At the time of accessing, those collected data must be the same data. If those devices produce different types of data that is not genuine, then it directs the application to the wrong function, so maintaining quality in data is mandatory for all transactions.

Authentication: Authentication is the process of identifying a source and destination for services that legitimate users have legal rights to access. Smart cities are scampering IoT devices via multiple network components along with some protocols used in them. All the data are transferred on networks and devices, so there is a possibility that anyone can intrude in between and do any type of attack to the flow. Hence the user identity verification at both ends, source and destination, confers authenticity of any transactions to avoid a meet in the middle (MITM) attack.

Integrity: Since data are collected and maintained by various manufacturers in different structures, integration of data from different sources is uneasy to perform in smart city applications combined with big data. For example collecting vehicle details, driver's license information, accident details, and insurance data all together is a tough process as these data are maintained by different authorities and their own structure format. Collecting particular data for a patient from various resources is important progress that has to be carried very fast from all sources. Integrity does this operation smoothly without affecting loyal information of data from any resources. No modification is possible when integrity works in the progress of data collection.

Confidentiality: It is the process of securing shrewd data and resources from unauthorized views and other access, many cryptographic functions are used to encrypt data on flow but still hacking tools can decrypt those data without modifying its originality. Due to the confidentiality rational data occurs during transactions. Intelligent service security application protocol applies security functions on its header files along with a digital signature through RSA, ECC, and a key management system (KMS) to do this confidentially between data transactions precisely [17].

Access Control: Evolving from communication technologies and ICT devices, the user has to involve themselves for accessing request and responses. Simply all users cannot engross the data on transactions since it violates authentication rights for accessing data and resources in illegitimate ways, so based on functions, each user is assigned roles.

Data Privacy: Privacy is the concern about data from users and resources that are to be used where, passed to whom, stored in which place, or used in the right application. Many applications are running with encryption technologies for protecting our data whereas privacy is applied on network protocols and communication mediums since data transactions happened in the middle of the network paths where security may get breached. People will not go to the new technologies for privacy until they face problems with data protections and security.

Geographic Data Augmentation: IoT-based smart city infrastructure contains a complex design of networks and device connectivity. From many parts of the city data has to flow towards the centric server. A collection of data from various resources has been done all the way through passing the data to many networking modes such as gateways, ad-hoc networks, sensors, and actuators. Proper network design with more security functions needs to be applied for this hierarchical network data flow.

14.6 Security Mechanism and Mitigations

Security of any functions or data through networks has been implemented with the confidentiality, integrity, and authenticity (CIA) triad under satisfaction of data confidentiality, integrity, authenticity, availability, and nonrepudiation. In most significant circumstances cyber-physical surveillance made for monitored and secure data transaction on all over smart infrastructure. Many security function standardizations, digital signatures, and specific protocol constitutions are responsible for the aforementioned CIA triads on access, control, and privacy.

14.6.1 Protocol Overhead Frame Size

In all the areas of a smart city, applications are embedded with small ICT devices that consume less power and little resources have accessing capacity such as CPU, memory, and a processor. It has the ability is to recognize entrenched protocols and cannot offer more storage space to carry much data from transactions. Formally a device's protocols used to hold 127 bytes on a physical layer level. Since each packet contains a frame overhead in a customary size of 25 bytes for some necessary functional requirements, the frame packet size is limited to 102 bytes in general. Then for applying security functions such as an Advanced Encryption Standard (AES) with a CCM (Counter with CBC-MAC) mode for data confidentiality and authentication, it takes 128 bits more for its computations. Since the AES-CCM function is applied on a protocol, still the frame size is reduced 21 bytes from 102, and it works with only 81 bytes. In the case of using an AES-CCM 32 would consume only nine bytes, leaving 93 bytes places to carry data transactions [18].

14.6.2 Authentication Management

In a general way for providing authentication and confidentiality all the users are assigned a username and password for accessing data or resources but this process is not safe while adversaries are also being assigned as legitimate. They can see, delete, update, or modify the original data on its flow. To provide CIA features for data transactions, key management functions are used, and keys are maintained in the public key management infrastructure. Sometimes simply the text authorization like a username and password fail in action. To enhance the security level, shared key management, digital signature, or biometric authentication processes are applied. A digital signature contains three main components including the global parameter (GK) for the private (PRK) and public key (PUK) of any length, signing parameter (SK), and verification parameter (VK). A signature is verified and accepted when SK is equal to the parameter VK which are derived from the global parameter GK. If SK and VK are not equal, then the signature is not valid and message is rejected [19].

14.6.3 Access Control Management

Nowadays, all smart devices have an advanced security mechanism to authenticate functions. Even human identification and recognition using sclera also was developed for authenticating the legal user [20]. Once authentication is verified and validated at the edge network's devices such as gateways, sensor networks, or at the local access control network the user is verified with an assigned role for accessing any functions. In a ubiquitous networks structure, every function is consigned with roles. To access those specific applications, each user is verified with role-based access control. Based on the activity, access control is split into three broad areas as listed next [21].

1. Discretionary access control (DAC): In which roles are assigned by an administrator to users in order to access resources.
2. Role-based access control (RBAC): In which roles are assigned to perform the specific task-based role activities.
3. Attribute-based access control (ABAC): In which rights are sanctioned to evaluate user rights attributes, handle resources requested, and identify that from where the request is made.

14.6.4 CoRE Network Digital Signature Management

A constrained RESTful environment (CoRE) is the network model for smart devices and IoT devices. Each device is activated as a node in the network by constrained application protocol (CoAP) functions. Since CoAP uses low-power consumption and low-memory consumption it is suited for CoRE devices networking models. IPV6 low power personal area networks (6LoPAN) is another constraint in IoT networks that ensemble smart devices such as energy management, fire sensors, home building automation and more. The CoAP protocol model is used in connection with smart devices on the same constrained networks or different constrained networks joined by the internet and it works on traditional IP network models [22]. CoAP and client to authentication protocol (CTAP) works along with domain transport layer security (DTLS) and TLS for providing security functions like authentication and confidentiality for its communications. The CTAP security function captures the constrained object signing and encryption (COSE) and JSON object signing and encryption (JOSE) algorithms for encrypting web files and JSON file information over network transactions. The internet assigned number authority (IANA) takes a part in CTAP to bind them into protective algorithms such as RSA, ECC, or the JSON web key elliptic curve for secured authentication operations.

COSE and JOSE employed a set of notations and conventions to interpret the security algorithm functions. The set of functional keywords appear in capital letters to specify its value and how much priority it needs to give on security algorithms. 'MUST', 'MUST NOT', 'SHOULD', 'SHOULD NOT', 'RECOMMENDED', 'NOT RECOMMMENDED', and 'OPTIONAL' are some of the example keywords pertained in JSON encryption techniques [23]. The RSASSA-PKCS1-V1-5 signature algorithm uses 2048-bit size keys for signing algorithms. A combination of RSA and secure hash algorithm (SHA) checks for the key format at both ends to ensure RSA key has been followed in creating a signature and verification [24]. The SecP256K1 algorithm enables standards for the efficient cryptography group (SECG), and the elliptic curve algorithm for web-oriented JSON data signature and verification. ECDSA signature algorithm along with the key size of 256 bits are used by JSON web key (JWK) and has implemented the elliptic curve cryptography (ECC) to confer more security on fewer numbers of key size than other security algorithms. This ensures the best performance of communication protocols on transportation over optimized data frame limits.

14.6.5 DoS/DDoS Resistance

Denial of Services (DoS) is one of the attacking methods in which the hacker used to send the continuous requests to the server and made the server inactive by means of acquiring its entire processing and power resources. Due to this unconditional flow of an infinite number of requests the targeted server won't be able to respond properly. Passing continuous request from various places to the targeted systems in order to make it busy or not to respond for a while is known as Distributed Denial of Service (DDoS). In a low-power communication networks, devices are in failure mode when the power goes off. In that particular moment, devices are unable to receive or respond to the requests and this circumstance leads to DoS so that target operations will not be available at the requested time. When requests are in the waiting queue, hackers can easily access information in this time duration. If devices are unable to receive packets, rerouting ispossible by the cybercriminal when packets are in queue with meaningful information. Host identity protocol (HIP) is used to avoid DoS or DDoS on smart device networks. The basis of IP addresses and DNS, HIP separates the end-to-end identifier and locator roles of a given IP address. It introduces the host identity (HI) name space in the public key management system to maintain authenticity of source and destination places.

14.6.6 Security in Database Authentication

The data stream management system (DSMS) and continuous authentication on data streams (CADS) mechanism is applied in security architecture for providing an authentication system at continuous data flow on databases or big data's table. In this process, a service provider takes continuous data from one or more owners of the transactions whilst it checks for authentication of all the transferred data from the entire user. DSMS system architecture cares about data streams and the confidentiality of concerned data based on a naming system, addressing methods and the specific structure in its architecture. Figure 14.5 shows the database outsourcing framework model.

Database outsourcing ensembles three essential key components such as service providers (SP), data owners (DO), and data clients (DC). Outsourcing has the benefits to assign roles and responsibilities for all the aforementioned three constraints. Outsourcing of the database framework model is shown in the Figure 14.5. A data owner doesn't want to request or contribute for databases to the clients or to the service providers whilst service providers are taking care of flexible and scalable database apportion to clients and owners and also assigning multiple owners to provide the best performance. These three constraints mainly serve for protecting data from a man-in-the-middle attack (MTMA). An adaptive version of CADS (A-CADS) updates the partition schemes to follow distributed database access [24]. A-CADS get the individual reports to process a query for avoiding the missing values and verifying the completeness of data processing on continuous data transactions to avoid the incurring of false transmission. This minimizes the processing and transmission overhead by elaborating indexing and virtual caching mechanisms.

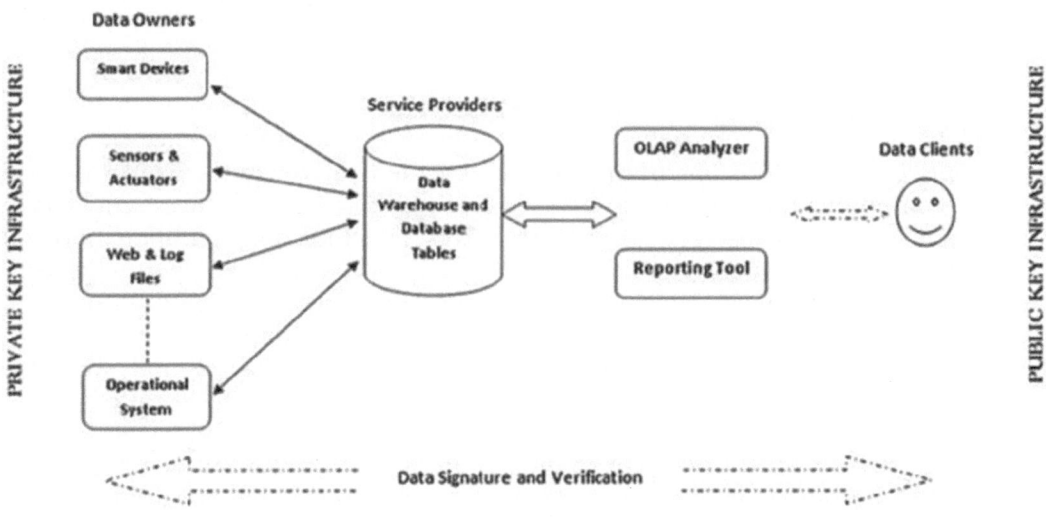

FIGURE 14.5
Database Outsourcing Framework Model..

14.6.7 Dynamic Distributed Key Management System

The entire network model in the smart city plan is distributed all over city places and those models enclosing many smart devices for regular and uninterruptible communications. Every network has to send and receive their sensible data to the centric server through a wired or wireless network medium based on the source and destination IP addresses. Newly established network and IP information needs to be update with the existing system information in the routing database table. To give authenticity for all the IP addresses of a routing table, public key infrastructure (PKI) is maintained and updated in a periodic manner [25]. Sometimes PKI goes wrong when the user is a criminal or a hacker's activities are found in between source and destination transactions. Asymmetric key perception is followed in PKI environment so that every user of a data has to possess two different kinds of keys called the private key and public key. In general, the public key is used for encryption, and the private key is used for decryption and sign verification progress. PKI functions and certifies distributions among a network's frame model as depicted in Figure 14.6.

The functional processes of different stages in Figure 14.6 are as follows:

Stage 1: Requisition for certificates

- Initialization for communication
- Verification of initiator's details
- Creation of certificates along with session ID
- Send the certificate to the initiator along with a request for acknowledgement

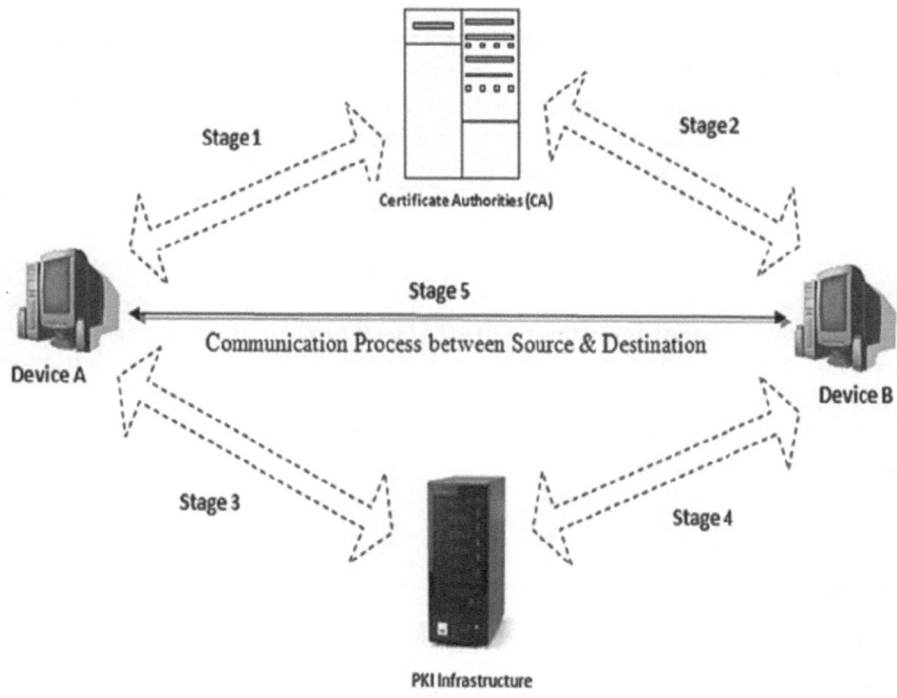

FIGURE 14.6
CA and PKI Functional Activities in Network Communication.

Stage 2: Passing acknowledgement

- Sending acknowledgement for received request
- Creation of a certificate and session ID for server-side verification
- Setting up the TTL and Port ID for communication devices and durability
- Sending authorized certificates and a copy to the server machine

Stage 3: Key generation and distribution

- Requesting for keys from KDA
- Verification of certificates and session IDs at KDA
- Generating keys for encryption and decryption
- Issuing the public key from PKI database for encryption and digital signature
- Issuing private key for decryption at both sides and signature validation

Stage 4: Signature verification and replying

- Extracting the digital signature from received messages
- Verification of source IDs and signature
- Verification of certificates and session IDs
- Acknowledging the received messages and replying

The key distributing authority (KDA) centers maintain and manage the two sets of keys in a PKI infrastructure for authentication purposes. Before commencing any of transaction each user has to register with certification authorities (CA) to get certificates for communication initialization and to access KDA to avail their private and public keys. Sooner than issuing certificates, CA checks for legal terms, originality, and trustworthiness of a user. Once the user identities are verified and validated then the user will be issued with their set of keys to accomplish encryption and decryption. Certification authorities verify user's details beside they will generate session logs and request a type then session ID for end communication devices. The initiator of a transaction should carry the payloads information on its IP header for opposite parties to verify the initiator's legacy information, time to live (TTL), validity, and security algorithms for signature and verification along with its private and public keys. As mentioned in the stages of Figure 14.6, both devices have to proceed their transactions based on their limitations and functionality requirements mentioned with their certificates and norms of the protocols to be used.

14.7 Network Layer-Based Security Management System

Since ICT devices used in smart city applications communicate through data exchange by connecting mediums such as wired or wireless connections, they are highly vulnerable to cyberattacks. The ISO/OSI model standardizes every protocol function by splitting

TABLE 14.3

Layer Model, Applications, Possible Attacks and Preventive Measures

S. No	ISO/OSI Layers	Features	Vulnerabilities	Security Mechanisms
1	Physical layer	Used to connect physical and electrical devices for data transfer. It acts like a communication medium between end devices	Intentional violence: Unplug the wire, cut off the connections, power off, natural disaster, and DoS or DDoS attacks	Biometric authentication mechanism, advanced locking systems, electro shield with magnetic effects, etc.
2	Data link layer	Making a data frame passed by physical layer to the upper layer	Spoofing, MAC identification, exploiting VLAN, and MAC addresses	MAC filtering, authentication verification, and firewall setup for Encryption
3	Network layer	Maintain and manage traffic flow of a data, routing, controlling, and addressing the packets	IP address spoofing, packet spoofing, and malicious attacks	Well-configured firewall settings, anti-spoofing mechanisms, and authentication verifications
4	Transport layer	Smoothing the data flow, error detection and corrections, fragmentation and defragmentation	Accessing of transmission protocols	Highly configured firewall set up for security mechanism
5	Session layer	Local and remote area application's interaction	Brute force attacks	Password encryption Process
6	Presentation layer	Conversion of standardize data formats to the different other formats in the readable mode	Injection attacks, malicious inputs on coding part	Cleaning before passing on to the networks
7	Application layer	End user data	Stealing of information, DoS, intrusions	Detection of malicious activities, intrusion detection, and preventive mechanism, malicious software detection

operations on seven different layers called the physical layer, data link layer, network layer, transport layer, session layer, presentation layer, and application layer. Table 14.3 shows the ISO/OSI layer list and the possibility of cyberattacks along with security mechanisms implied.

14.7.1 Protocol Analysis

In the ISO/OSI model seven layers are configuring and controlling entire networks and data flow functions through a set of ruling decrees called protocols. These protocols are the specific functions with source and destination identifications, encryption and decryption algorithms, routing information, and payload overhead based on the layer level. Some of protocols and functional notations are described in Table 14.4.

TABLE 14.4

ISO/OSI Layer's Protocol Suits

S. No	Category	Network Layers	Protocols	Protocol Components	Applications
1	Application sets and function	Physical layer	CAN, ISDN, PON, OTN, Digital Subscriber Line (DSL), IEEE802.03, IEEE802.11, Bluetooth, etc.	1) Version 2) Protocol 3) Router ID 4) Source and destination IP address 5) Session ID	Activate, deactivate and maintaining the physical connections, data encoding and data conversion to voltages and pulses, etc.
2		Datalink layer	CSLIP, PLIP, SDLS, PPP, X-25, ARP, HDLC, etc.	6) Source and destination port ID	Frame synchronization, LLC, MAC functions.
3		Network layer	TCP/IPV4 and V6, ICMP, IGMP, IPX, EGP, EIGRP, NAT, IPSEC, HSRP, VRRP, etc.	7) Data length 8) Data type 9) Checksum value 10) ACK and NACK	Data transfer form node to node.
4	Transport sets and function	Transport layer	DCCP, NBF, SCTP, SPX, NBP, TCP, UDP, TUP, NetBIOS, etc.	11) Priority ID 12) Flags 13) Time to live 14) Time wait 15) Closed 16) Finished	Identifying the reliable communication medium between end devices.
5		Session layer	NCP, PAP, RPC, RTCP, SDP, SMB, SMPP, ZIP, SOCKS, iSNS, NetBEUI, etc.		Session management, critical operation management, dialog control.
6		Presentation layer	FTP, SSH, Telnet, TLS, IMAP, etc.		Translation of data formats, encryption, and decryption.
7		Application layer	HTTP, HTTPS, POP3, PGP, NFS, FTP, IRC, SSH, AMQP, SNMP, SMTP, DHCP, SOAP, etc.		End user applications such as browsing web pages and e-mail transactions, etc.

14.7.2 Protocol Security Mechanism

Protocols are the main tools for making perfect communication. These are working like a medium to make a data transaction. Through this connecting medium any unwanted interruption from illegitimate users will disturb communication or eavesdrop on sensitive data. To avoid this kind of interruption network devices have to be implementing with an intrusion detection system (IDS) mechanism [26]. Internet protocols (IPV4/IPV6) offer security by authentication header (AH) and encapsulation security payload (ESP). Applying these security parameters is different in transport and tunnel mode. In transport mode, AH or ESP is inserted in between the IP header and data packet payload, whereas in tunnel mode the entire IP packet is encapsulated with the AH or ESP. On the other hand, an IP packet is covered by another IP packet. Identical fields for IP packet are as follows [27]:

Differentiated Service Code Point (DSCP): This field is about intermediary router characteristics. It controls and manages simplified or scalable network groups information. It provides quality of service at the layer3 IP networks in modern networks by its precedence values.

Explicit Congestion Notification (ECN): ECN maintains the lossless packet transmission between end-to-end communications. Due to congestion raises on the network path, sometimes TCP/IP drops their packets before delivering it to the targeted system. To avoid this kind of packet loss, ECN checks for clogging of end devices during transmission time.

Data Fragment (DF): Data fragmentation is available when a larger data transaction takes place. DF is applicable only on tunnel mode, whereas in transport mode the IP header encapsulates only the IP header not the entire data packet. This bit field is set for making data fragmentation.

Fragment Offset (FO): Wherever data fragment is happening, at those places along with header information fragmentation offset value also set to identify fragmented data for data integrity. Otherwise simply the offset value 'ZERO' is set for non-fragmented data packet notifications.

Time to Live (TTL): In the smart city application or in any other network model, a data packet has to cross many intermediary routers for better and faster communication. In such situation, when the data crosses any intermediary, routers then by default decrement the value of one from its count.

Checksum: Checksum is the value used for recalculation of received information and integrity property. If any values got changed or deleted, then the outcome of this checksum mismatches the original checksum values.

14.7.3 Lightweight Cryptographic Mechanisms

A smart city network holding higher numbers of IoT devices together form a network, and those devices contains less resource powers such as RAM, power management, computation processes, and hardware resources due to less resource management administrators not using highly computational algorithms to provide security functions. The lesser power processors will not be able to tackle many complex calculations. Instead of boosting up the resources, it is possible to implement lightweight computational security functions into the network devices. For example, implementing elliptic curve cryptographic (ECC) security functions are not feasible due to its multifaceted calculations. As a replacement for ECC, devices can be embedded with AES and DES security algorithms in order to reduce calculative burdens [28]. The tiny encryption algorithm (TEA) uses the international data encryption algorithm (IDEA) principles for applying cryptographic functions to provide security for the data. Since the IDEA algorithm uses addition, XOR addition, and multiplication along with shift operation, it is very easy to implement in mini-processor devices. Many of the smart device networks run this TEA algorithm in it for data security. TEA exercises on 8-bit values in CPU and this could be easily processed on any smart device's processor. A scalable parameter needs to be used when devices want to perform slightly more complex computations. The scalable encryption algorithm (SEA) is parameterized in order to allocate processors based on the sizes of plain text and keys.

Microcontrollers such as AVR and 8051 require 0.81 seconds to perform a point multiplication operation on a ECC single curve function GF(p) and 11 seconds are needed to do the same security function implementations with the size of 1024 bits using RSA algorithm.

TABLE 14.5

Area Optimization and Lightweight Cipher Implementation Comparison

Module	Speed Optimized	Area Optimized	Algorithm	Block Size	Key Size	Area #GE	Speed kbps@ 100KHz	Logic Process
64-bit data register	384	192	XTEA	64	128	3490	57.1	$0.13\,\mu m$
Key addition	87	87	HIGHT	64	128	3048	188.2	$0.25\,\mu m$
S-box layer	174.8	174.8	mCrypton	64	128	2500	492.3	$0.13\,\mu m$
P layer	0	0	DES	64	56	2300	44.4	$0.18\,\mu m$
32-bit XOR	87	87	DESXL	64	184	2168	44.4	$0.18\,\mu m$
80-bit key register	480	212	KATAN	64	80	1054	25.1	$0.13\,\mu m$
S-boxes (key schedule)	43.7	30	TANTAN	64	80	688	25.1	$0.13\,\mu m$
5-bit constant XOR	13.5	13.5	PRESENT	64	80	1570	200	$0.18\,\mu m$
Control logic	50	70	LBlock	64	80	1320	200	$0.18\,\mu m$
Sum	1320 GE	866.3 GE (with RAM)	–					

Wenling Wu and colleagues have done research on reducing the complexity of calculations on block cipher over smart devices [29]. They presented their thoughts and implementation on lightweight block cipher at ACNS. The result after their research implementation is represented in Table 14.5.

Lightweight block cipher and stream cipher implementations on smart devices are helping us to reduce the power resources, RAM, CPU, processor utilization for making simple and complex computations as well. Each and every operation for computations takes a single cycle in the processor associated with clock cycles. These mechanisms are all simply to perform encryption and decryption operations based on symmetric (single for both encryption and decryption techniques) and asymmetric key (more than one key is used for encryption and decryption) methods.

14.7.4 Penetration Testing Analysis

These are all few mechanisms for securing data during transfer between two end devices. Apart from these few manual testings also to be carried out on the network to find vulnerabilities in it. Network vulnerabilities are easily leading hackers inside the system to take control over devices, machines, or networks. Penetration testing has to be done manually by the ethical hacker to find the loophole of any vulnerability. Once the susceptibility of the systems is found, then the required action will be taken on the system's places. The following areas are the feeble places of vulnerability:

- Devices on the networks
- Connecting gateways
- Cloud data centers

Reconnaissance Areas of Penetration Analysis

- Hardware level
- Network level
- Firmware level
- Web application level
- Cloud storage level

Pentesters carry out their testing in five different levels of places to analyze the weakness of a networks and devices. Reconnaissance consigns of pentesters are hardware, network, firmware tools, web applications, and cloud data access. Once the penetration testing is over, then pentesters attack the systems and devices on reckoned places with the help of hacking tools to give remedial actions against the attacks. Most hackers try man-in-the-middle attacks and brute force attacks on networks to exploit resources or data. Most of the pentesters with knowledge of Windows and Linux operating systems based on the protocols TCP, UDP, and FTP are doing common types of analysis whereas a smart application belonging to the MIPS architecture runs on the protocols of Bluetooth, Zigbee, and NFC. A drawback of most of the pentesters is not having knowledge on smart device-based architectures and protocols.

14.8 Future Direction

As discussed in the previous sections, many security issues have been found in all of the internet communication process. AI is an important concept. Systems learn about attacking methods from various sources and protecting users and systems, as well from those assaults crafted by hackers. ML is a branch of AI and it has two approaches to make a system learn the real-time occurrences such as supervised and unsupervised learning. On the basis of analyzing the factors of all attacking methods, those factors are feeding to the machine as an input to take necessary action against attacking methods found on the networking path. Intrusion detection and prevention (IDS and IPS) tools are working on principles of analyzing the attacking factors and prediction of attacking methods. Blockchain maintains the digital ledger of all the user transactions with a secured manner. A hash function is applied in the security function and once the hash code is generated for any user then automatically all the connected users' hashing values are updated through the blockchain concept. Since blockchain uses hash function, changing a single bit value of any data in the blockchain yields changes in multiple bits of an output. By means of an AI method, blockchain uses the security function over network. In today's digital era IoT, AI, and blockchain technologies are functioning through networks towards responding to requests, so keeping them more secured is mandatory for protecting data and system resources with the help of all updated security functions and software patches.

14.9 Conclusion

Apart from many security issues a smart network must be active on all critical situations like system failure, power failure, and natural disaster; even a minor significant fault in the devices or a system could make a communication failure over a network or lead to a perform malfunction. Smart application frameworks and infrastructures must be able to recover a communication link, and system failure in some kind of critical circumstances. Managing sensors and actuators in case of a failing situation is mandatory to maintain data transmission. Data security and cryptographic functions on the smart networks are for providing the CIA triads with optimized progress improving the ability and performance of the better transaction. In the future smart city application networks and their infrastructure are to be configured and controlled by the automation process by ML for service continuity and uninterrupted services.

References

1 https://www.indiatoday.in/india-today-insight/story/why-the-smart-cities-mission-will-miss-its-deadline-1574728-2019-07-29

2 http://smartcities.gov.in/upload/uploadfiles/files/Appbased%20Projects_60%20cities.pdf

3 http://smartcities.gov.in/upload/smart_solution/58df96e1ac038Pan_Solutions_Components_1 3Fasttrackcities.pdf

4 https://www.softwaretestinghelp.com/iot-devices/

5 Fatmasari Rahmana, Leila, et al. 2018. "Understanding IoT Systems, Procedia Computer Science." *Procedia Computer Science* 130: 1057–1062. doi: 10.1016/j.procs.2018.04.148.

6 https://www.leanix.net/en/blog/iot-devices-sensors-and-actuators-explained

7 Strohbach, Martin, Holger Ziekow, Vangelis Gazis, and Navot Akiva. 2014. *Towards a Big Data Analytics Framework for IoT and Smart City Applications*, edited by Fatos Xhafa, Leonard Barolli, Admir Barolli, and Petraq Papajorgji. 2196–7326. Springer.

8 Stonebraker, M. 2012. "What Does 'Big Data' Mean? (Part 3)." *BLOG@ACM.* http://cacm.acm.org/blogs/blog-cacm/157589-what-does-big-data-mean-part-3/fulltext

9 Manjunatha, and B. Annappa. 2018. *"Real Time Big Data Analytics in Smart City Applications." International Conference on Communication, Computing and Internet of Things (IC3IoT).* doi: 10.1109/ic3iot.2018.8668106.

10 Waqas, Ahmad, Abdul Waheed Mahessar, and Nadeem Mahmood. 2015. "Transaction Management Techniques and Practices in Current Cloud Computing Environments: A Survey." *IJDMS* 7 (1): 41–59.

11 https://www.guru99.com/dbms-concurrency-control.html

12 Fataniya, B. 2017. "A Survey of Database Security Challenges, Issues and Solution." *IJARIIE-ISSN(O)* 3 (5): 2395–4396.

13 Zou, G., J. Wang, D. Huang, and L. Jiang. 2010. "Model Design of Role-Based Access Control and Methods of Data Security." *International Conference on Web Information Systems and Mining.* doi: 10.1109/wism.2010.100.

14 Lima, Chiehyeon, Kwang-Jae Kimb, and Paul P. Maglioc. 2018. "Smart Cities with Big Data: Reference Models, Challenges, and Considerations." *Cities* 82: 86–99.

15 Atlam, Hany F., et al. 2018. "Developing an Adaptive Risk-Based Access Control Model for the Internet of Things." *IEEE Xplore.* doi: 10.1109/iThings-GreenCom-CPSCom-SmartData.2017.103.

16 Liu, Jing, Yang Xiao, and C. L. Philip Chen. 2012. "Authentication and Access Control in the Internet of Things." *IEEE Xplore.* doi: 10.1109/ICDCSW.2012.23.

17 Abdukhalilov, S. G. 2017. *"Problems of Security Networks Internet Things." 2017 International Conference on Information Science and Communications Technologies (ICISCT).* doi: 10.1109/icisct.2017.8188588.

18 Maple, Carsten. 2017. "Security and Privacy in the Internet of Things." *Journal of Cyber Policy* (2) 2: *The Internet of Things.* 155–184.

19 Kittur, A. S., A. Jain, and A. R. Pais. 2017. "Fast Verification of Digital Signatures in IoT." *Security in Computing and Communications*. 16–27. doi: 10.1007/978-981-10-6898-0_2.

20 Vijayalakshmi, S., and V. Gokul Rajan. 2018. "A Novel Approach for Human Identification Using Sclera Recognition." *International Journal of Computer Sciences and Engineering* 6 (4): 228–235.

21 Heer, T., O. Garcia-Morchon, R. Hummen, S. L. Keoh, S. S. Kumar, and K. Wehrle. 2011. "Security Challenges in the IP-Based Internet of Things." *Wireless Personal Communications* 61 (3): 527–542. doi: 10.1007/s11277-011-0385-5.

22 Jones, Michael B. 2019. http://self-issued.info/docs/draft-jones-cose-additional-algorithms-00.html

23 Jones, M. 2017. "Using RSA Algorithms With CBOR Object Signing and Encryption (COSE) Messages." *RFC 8230*. doi: 10.17487/RFC8230.

24 Papadopoulos, Stavros, Yin Yang, and Dimitris Papadias. 2010. "Continuous Authentication on Relational Streams." *The VLDB Journal* 19: 161–180. doi: 10.1007/s00778-009-0145-2.

25 Laurence Boren, Stephen, et al. 2015. "Dynamic Distributed Key System and Method for Identity Management, Authentication Servers, Data Security and Preventing Man-in-the-Middle-Attacks." Patent No.: US 9,166,782 B2.

26 Rashmi, R., et al. 2018. "*IDS Based Network Security Architecture with TCP/IP Parameters Using Machine Learning.*" In *2018 International Conference on Computing, Power and Communication Technologies (GUCON 2018)*. IEEE.

27 Perez, André. 2017. *Implementing IP and Ethernet on the 4G Mobile Network*. Imprint ISTE Press. Elsevier.

28 Eisenbarth, T., S. Kumar, C. Paar, A. Poschmann, and L. Uhsadel. 2007. "A Survey of Lightweight-Cryptography Implementations." *IEEE Design & Test of Computers* 24 (6): 522–533. doi: 10.1109/mdt.2007.178.

29 Wu, Wenling, and Lei Zhang. 2011. "LBlock: A Lightweight Block Cipher." *ACNS LNCS* 6715: 327–344.

15

Security of IoT in Healthcare

Supriya Khaitan
Galgotias University, Greater Noida, Delhi-NCR, India

Rashi Agarwal
Galgotias College of Engineering & Technology, Greater Noida, Delhi-NCR, India

T. Poongodi, R. Indrakumari, and A. Ilavendhan
Galgotias University, Greater Noida, Delhi-NCR, India

CONTENTS

15.1 Introduction

The Internet of Things (IoT) is a connection of interwoven devices that are used to send information without the interaction of a human or a computer. IoT devices may or may not be wearable. IoT is a booming field that many people are moving towards. With increases in internet usage, there is a huge demand for fitness bands and online health applications to keep track of activity and health. IoT in healthcare unleashes the quick and immediate flow of information that helps in dealing with chronic illness well in time even from a remote location.

For elder people and patients living alone these devices enable a constant capturing of data. In case of an emergency, it may also contact the family and paramedics of patients.

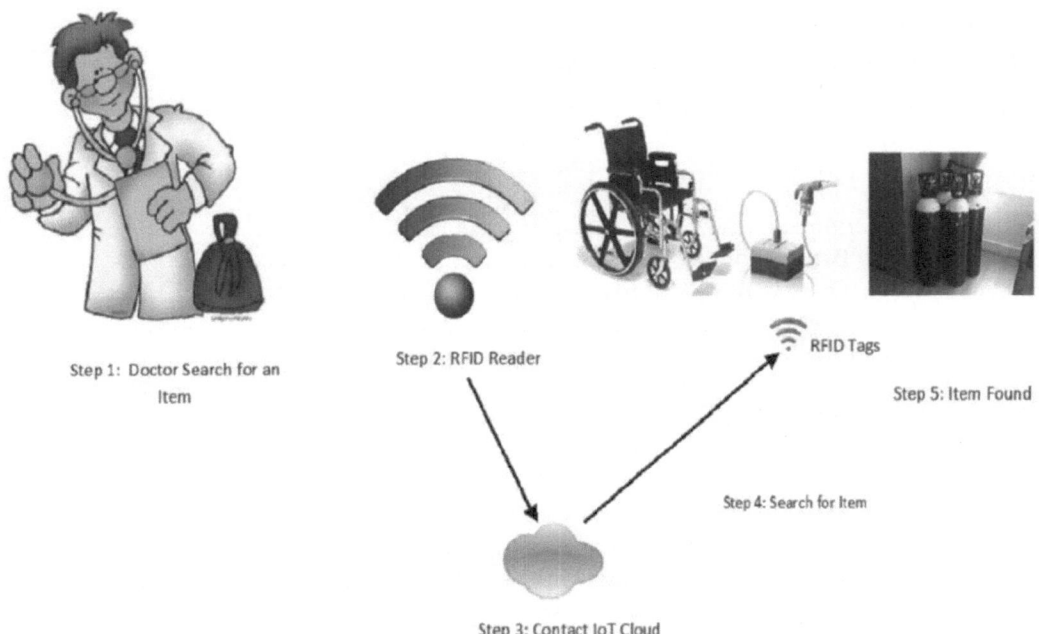

FIGURE 15.1
IoT for Hospitals.

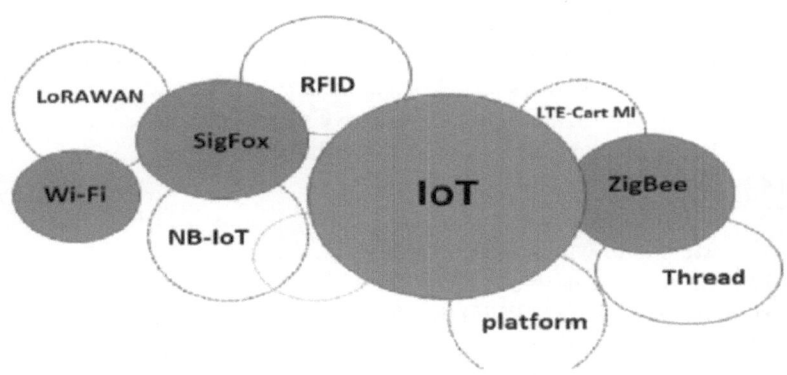

FIGURE 15.2
Technologies Used in IoT.

For doctors these IoT devices help them for better understanding of the health of patients in a timely manner. This leads to a proactive approach in case of any medical emergency [1]. For hospitals sensors in medical equipment like wheelchairs, oxygen cylinders, and nebulizers can be used for scouting them in better manner. IoT devices may also help in deployment of medical staff in an efficient and effective manner. Figure 15.1 shows the use of IoT for hospitals. Figure 15.2 gives the different technologies used in Internet of Things.

Hygiene maintenance IoT devices are also becoming popular to help in prevention of spreading infection. Insurance companies can benefit from the data captured by the health monitoring system. It is easier to detect fraud and apply claims as all data is in real time. The real-time transfer of data for millions of patients takes place and all the data gets stored on the cloud. With this data comes the need of securing this data. Also, it is important to provide integrity, accuracy, confidentiality, availability, and authorization of data [2]. Classical methods of security cannot be implemented in the aforementioned scenario. The protocols stack used by IoT is different than these traditional methods. The Gartner report forecasts the IoT market will grow to 5.8 billion endpoints in a year out of 4.8 billion will be in use. In 2020 that is a 21% increase from the year 2019.

15.1.1 Benefits of IoT

- Real-time and simultaneous checking on patient health data can save lives.
- End-to-end association and availability of data in an effective manner with the help of the next generation of technologies.
- IoT-based healthcare services enable a better and easier human services investigation and an information-driven experiences to expedite basic leadership and are less inclined to make mistakes.
- IoT gadgets accumulate on time-critical information and transfer it to specialists for constant follow-up through versatile applications and other connected gadgets.
- In crisis and emergency situations, patients can contact a specialist who is far away with savvy versatile applications.

15.1.2 Disadvantages of IoT

Massive implementation of IoT in the healthcare industry also comes with some downsides like the following:

- *Privacy can be potentially undermined:* All the devices used in IoT-enabled healthcare systems can communicate to each other through internet only. As per the security issue is concerned, devices can be hacked. So a lot of security measures and attention is required in the usages of such systems.
- *Unauthorized access to centralization:* There can be a problem within the system that some dishonest interlopers access data with some cruel intentions and harm humanity.
- *Global healthcare regulations:* Some guidelines that are already issued by the international health administration regarding the working principle of IoT integrated devices. These restrictions may put some restrictions on the capabilities of new technologies to some extent.

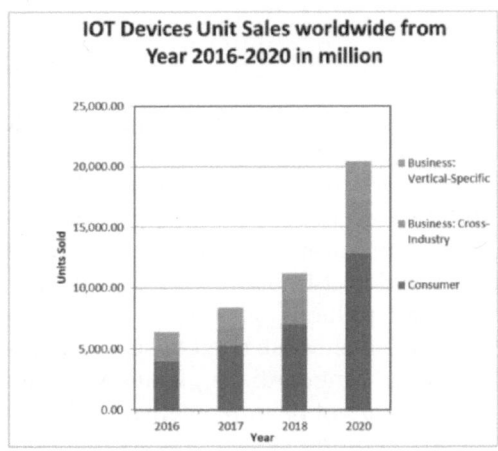

FIGURE 15.3
IoT Device Unit Sales in World.

15.2 Key Qualities of IoT

Users can achieve deeper automation analysis in IoT-embedded technology systems. A modern attitude towards technology, prices fall on hardware devices, and advances in software brought major changes in the services, development, and delivery of products. Figure 15.3 gives the sales of IoT devices from year 2016 to year 2020. The major contribution of technology in IoT-embedded systems is towards AI, connectivity, and it enhances every aspect of life [3] to be 'smart' (such as if refrigerators could know about the quantity of milk in its sensors) with active engagement and use of smart devices.

- AI: IoT or if your cabinets can give userts information about the quantity of cereal and can place an order for users.
- Connectivity: Connectivity not only refers to creating a connection within a major network but also creates a small and cheaper networks in between the systems.
- Sensors: Sensors are the devices that are able to convert passive objects in an IoT-integrated system into active objects to achieve real-world integration within the system.
- Active Engagement: An introduction of IoT in technology enables users to interact with active objects like content, product, and service engagement.
- Small Devices: The size of the devices becomes small over time. Also, cheaper devices are specially designed for the specific purpose to deliver services to the user.

15.3 Applications of IoT

IoT is a network of technology-embedded objects that makes them communicate and interact with each other and with the outside environment. This makes all objects intelligent to make the life comfortable. It uses live and rising expertise and equipment for converting the objects into more intelligent ones so that they can behave as robots. Because of this reason IoT has many applications in healthcare.

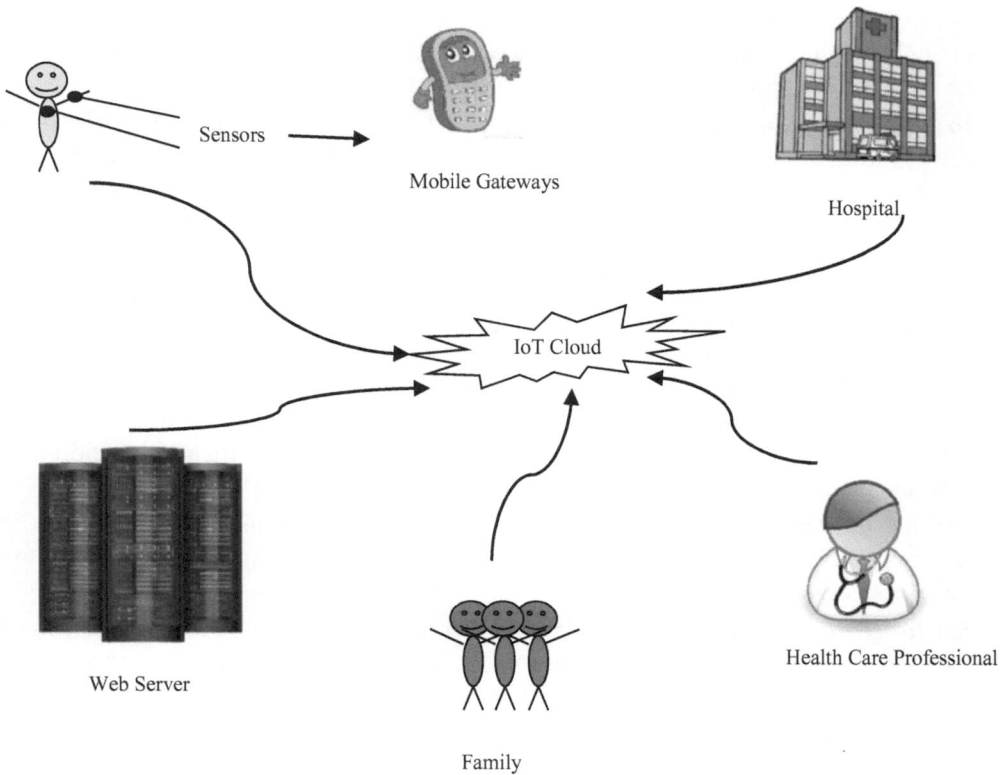

FIGURE 15.4
Applications of IoT in health Care.

Some of the applications are shown in Figure 15.4. With a huge range of applications comes a lot of challenges and security issues in an IoT-based healthcare system. IoT benefits include the following:

- A better understanding of patient's condition
- An advanced healthcare system based on modern healthcare delivery design
- A decision support system of clinical information
- A program designed by a team of experts to cater patients' needs

Information captured through the IoT-enabled devices can be useful to a range of processes applicable to different stakeholders like hospitals, insurance companies, doctors, patients, and relatives.

- *IoT for Patients:* The invention of IoT in medical equipment has changed the thoughts and lifestyle of all people. People are becoming more sensitive to know about their health and fitness. They are keen to do many activities that can lead their lives to better health. Constant tracking of blood sugar, heart rate, and glucose levels encourages patients to motivate themselves to do more energetic and regular exercises. Devices like fitness bands and wireless devices like a glucose meter and a blood pressure measuring device provide patients with personal attention. They do not want any appointment for such regular observations. Also calorie counters and exerciser checkers help them to make the right choices with their food and motivate them to change food habits if required.

- *IoT for Physicians:* Nowadays patients and families are becoming knowledgeable after getting enough online information. They worry about the health of the patients. Seeing the severity of critical problems, healthcare professionals need to be more connected to patients online. Through IoT-enabled devices, physicians can connect to the patient and can get an expected outcome. In case of an emergency, doctors can effectively change their treatment and provide immediate attention.

- *IoT for Hospitals:* The invention of IoT in the health industry also changes our hospitals. Now hospitals are using equipment like wheelchairs, monitoring devices, and nebulizers that are tagged with sensors to track their location. Medical staff can also be tracked in the hospitals. Devices used in pharmacy inventory control are used to efficiently manage information about drugs. Temperature control and humidity control sensors are used well in hospitals.

- *IoT for Health Insurance Companies:* Health insurance companies are using IoT devices to detect fraud claims; this brings transparency among insurers and customers in risk assessment processes.

- *IoT for Healthcare Insurers:* Keeping track of routine activities, precautionary health measures, and treatment plans of people, insurers can give some offers or incentives using IoT data. Also insurance companies can validate claims using data.

15.4 Types of IoT-Enabled Devices

IoT-enabled healthcare devices can be categorized into three different types:

- *Life-critical devices:* The role of such devices is very critical as they are used to monitor patients and to transmit data required. They are essential in life or death care. If functioning of such devices fails to fulfil the requirement, then the patient's life will be at risk (e.g., pacemakers and ventilators devices).

- *Noncritical monitoring devices:* The role of these devices is to record and transmit data only if required.

- *IoT health and wellness devices:* The roles of such devices are not typically related to medical problem by any means (e.g., use of smart watches and Fitbits). People are increasingly more interested in personal health and fitness data. They want to keep track of their activity levels, pulse rate, dietary habits, sleep patterns, and to self-monitor personal health and wellness.

15.5 IoT-Integrated Healthcare Devices

The evolution of IoT has changed the lifestyle of all people as they are now aware of their medical health issues and want to do better with their lives. Seeing people's interest towards their health, all the medical equipment is now replaced or integrated with technology for increasing their functioning. These devices are changing the era of medical devices in terms of efficiency, cost saving, and quality of life.

- Efficiency: Physicians are able to keep track of their patients remotely and regularly. It reduces time to manage the long running queue. Therefore, these devices are efficiently managing the time of both patients and physicians.

- Cost saving: The early detection of the problem reduces appointments with doctors in less time with less cost. Also insurance companies can give their best offers as per their online health records. Patients can be monitored remotely if required, reducing the billing of hospitals.

- Quality of life: Sleep tracking record, heart rate records, and other statistics of health motivate a person to change their eating habits. People are interested regularly exercising. Hence, quality of life is improved.

15.5.1 Examples of Existing IoT-Enabled Devices

15.5.1.1 Pulse Oximeters

Pulse oximetry is a painless test used to measure oxygen saturation level and pulse rate in the blood. According to www.nonin.com/company-history, Nonin is known as the first one to develop oximetery equipment for personal and professional purpose. Monitoring devices are used by physicians for increasing the efficiency of the treatment during worse conditions. Following are the common applications of oximetry:

- To measure the function of lungs
- To evaluate breathing process
- To check the body functions with ventilator
- To check oxygen levels in the body before and after surgical operation
- To know the effectiveness of any new supplementary oxygen therapy
- To measure the tolerance ability during physical activity
- To study sleep apnea cases

An oximeter is small, lightweight, easy to use and non-invasive tool attached to fingertip, toe, or earlobe to measure oxygen. A small beam of light passes through the blood. Changes in the light beam are because absorption denotes the oxygen level in the body [4]. It is very easy to use as shown in Figure 15.5.

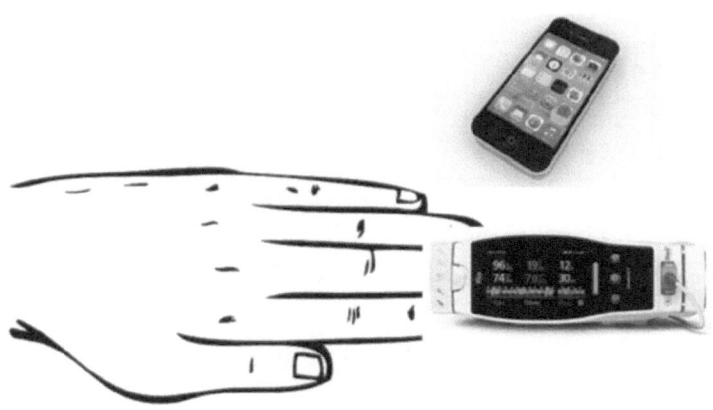

FIGURE 15.5
Pulse Oximeter.

Accurate pulse rate and SpO$_2$ readings are very important for patients having heart and breath problems. Frequent drops in oxygen level can be easily monitored under physician guidelines.

SpO$_2$ is the measure of the oxygen level in the blood. A reading of 95 percent or above is considered normal. If oxygen level is 92 percent or lower, then it means poorly saturated blood, which causes chest pain, increases heart rate, and breath shortness-like problems.

15.5.1.2 Treatment of Cancer

For patients with head and neck cancer, a sensor and mobile technology are used to monitor them from remote locations. These patients are receiving radio therapy. A cyber-infrastructure for comparative effectiveness Research (CYCORE) technology is used. The CYCORE uses a sensor and port used for tracking. A study on 357 patients was performed, out of which it was seen that people using CYCORE technology [5] were much better. The problems encountered during research were poor patient involvement, effects of environmental impact, inadequate knowledge about patient living style, and the limited capability to handle and evaluate a huge number of variables.

All the data generated by sensors was monitored in real time by doctors and the care to patients was provided well in time. Figure 15.6 gives an abridged view of CYCORE. It shows how all collaborators interact with the system. The stakeholders may be a doctor, researcher, family members, or any community member. The main aspiration of CYCORE is to provide quality assurance and timely data to researchers and other stakeholders of cyber-infrastructure. Its main concern is to address security and privacy of the system. The data generated by the system can be mishandled by an imposter.

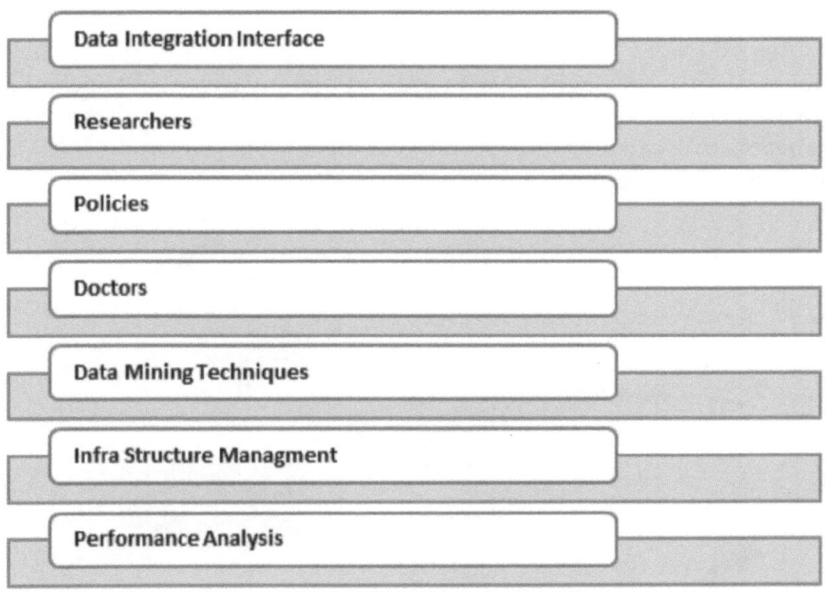

FIGURE 15.6
CYCORE Components.

15.5.1.3 Mood Enhancers

Mood enhancers are devices that are used to change your mood and make you feel happy. It is still under research and used by neuroscientists for the same. The mood enhancers use sensors to measure emotion; they use different biomarkers like heart rate, sudden jump in pulse rate, diminished sleep, and more perspiration [6]. A head-mounted device is used that passes on a low ferocity current in the brain to boost the mood of a person. On the basis of the aforementioned data the decision of the amount of current is passed. An application may also be used that can push patient to call family or friends when they are not performing their regular tasks. It may also ask to go for a walk if movement is not registered for long. All this information is sent to the doctor, however if this information is hacked then someone may use the situation for their own benefit.

15.5.1.4 Continuous Glucose Monitoring System

It is a device attached to a minuscule sensor on the stomach or arm. The sensor takes reading of the glucose level every few minutes. The device contains a transmitter that sends a signal to monitor; that may be an insulin pump or a remote device. It helps in regularly monitoring glucose levels. If a level is found high or low it raises an alarm. It also helps in finding out the glucose trends; if the diet is also monitored then it can help in predicting the reason behind high or low glucose levels. However, the glucose monitoring system (GMS) is not a fully reliable system, one need to use the other methods of checking glucose level before changing the dosage of insulin. Figure 15.7 shows the working of GMS. It is more beneficial for low glucose patients [7].

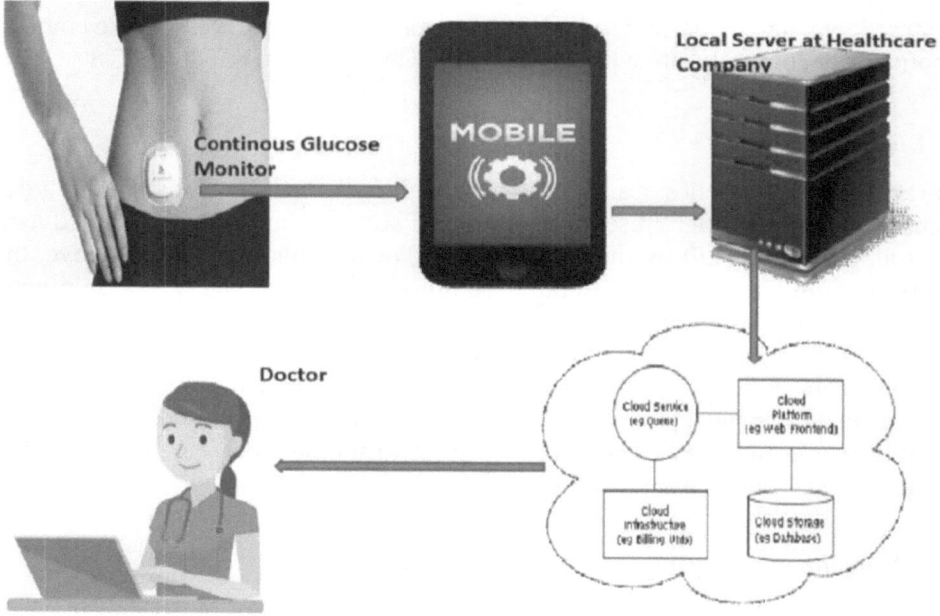

FIGURE 15.7
IoT in Healthcare Systems.

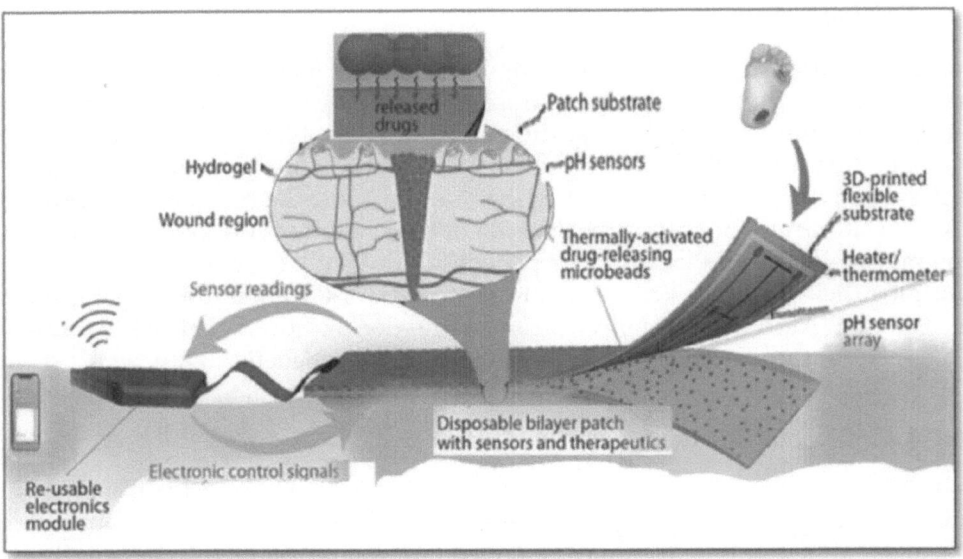

FIGURE 15.8
Smart Bandages.

15.5.1.5 Smart Bandages

A bandage is designed by Tufts University to monitor chronic, deep rooted wounds, laceration, and burns for any skin-related trauma. It uses sensors to monitor the pH level of skin and temperature sensors to check any kind of infection and inflammation. It can sense any kind of changes on skin and can release drugs for oxygenating the skin. The patch can be charged by using a detachable cable [8]. Sensors and microprocessors are embedded in the bandage to communicate with handheld devices as shown in Figure 15.8.

15.5.1.6 Smart Beds

There are different varieties of smart beds available in the market. These beds have special sensors to perform a different set of operations. Patients who need to lie in bed for a protracted length can benefit by these smart beds. For patients who cannot move, this bed provides continuous rotational therapy (i.e., moving patients every few minutes). This helps patients who have some movement-related or pulmonary problems that make them stay in bed for a long time. The bed has an alarm system incorporated that can be used if the patient is not resting ain a particular position or if movement of some kind is required. The smart beds also have inbuilt cooling systems that prevents patients from getting bed sores. It is easy to move these beds from one place to another; locating these beds in case of emergency is smooth as shown in Figure 15.9.

15.5.1.7 Healthcare Charting

Audemix, a voice-driven IoT device, can be used to store records of patients in more comfortable and easy ways rather than a manual system. It is an easy way to access and inspect patient data and manage things in a better way.

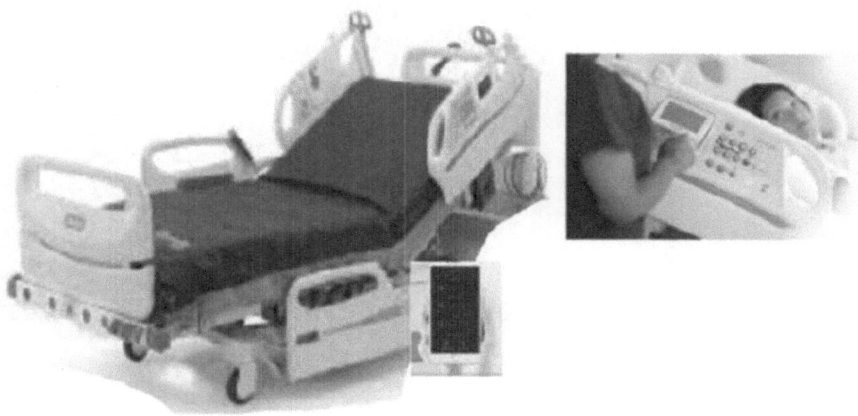

FIGURE 15.9
Smart Bed.

15.6 Security in IoT

It is important to make the system secure by use of strong passwords, firewall, and by adopting a secure culture. Some of the security features are shown in Figure 15.10. As per the Gartner survey almost 20 percent of companies have been attacked at least once on

FIGURE 15.10
Security Features in IoT.

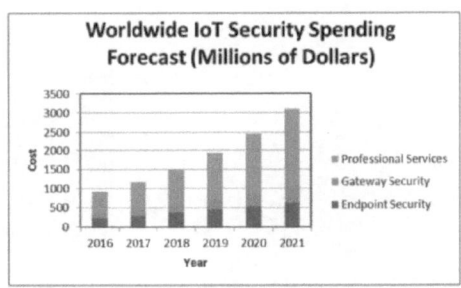

FIGURE 15.11
Spending on IoT Security.

IoT-based devices. Most of the companies are spending billions of dollars to prevent these attacks and the market is likely to spend above 3.1 billon U.S. dollars in the year 2021. That is more than double of what was being spent in 2019. The requirements in the healthcare industry and the challenges made it important for healthcare providers to provide quality and standardized services. Figure 15.11 gives the worldwide IoT security spending forecast from 2016 to 2021.

It is observed that only 30 percent have knowledge about IoT on their network. Only 44 percent know about the different policies.

15.7 Security Challenges in IoT-Based Devices

- *Confidentiality:* It is important to maintain the secrecy of patient records. Any unauthorized access of health records can lead to exposure of sensitive information.
- *Authentication:* Every device has a unique identification number used to identify it on the internet. Before entering any network, the devices have to be recognized and validated. Authentication is the process of recognizing the devices.
- *Fault tolerance:* Even in the case of device and system failure, the system can still support. All the nodes in the IoT network must deliver the lowest security required even in case of any kind of failures. It is also important to have resilience of data to prevent any attack.
- *Updated data:* In the case of remote monitoring of a patient, the data used must be the latest and the last updated data. For example, in case of glucose monitoring the latest sugar level is required for accurate diagnosis and treatment.
- *Memory constraints:* Most IoT devices are small and embedded inside a system. Due to this reason the memory of an IoT device is limited and they have to rely on the cloud to store data.
- *Speed:* The computational speed of an IoT device is low as they have low power processors and small devices. The tasks of the processor are manage, sense, analyze, and communicate with the limited power and memory, this makes it much slower.

- *Power consumption:* Most IoT devices used have nano sensors embedded in them, so they have low battery capacity. Because of this they enter the battery saving mode, so these devices cannot perform security protocols all the time.

- *Communication:* The IoT-based devices work mainly on wireless networks. These networks are not secure. It is an important and difficult task to find a secured protocol to communicate.

- *Frequent update:* The security update needs to be done frequently to monitor against any kind of security breach.

15.8 Security Architecture

A system is secured if it is furnished with three security goals: confidentiality, integrity, and availability. The need of security is high as many crucial areas are involved. The lowest level of IoT security architecture is the perception layer; it is a layer that co-assembles all information through equipment like RFID readers, sensors, and GPS. The next level is the network layer; the main work of this layer is to transfer data across networks with the help of wireless networks [9,10]. The application level is the topmost level that illustrate the use of a device according to the need of a user. Every level has their importance as illustrated in Figure 15.12.

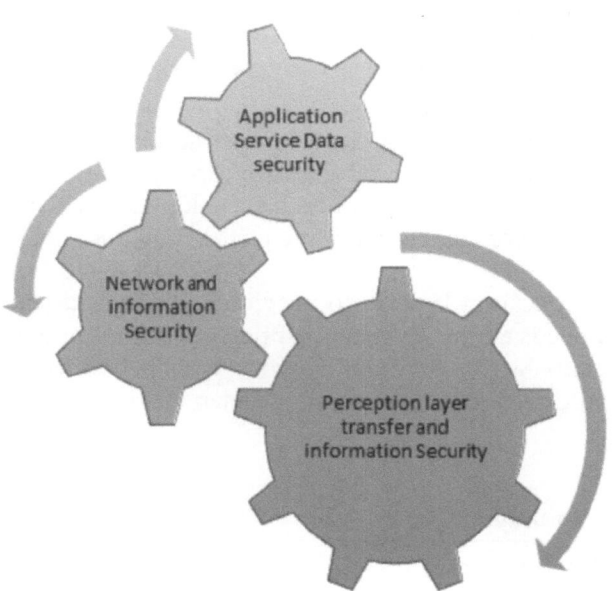

FIGURE 15.12
IoT Security Architecture.

TABLE 15.1

Perception Layer Security

Perception Layer Security			
RFID Security	**WSN Security**	**RSN Security**	
Protocol security	Routing protocol security	Fusion security	GPS Technology Security
Base station security	Cryptographic algorithms	Sensor and RFID security reader · RFID and WSN security	
Reader security	Key management	Sensor and tag security	
Tag counterfeit security	Node trust management		
Tag encode security			

- *Perception Layer:* This layer consists of sensors; most of the sensors are small and have the disadvantage of power and storage capacity. It is challenging to apply security algorithms. It is important to authenticate access to devices to protect them from any unauthorized access. The data transfer between any two nodes needs to be in a confidential manner. Due to the drawback of low power a lightweight encryption technology is required to protect data. The detail pf perception layer security is given in Table 15.1.

- *Network Layer:* This layer makes communication across the network possible. It is responsible for secure transmission of data from a perceptual layer initially processing data. This includes access to network security, core network security, and LAN security. Detailed are listed is in Table 15.2. It is challenging to apply existing security mechanisms in the current layer; however, it is important to apply confidentiality and integrity. Also the threat to availability and a DDoS-distributed DoS attack is the matter of concern, as it is severe for IoT. The network layer is responsible to resolve a DDoS attack for vulnerable nodes.

- *Application Layer:* Above the application layer and below the network layer lies a support layer. The purpose is to draw an appropriate platform to support different applications provided by the application layer. Cloud security, middleware security, and information development security are handled by this sublayer [11].

The purpose of the application layer is to provide personalized services mentioned in Table 15.3. This layer needs all the strong encryption algorithms and antiviruses. The concept of key authentication across different heterogeneous networks is the key to securing this layer. Another important concept is password management.

TABLE 15.2

Network Layer Security

Network Layer Security		
Access Network	**Core Network**	**Local Area Network**
Ad-hoc network security	Internet security	LAN security
GPRS security	3G security	
Wifi security		

TABLE 15.3

Application Layer Security

Application Layer Security	
IoT Application	**Application Support Layer**
Intelligent logistic security	Middleware security
Smart home security	Cloud security
Smart grid security	Service support security
Smart healthcare device security	Information development security
Environmental monitoring	

15.9 Vulnerabilities in IoT Devices

With the increase in usage of IoT medical devices more security issues are arising. Researchers have found vulnerabilities in many devices like pacemakers and insulin pumps where a KRACK key reinstallation attack was made that allowed the attacker to modify patient records. There are many issues associated with manufacturing and the security of IoT devices. Some of the vulnerabilities are listed next and it is depicted in Figure 15.13.

15.9.1 Feeble Passwords

The choice of a password by most of users depends on common dictionary words, relative name, repetitive words, and more. These passwords are easy to crack and can be compromised. An effortless exhaustive search can easily crack the system. The high technology, IT-savvy people make use of a backdoor in software to crack these devices. Any intimate information needs to be dealt in a secured way; permission is the keyword here. Only an authorized person can access the sensitive information.

15.9.2 Transfer of Data on an Unsecure Channel

Nowadays the internet is the medium to send data; this channel of medium is highly unsecure. Not only secure storage but also the transmission of data in a secured manner to keep confidentiality and the integrity of data intact is important. Many researchers are working towards achieving this goal.

15.9.3 Inadequate Measures of Device Management

Inadequate supports of device management includes asset management, update management, secure decommissioning, systems monitoring, and response capabilities. Small IoT devices may be produced in bulk; however, it is crucial to manage even the smallest IoT devices.

15.9.4 Devices with Default Settings

Many devices have default settings if vendors cannot change them. This makes the devices more unsecure.

15.9.5 Unsure Interfaces

The API interfaces are the power of IoT; IoT solutions are the strength of all the technology around us be it big data, cloud, or mobility. APIs are interconnectors that combine internet

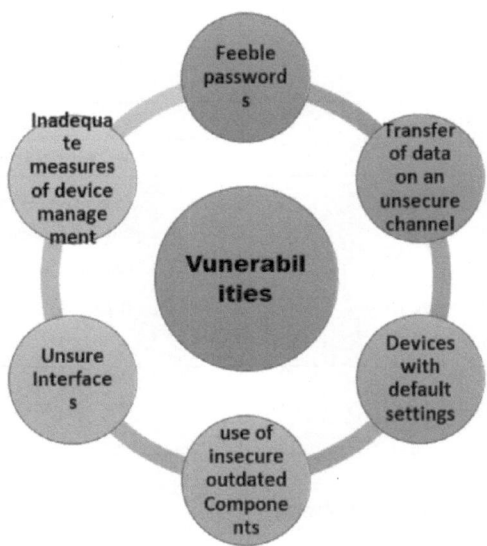

FIGURE 15.13
Vulnerabilities in IoT Network.

with things. Creating an API that deals with critical mass is very challenging. The knowledge of protocols and requirements is needed to create an API for a device. Achieving this kind of information and knowledge itself is a challenge. Threat-Challenge-Opportunity of various features is given in Table 15.4.

The following medical devices are found most vulnerable to attacks:

- *Insulin pumps:* It is one the most used medical IoT devices that can be exploited by an attacker who abuses the available capacities that connect drug delivery systems and patient records.

TABLE 15.4

Threat-Challenge-Opportunity

Feature	Threat	Challenge	Opportunity
Interdependence	Bypass static defenses, overprivileged	Access control, privilege, distance limitations	Context-based permission, next-gen packet optimal platforms
Diversity	Insecure protocols	Fragmented	Intrusion detection system
Restraint	Insecure system	Lightweight defense protocols, capacity limitations	Combination of physical and biological characteristics, coherent optics
Myriad	IoT botnets, DDoS	Intrusion, detection, and prevention	Intrusion detection system (IDS)
Neglected	Remote attack	Remote verification	Lightweight, trusted light weight, trusted execution
Confidentiality	Confidentiality attack	Privacy protection	Anonymous protocols, homographic encryption
Mobile	Malware propagation	Cross domain identification	Dynamic configuration
Pervasive	Insecure configuration	–	Safety consciousness

- *Smart pens:* Data stored in smart pens is an easy target to exploit. Researchers are using these devices to store patient medical records.
- *Cardiac devices:* Many cardiac devices are implantable, like a pacemaker. A denial of service (DoS) attack can lead to dire consequences and may lead to death.
- *Wireless vital monitors:* The devices that can transmit heart rate, insulin levels, and other vital information of a patient via Bluetooth or other wireless devices is comfortable. It is important to interface these monitors in a secured and confidential way.
- *Temperature sensors:* A casino in the year 2018 was hacked using a fish tank's smart thermometer. Technology is affordable today, however with it comes many risks. If used on patients, it may lead to leakage of their health record.

15.10 Security Attacks towards IoT and WSN

The attacks on different layers of IoT security architecture is mentioned in Table 15.5. Jamming DoS is a kind of DoS attack where the malicious node can interrupt a signal by performing transmissions at the same frequency. It creates noise in the carrier by transmitting signals continuously in a region that degrades or thwarts the communication process among all nodes. Jamming can also be performed at certain time intervals that still affect the transmission process.

In spoofing, an intruder node identifies the addresses of a victim node and exploits those identities anywhere in the network. The attacker starts transmitting messages in the network by spoofing the addresses. The objective of this spoofing attack is to associate the spoofed identity with the other original addresses and to redirect the traffic meant for the legitimate node to be redirected to the intruder instead.

In traffic analysis the information about the network topology can be obtained by analyzing traffic patterns. In WSN the nodes that are nearer to the base station transmit more packets than the nodes located far from the base station. In particular the cluster nodes are much busier when compared to other normal nodes in the network. Detecting cluster heads are highly preferable for adversaries since DoS attacks will have a huge impact on these nodes. Moreover if there is a sudden increase in normal traffic flow, it may indicate the beginning of a deliberate attack.

In a sybil attack, a single node can provide multiple identities to create confusion among all remaining nodes in the network. This results in conflicting network routing paths that automatically decrease the impact of fault-tolerance approaches and causes some potential

TABLE 15.5

Attacks on Different Layers of IoT Security Architecture

Layer	RFID Attack	WSN Attack
Perception	Replay, jammer, sybil, selective forwarding, synchronization	Jamming, replay, destruction of RFID reader
Network	False routing, selective routing, session flooding, snooping	Cloning, spoofing, masquerading, network protocol attack
Application	Injection, buffer overflow	Injection, buffer overflow, tag modification
Multi-layer	Replay, side channel attack, traffic analysis	Replay, side channel attack, traffic analysis

threats to routing protocols [12]. Moreover these services affect the performance of various schemes such as data aggregation, anomaly detection, and distributed storage.

In a black hole attack, the suspicious node drops all packets that are forwarded to it. This attack becomes more effective when the node acts as a sinkhole. In a sinkhole, an adversary node broadcasts itself to the neighboring nodes that it is the best route for transmitting packets to reach the destination. Once the node becomes a sinkhole that will receive all the packets transmitted to the base station, the sinkhole nodes do not drop packets for the data traffic pass is directed to the single node in the network. Moreover the main goal of this attacker node is to remain undetected, which eventually sinks the data it receives.

A gray hole is a selective forward attack in which it does not forward all packets it receives instead it selectively drops the packets. In this way, an attacker node remains undetected by dropping only a few packets while forwarding all remaining packets. The participating nodes in the sensor network generally forward the messages as they are received. The adversary node in a selective forward attack that may drop certain packets without forwarding them.

In a wormhole attack a special tunnel is created between nodes for the rapid transmission of packets. The adversaries who are located apart will announce their locations to attract the data traffic [13]. An intruder node obtains the data packets and transfers to another intruder node for replaying it. The nodes involved in the transmission process believe that the packets are received directly from the first intruder node. Wormhole attacks are usually launched through a faster channel; this attack is complex to identify and affects various network services such as localization, time synchronization, and data fusion.

A node-replication attack is also known as a cloning attack in which an attacker purposely places duplicates of compromised nodes in various places on the network to maintain inconsistency. An attacker can divert the network behavior by exploiting some copies of previously attacked nodes.

6LoWPAN is an internet protocol introduced as the extended version of IPv6 to be used by the smart things for routing packets in WSN.

A fragment duplication attack is a 6LoWPAN attack where intruders can place their own fragments in the chain. The significant benefit of this fragment duplication attack is that the destination node cannot identity that the fragment has been originated from the same source node. There is no specific authentication mechanism available at the destination node to verify whether the received fragment is a spoofed one or the original. The receiver cannot differentiate the legitimate one from the spoofed fragments. Instead, the received fragments are processed that seem to appear as received from the same source based on verifying the datagram tag and the source node MAC address. Moreover, there are more chances to be engaged in subsequent attacks such as Denial of Service (DoS) attacks [14,15].

Session hijacking is an attack commonly referred to as 'tampering with' and 'exploitation of' a legitimate communication session (known as a session key) to obtain the unauthorized access on the available resources or services of a particular system. More specifically, the session hijacking of a set of TCP messages significantly affects and creates more trouble in the IoT network.

In a SYN-flooding attack, a malicious node attempts to exhaust the memory and the energy of a particular node by flooding with surplus messages. It can be succeeded by sending multiple requests to establish a connection, rather that obtaining the connection it eventually overwhelms the buffer, which causes the node to be exhausted [16]. Moreover an adversary in a TCP SYN flooding attack transmits multiple TCP requests to establish a connection without having any intention of establishing it. Last the target automatically gets overwhelmed or exhausted [17].

A constrained application protocol (CoAP) exploit is an attack that occurs in the application layer protocol where the protocol was introduced as the replica of HTTP for the tiny IoT devices to facilitate the communication among themselves. There are several security challenges posed for multicasting messages in IoT applications with this protocol [18,19].

A false data infection attack leverages the overall performance of a reading or a measurement, where the intruder nodes intentionally inject bogus data in the network. Thus it is concluded that the false data injection attack occurs at a semantic level, not at a logical level. Sensor overwhelming is an attack that modifies the measurement of a sensor node, hence the targeted sensor node is overwhelmed with false messages for redirecting them with fake stimuli.

15.10.1 Countermeasures against Various Attacks towards IoT and WSN

Routing protocols are designed in such a way that an adversary cannot make it to malfunction by compromising the nodes. Several preventive measures are encountered to fight against known threats that may not be successful against unknown threats. Threat detection schemes are designed in a generic way to handle malfunctioning or misbehaving nodes. The strategies and techniques to defend against attacks towards IoT and WSN are discussed next:

- Link-layer encryption can prevent outside attacks such as eavesdropping and information spoofing; hence the attack towards WSN can be avoided by incorporating link-layer symmetric key techniques using shared keys, which are shared globally. Secure network encryption protocol (SNEP) is a commonly known encryption protocol specifically designed for WSN [20–22], which can prevent spoofing activities.

- A SensorWare communication multicast model is proposed that provides link-layer encryption using RC6 algorithm. This algorithm is robust because of choosing the number of rounds parameter. The key selection process is performed in a pseudo-random manner; every node exploits the same seed of randomized function [23].

- In [24], two protocols are presented to defend against a sybil attack. In the first protocol, the 'radio resource testing' approach is employed where each sensor node shares a unique channel to each of their neighbor nodes. Then it is tested if the communication happens via the pre-declared channels. Sending and receiving data accomplished simultaneously with the available radio circuitry of a sensor node. The other protocol implements 'ID-based symmetric keys' to prevent malicious activities. The pool of keys is maintained where every sensor node is loaded with a key linked to its ID. The ID of a malicious sensor node is investigated by the witness node by comparing the keys shared among the witness sensor node and suspicious sensor node.

- A rule-based anomaly detection system (RADS) is presented [25] to monitor and identify the sybil attack that occurs in the sensor networks. The system follows an ultra-wideband ranging detection algorithm and it is executed in a distributed manner where each node can trigger an alarms if any suspicious activities are detected. A scheme proposed in [26] detects sink hole attacks in WSN using a redundancy mechanism. In order to identify the sinkholes, messages are sent to the malicious nodes via multiple paths. Examining the response messages from the malicious nodes assists in confirming the attacker nodes.

- A cross-layer security scheme is proposed [27] as 'swarm intelligence' to mitigate and detect a jamming DoS attack against WSN. It affords a mapping protocol to identify the jammed area in a WSN. The identified jammed region can be avoided by re-routing the packets [28].

- A wormhole technique is presented in [29] that effectively handles a jamming DoS attack and provides an effective solution for threats against WSN.

- Anomaly detection IDS is introduced to defend against blackhole and gray hole attacks against WSN, which functions on each node and is more effective on the network layer [30].

- A REWARD scheme [31] is proposed by introducing a routing algorithm to detect and avoid blackhole attacks against WSN; hence the normal operation can be resumed in WSN in this scheme.

- Findings in [32] quoted that mesh topology is highly resilient on blackhole attacks in WSN when compared to star and tree topological structures.

- A threshold-based dynamic defense scheme is introduced, the contextual behavior of the operational network scenario is analyzed to detect the node's misbehavior. It is investigated to identify the node's abnormal behavior and mitigate the impact of it [33].

- A watchdog mechanism is proposed for detecting blackholes and gray holes (selective forwarding) attacks. Sensor nodes in each cluster monitor their neighbors and inform the cluster head if any misbehavior is detected [34].

- Localized Secure Architecture for MANET (LSAM) protocol is proposed to detect the blackhole attack. Secure monitoring nodes are activated only if the fixed threshold value is exceeded and those will be triggered to remove the malicious node from the proximity region [27].

- Multi-detection routes are created in the ActiveTrust [35] method; hence the detection routes will not be disclosed to the adversaries. The behavior and location of an adversary can be detected using the ActiveTrust method that assists in avoiding blackhole attacks.

- A technique [36] is proposed to detect and prevent a blackhole attack in a clustered wireless sensor network; a centralized coordinator node will be selected by the nodes based on the election criteria defined to become a cluster head. Once an attacker is detected in the cluster, then the cluster head is responsible to remove and terminate the communication with the malicious node.

- A resistive methodology is proposed as a lightweight RSDA protocol [37] that detects malicious nodes that are trying to launch a selective drop attack. It provides reliability by authenticating the node and disabling the link using the elliptic curve digital signature algorithm.

- An algorithm is presented to detect a sinkhole attack by maintaining a list that consists of suspected nodes and the opinion about the suspected node will be collected from its neighbors that help in making a decision [38].

15.11 Security Attacks towards RFID in IoT

IoT devices have low security that may lead to give unauthorized access to connected devices therefore hampering the confidentiality of the user. Radio frequency identification (RFID) is the most common technology in IoT; it consists of tags, RFID readers, and a database server. They are costly and are tagged or embedded onto devices [39]. There are two types of RFID tags:

- **Active Tags:** They are expensive and need the power of an external source. The storage and broadcast capacity are much higher as compare to a passive tag.
- **Passive Tags:** They use the energy transmitted by the reader; the range of passive tags is shorter and they consume much less energy [40]. They may work within a range of 30 centimeters to 1.5 meters and are categorized into 3 types: low frequency, high frequency, and ultra-frequency tags.

15.11.1 Limitations of RFID

- It is not very secure.
- Very few security protocols are implemented on RFID.
- Limited storage space; therefore, it is difficult to implement security protocols.

15.11.2 Security Issues in RFID

- **Jamming:** A process of stopping the harmony between the tag and RFID reader. This leads to stopping communication between the tag and reader. A radio noise is generated in the frequency used by RFID to prevent this communication [41].
- **Eavesdropping:** The attacker uses the fake reader, as RFID text is not encrypted it is easier for the attacker to retrieve information [42].
- **Replay attack:** The attacker records the details and replicates the same details later on the RFID system.
- **Deactivation:** The attacker deactivates the RFID tag and makes it useless for the attacker.
- **Spoofing:** The attacker uses the RFID security protocol to write data with the same format to RFID attack.
- **Man-in-the-middle attack:** The attacker places a fake reader in between the tag and RFID reader. All information that is transferred between the reader and tag is received by a fake reader and can be manipulated.

15.12 Conclusion

With the huge demand of IoT in the near future and their use in critical situations, one may accept that these gadgets are ensured with a similar degree of security as the average system server; this isn't the situation. IoT gadgets present some significant security concerns that have been mentioned in this chapter. Typically IoT gadgets are created for a particular use and have less power and speed, and frail hardware; this leads to many vulnerabilities that attackers could exploit. So it has become important to safeguard the data generated by these devices. The manufacturers and users must come together to implement better practices so that future damage can be avoided.

References

[1] Meinert, Edward, Michelle Van Velthoven, David Brindley, Abrar Alturkistani, Kimberley Foley, Sian Rees, Glenn Wells, and Nick Pennington. 2018. "The Internet of Things in Health Care in Oxford." *Protocol for Proof-of-Concept Projects* 20. doi: 10.2196/12077.

[2] Musonda, Chalwe. 2019. *"Security, Privacy and Integrity in Internet of Things – A Review."* *Proceedings of the ICTSZ International Conference in ICTs*, Lusaka, Zambia, pp. 148–152.

[3] Patel, Keyur, Sunil Patel, P. Scholar, and Carlos Salazar. 2016. "Internet of Things-IOT: Definition, Characteristics, Architecture, Enabling Technologies, Application & Future Challenges." doi: 10.4010/2016.1482.

[4] Van de Louw, A., C. Cracco, C. Cerf, A. Harf, P. Duvaldestin, and F. Lemaire. 2001. "Accuracy of Pulse Oximetry in the Intensive Care Unit." *Intensive Care Medicine* 27: 1606–1613.

[5] Patrick, K., L. Wolszon, K. M. Basen-Engquist, et al. 2011. "CYberinfrastructure for COmparative effectiveness REsearch (CYCORE): Improving Data from Cancer Clinical Trials." *Translational Behavioral Medicine*: 83–88. doi:10.1007/s13142-010-0005-z.

[6] Young, Simon N. 2007. "How to Increase Serotonin in the Human Brain Without Srugs." *Journal of Psychiatry & Neuroscience* 32 (6): 394–399.

[7] Nguyen gia, Tuan, Mai Ali, Imed Ben Dhaou, Amir M. Rahmani, Tomi Westerlund, Pasi Liljeberg, and Hannu Tenhunen. 2017. "An IoT-Based Continuous Glucose Monitoring System: A feasibility Study." *Procedia Computer Science* 109: 327–334. doi:10.1016/j.procs.2017.05.359.

[8] Mostafalu, Pooria, Ali Tamayol, Rahim Rahimi, Manuel Ochoa, Akbar Khalilpour, Gita Kiaee, Iman K. Yazdi, et al. 2018. "Smart Bandage for Monitoring and Treatment of Chronic Wounds." *Nano Micro Small* 13 (33): 1–9.

[9] Ngai, E. C., J. Liu, and M. R. Lyu. 2006. *"On the Intruder Detection for Sinkhole Attack in Wireless Sensor Networks."* In *Communications. ICC'06. IEEE International Conference on*. IEEE, Vol. 8, pp. 3383–3389.

[10] Dewangan, Kiran, and Mina Mishra. 2018. "Internet of Things for Healthcare: A Review."

[11] Borgohain, Tuhin, Uday Kumar, and Sugata Sanyal. 2015. "Survey of Security and Privacy Issues of Internet of Things."

[12] Gupta, H. P., S. Rao, A. K. Yadav, and T. Dutta. 2015. "Geographic Routing in Clustered Wireless Sensor Networks Among Obstacles." *IEEE Sensors Journal* 15 (5): 2984–2992.

[13] Hu, Y.-C., A. Perrig, and D. B. Johnson. 2003. *"Packet Leashes: A Defense Against Wormhole Attacks in Wireless Networks."* In *INFOCOM 2003. Twenty-Second Annual Joint Conference of the IEEE Computer and Communications*. IEEE Societies, Vol. 3, pp. 1976–1986.

[14] Pongle, P., and G. Chavan. 2015. *"A Survey: Attacks on RPL and 6LoWPAN in IoT."* In *Pervasive Computing (ICPC) International Conference on*. IEEE, pp. 1–6.

[15] Hummen, R., J. Hiller, H. Wirtz, M. Henze, H. Shafagh, and K. Wehrle. 2013. *"6LoWPAN Fragmentation Attacks and Mitigation Mechanisms."* In *Proceedings of the Sixth ACM Conference on Security and Privacy in Wireless and Mobile Networks*. ACM, pp. 55–66.

[16] Wood, A. D., and J. A. Stankovic. 2002. "Denial of Service in Sensor Networks." *Computer* 35 (10): 54–62.

[17] Raymond, D. R., and S. F. Midkiff. 2008. "Denial-of-Service in Wireless Sensor Networks: Attacks and Defenses." *IEEE Pervasive Computing* 7 (1): 74–81.

[18] Frank, B., Z. Shelby, K. Hartke, and C. Bormann. 2001. "Constrained Application Potocol (coap)." In IETF-Draft. *IETF*.

[19] Rahman, R. A., and B. Shah. 2016. *"Security Analysis of IoT Protocols: A Focus in COAP."* In *Big Data and Smart City (ICBDSC), 2016 3rd MEC International Conference on*. IEEE, pp. 1–7.

[20] Karlof, C., and D. Wagner. 2003. "Secure Routing in Wireless Sensor Networks: Attacks and Countermeasures." *Ad Hoc Networks* 1 (2): 293–315.

[21] Perrig, A., R. Szewczyk, J. D. Tygar, V. Wen, and D. E. Culler. 2002. "Spins: Security Protocols for Sensor Networks." *Wireless Networks* 8 (5): 521–534.

[22] Deng, J., R. Han, and S. Mishra. 2003. "A Performance Evaluation of Intrusion-Tolerant Routing in Wireless Sensor Networks." In *Information Processing in Sensor Networks*. Springer, pp. 552–552.

[23] Slijepcevic, S., M. Potkonjak, V. Tsiatsis, S. Zimbeck, and M. B. Srivastava. 2002. "On Communication Security in Wireless Ad-Hoc Sensor Networks." In *Enabling Technologies: Infrastructure for Collaborative Enterprises*. IEEE, pp. 139–144.

[24] Newsome, J., E. Shi, D. Song, and A. Perrig, 2004. *"The Sybil Attack in Sensor Networks: Analysis and Defenses."* In *Proceedings of the 3rd International Symposium on Information Processing in Sensor Networks*. ACM, pp. 259–268.

[25] Sarigiannidis, P., E. Karapistoli, and A. A. Economides. 2015. "Detecting Sybil Attacks in Wireless Sensor Networks Using Information." *Expert Systems with Applications* 42 (21): 7560–7572.

[26] Zhang, F.-J., L.-D. Zhai, J.-C. Yang, and X. Cui. 2014. "Sinkhole Attack Detection Based on Redundancy Mechanism in Wireless Sensor Networks." *Procedia Computer Science* 31: 711–720.

[27] Muraleedharan, R., and L. A. Osadciw. 2006. *"Cross Layer Denial of Service Attacks in Wireless Sensor Network Using Swarm Intelligence."* In *Information Sciences and Systems, 2006 40th Annual Conference on*. IEEE, pp. 1653–1658.

[28] Wood, A. D., J. A. Stankovic, and S. H. Son. 2003. *"Jam: A Jammed Area Mapping Service for Sensor Networks."* In *Real-Time Systems Symposium. RTSS 2003. 24th IEEE*. IEEE, pp. 286–297.

[29] Cagalj, M., S. Capkun, and J.-P. Hubaux. 2007. "Wormhole-Based Antijamming Techniques in Sensor Networks." *IEEE Transactions on Mobile Computing* 6 (1): 100–114. doi:10.1109/TMC.2007.250674.

[30] Liu, Y., Y. Li, and H. Man. 2005. *"Mac Layer Anomaly Detection in Adhoc Networks."* In *Information Assurance Workshop. IAW'05. Proceedings from the Sixth Annual IEEE SMC*. IEEE, pp. 402–409.

[31] Karakehayov, Z. 2005. "Using Reward to Detect Team Black-Hole Attacks in Wireless Sensor Networks." *Wksp. Real-World Wireless Sensor Networks*, pp. 20–21.

[32] Krishnan, S. N., and P. Srinivasan. 2016. "A QOS Parameter Based Solution for Black Hole Denial of Service Attack in Wireless Sensor Networks." *Indian Journal of Science and Technology* 9 (38).

[33] Poongodi, T., M. Karthikeyan, and D. Sumathi. 2016. "Mitigating Cooperative Black Hole Attack by Dynamic Defense Intrusion Detection Scheme in Mobile Ad Hoc Network." 15 (23): 4890–4899.

[34] Tiwari, M., K. V. Arya, R. Choudhari, and K. S. Choudhary. 2009. *"Designing Intrusion Detection to Detect Black Hole and Selective Forwarding Attack in WSN Based on Local Information."* In *Computer Sciences and Convergence Information Technology. ICCIT'09. Fourth International Conference on*. IEEE, pp. 824–828.

[35] Poongodi, T., and M. Karthikeyan. 2016. "Localized Secure Routing Architecture Against Cooperative Black Hole Attack in Mobile Ad Hoc Networks." *Wireless Personal Communications* 90 (2): 1039–1050.

[36] Liu, Y., M. Dong, K. Ota, and A. Liu. 2016. "Activetrust: Secure and TrusTable 15. Routing in Wireless Sensor Networks." *IEEE Transactions on Information Forensics and Security* 11 (9): 2013–2027.

[37] Wazid, M., A. Katal, R. S. Sachan, R. Goudar, and D. Singh. 2013. *"Detection and Prevention Mechanism for Blackhole Attack in Wireless Sensor Network."* In *Communications and Signal Processing (ICCSP), 2013 International Conference on*. IEEE, pp. 576–581.

[38] Poongodi, T., Mohammed S. Khan, Rizwan Patan, Amir H. Gandomi, and Balamurugan Balusamy. 2019. "Robust Defense Scheme Against Selective Drop Attack in Wireless Ad Hoc Networks." (7): 18409–18419.

[39] Dass, Prajnamaya, and Hari Om. 2016. "A Secure Authentication Scheme for RFID Systems." *Science Direct* 78 (1): 100–106.

[40] Mohite, Sangita, Gurudatt Kulkarni, and Ramesh Sutar. 2013. "RFID Security Issues." *International Journal of Engineering Research & Technology (IJERT)* 2 (9): 746–748.

[41] Desai, Nidhi, and Manik Lal Das. 2015. "On the Security of RFID Authentication Protocols, Electronics, Computing and Communication Technologies (CONECCT)."

[42] Li, Tieyan, and Guilin Wang. 2007. "Security Analysis of Two Ultra-Lightweight RFID Authentication Protocols." *IFIP International Federation for Information Processing* 232: 109–120.

16

Healthcare Internet of Things – The Role of Communication Tools and Technologies

K. Lalitha
Kongu Engineering College, Erode, Tamil Nadu, India

D. Rajesh Kumar
Galgotias University, Greater Noida, Delhi-NCR, India

C. Poongodi
Kongu Engineering College, Erode, Tamil Nadu, India

Jeevanantham Arumugam
Kongu Engineering College, Erode, Tamil Nadu, India

CONTENTS

16.1 Introduction

In the marketplace of 21st century era, the number of devices attached to the internet is emerging exponentially and as the number of devices increases day by day, the need to implement the interaction between devices with machine-to-machine (M2M) communication is also much higher. No one could predict the impact of Internet of Things (IoT) or connected devices in our day-to-day lives. IoT is creating a new world in a quantifiable and measurable manner in which people can manage their lives and businesses in a better way. The emergence of the internet and IoT drive produce the significant theoretical and practical improvements in daily life of people by helping them to make wise and timely decisions when they come across any kind of need or emergency situations.

IoT is one of the technologies where physically located devices are connected to each other and provide the facility to interact with each other without any human intervention as shown in Figure 16.1. It is an extension of the connectivity through internet where the devices will interconnect with one another, share the data, and based on the analytics, the device itself can take action with respect to the external stimuli. In addition to the internet connectivity, devices get connected with electronics and other hardware like sensors. In simple terms, 'Anytime, anywhere and anything in connectivity' can be accessible [1,2]. Without the intervention of a human role, IoT devices can collect data or create information about human behaviors, analyze it, and may take action on their own.

16.1.1 Healthcare IoT

IoT is the most commonly identified prospective key to release the burdens on healthcare systems, and also it is been focused widely in recent research. Anything will be connected and managed with the virtual world with the emergence of IoT in the real world. It should be identified literally as a collection of networked devices interacting with each other without human intervention.

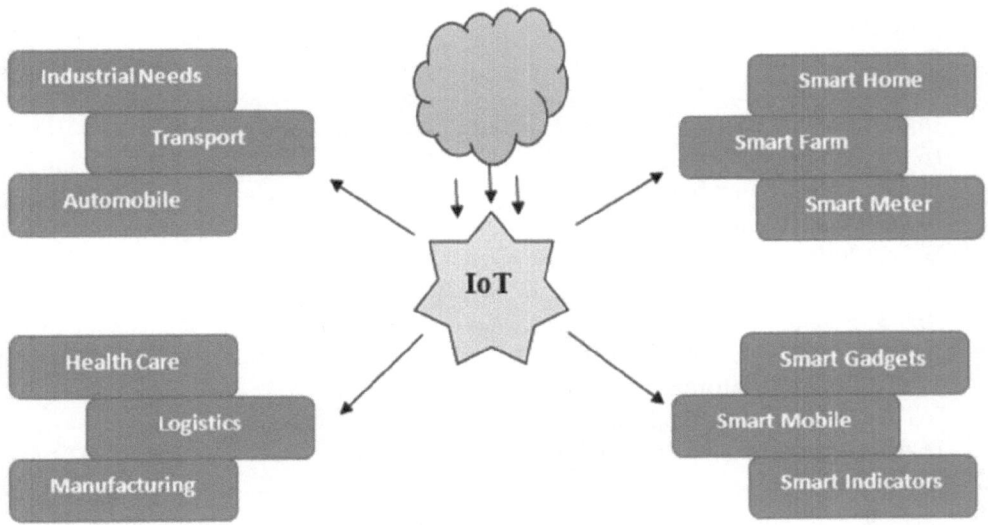

FIGURE 16.1
IoT Will Revolutionize the World.

The healthcare monitoring system could not be imaginable without IoT nowadays. A few years ago the healthcare system could rely on maintaining patient data in computers. The data physically collected from patients are documented as well as maintained in computer databases with ID. Generally, the healthcare system followed in India is reactive; solution are provided when the problem is identified with the patient. The issue may be identified at the last stage, probably in the worst-case scenario. From an accessibility point of view, healthcare facilities may not be uniformly accessible in all locations. If in case, any therapy is required, qualified physicians may not be available in all the locations. The phenomenon gets changed rapidly with the potential use of IoT in a healthcare monitoring system, but it is still in the embryonic stage as shown in the Figure 16.1. IoT is at the core of digital transformation. Improvements have been seen in agriculture, smart homes, smart cities, medical equipments, and therapies given for various illnesses especially gene therapies and body part transplantation are the few clear-cut samples where in IoT is strongly established.

16.2 IoT Tools and Technologies

In day-to-day life, technologies are used in manifold; from morning wake up to late night sleep including water supply, vegetables and groceries supply to home, cooking, driving a vehicle, locating a person or a location, or ordering food items. The work may not be fulfilled without Google assistance as per the current scenario. Since IoT is a quintessential requirement in the smart world, technologies that are being used in IoT devices gain much attention in the community.

IoT will demand an extensive, broad range of new technologies especially in the healthcare domain. Since IoT is going to deal with the predictions of many people's health issues, it is quintessential that a technology should produce the proper decision system. Immaturity of technologies and software/services may pose a significant challenge and create risk in all sectors [1]. An organization should have a good impact on technologies such as IoT architecture, network design, and security, and also may expect a wide range of coverage like risk management.

Coming to the driving force point of view, there are eight main components used in IoT that make revolutionary changes in the world. The components are:

- Define purpose and requirements of IoT system for healthcare
- Design the system interactions
- Explore domain model specifications
- Identify the microcontrollers
- Define communication options of IoT
- IoT device management
- IoT data storage and analytics
- IoT security

This chapter focuses on five main components of healthcare IoT System as given in the Figure 16.2 (Numbered as 4,5,6,7 and 8).

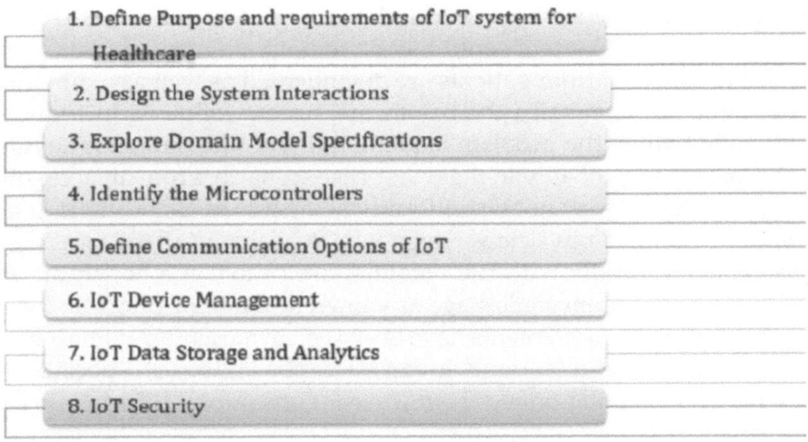

1. Define Purpose and requirements of IoT system for Healthcare

2. Design the System Interactions

3. Explore Domain Model Specifications

4. Identify the Microcontrollers

5. Define Communication Options of IoT

6. IoT Device Management

7. IoT Data Storage and Analytics

8. IoT Security

FIGURE 16.2
Roadmap of the Components of Healthcare IoT System.

16.2.1 Microcontrollers and Operating Systems

16.2.1.1 Raspberry Pi

The Raspberry Pi is a tiny sized minicomputer powered by SoC based ARM11 and identified with Broadcom BCM 2835. Since Raspberry Pi has inbuilt GPU along with audiovisual proficiencies, it can be effortlessly plugged into a monitor. Raspberry Pi extended its service which includes internet surfing, a letter preparation with a word processor, and email sending. Raspberry Pi is a highly recommended device recommended for the learning aspirants and field scientists since it is available at an affordable cost, hard to break, and highly controllable according to the need. Features supported by this tiny device are enormous like high-quality audio streaming, and playing HD quality videos and audios with playback options along with the gaming support [2].

RASPBMC, RASPBIAN, OPENELEC, PIDORA, ARCH LINUX, and RISC OS limited software is used. Altogether, the previously said software is taken up from the official forum that falls in the NOOBS category. To write the functions of the device and coding, Python programming choses as the primary programming language. In addition to Python, it supports C, BASIC, C++, Perl, and JAVA and Ruby languages. The picture given below (Figure 16.3.) shows the different types of microcontrollers.

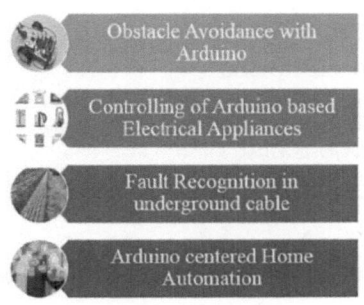

Obstacle Avoidance with Arduino

Controlling of Arduino based Electrical Appliances

Fault Recognition in underground cable

Arduino centered Home Automation

FIGURE 16.3
Microcontrollers Raspberry Pi, Arduino, and ESP 2866.

Some of the applications include but not limited to:

- Applications with playoffs: Raspberry pi has the skill of supporting playoffs like 3D games and HD videos since it has the capability to do graphics, and multimedia applications.
- Pi in the sky: This device is designed for tracking great altitude with balloon flights, provisioned with a GPS receiver as well as radio transmitter.
- R2D2 powered by Rasperry Pi: R2D2 tracks expressions and gesture and object movement is recognized with motorized capability and it is able to react to voice directions.
- Animated effects on devices: A camera product like Otto captures images with different views and perspectives. Animated GIFs with huge effects can be created by conversing the captured images.
- Live bots: This scheme allows users to monitor and control various robots centered on Raspberry Pi through the internet.
- Medical application: Heartfelt Technologies and NuGenius are revolutionizing medicine and health with Pi-powered products.

16.2.1.2 Arduino

The Arduino board is an open source, freely available development microcontroller that is capable enough to manage various communication protocols to suit IoT devices with different types. The cost of the board is lesser and accompanied with rich features. The accessibility of a range of daughter boards with rich features have an incredible assembling feature to the main motherboard. To bring out rapid prototyping and to do programming in an easier way with devices, low-power Arduino shields facilitate the availability of Ethernet and Wi-Fi communication options.

Applications

- The robot operated based on obstacle avoidance with Arduino
- Controlling of Arduino-based electrical appliances via IR
- Fault recognition in underground cable by means of applying the Arduino board
- Arduino-centered home automation

16.2.1.3 ESP-8266

The ESP-8266 is a Wi-Fi-enabled system on chip (SoC) firmware mostly used with IoT applications. This Wi-Fi board is integrated with a full stack that will provide access to any microcontroller in the network. The board is skillful enough to host any application with discharging various Wi-Fi networking functionalities. Also this is utilized as a preferable sensor node to collect data through any object that is connected wirelessly and gathered data are transferred to the chief server.

This system is commonly used in numerous projects given like [3] with the support of Wi-Fi proficiency, however the main applications are listed as the following:

- Wireless web server
- Geolocation using ESP8266

FIGURE 16.4
Structure of RTOS.

- Pressure sensors on railway tracks
- Temperature logging system
- Wi-Fi controlled robot
- M2M using ESP8266

16.2.1.4 RTOS

A set of software that handles the operations of computer hardware constitutes an operating system. RTOS is an operating system dedicated to support real-time operations and is implemented in embedded systems (see Figure 16.4). Embedded systems used today are a complex combination of software and hardware. RTOS are multitasking operating systems that not only depends on logical correctness but also upon the application delivery time. Overall system performance measurement and the set of tools and methodologies to solve any task are highly depends on the RTOS.

The major tasks of an RTOS operating system consists of four different categories: interprocess communication, process management, memory management, and synchronization and I/O management. Some of the popular RTOS are quantitatively and qualitatively analyzed. RTOS is broadly classified in to three categories: hard, soft, and firm.

In any hard real-time systems, the issue assigns the request to complete the task is finished within the given time limits, and the request not met by the deadline is allowed only in a few firm real-time systems. This scenario is entirely changed in soft real time systems that meet the deadline properly but are not tightly bounded with time. The performance of an RTOS is determined by the set of applications and the suitable parameter selection [4].

16.3 Low-Power Embedded Systems

Electronic system design plays a very vital role when the performance of a device is being concentrated, such as less energy consumption or high performance during the operational view of electronic devices. So the embedded systems should be aimed in such a mode that battery consumption will be minimal. Since the size of the data collected through IoT devices from a field is massive and it has to be stored in a storage location reliably, cloud computing plays the major role. This huge amount of data is to be processed and it

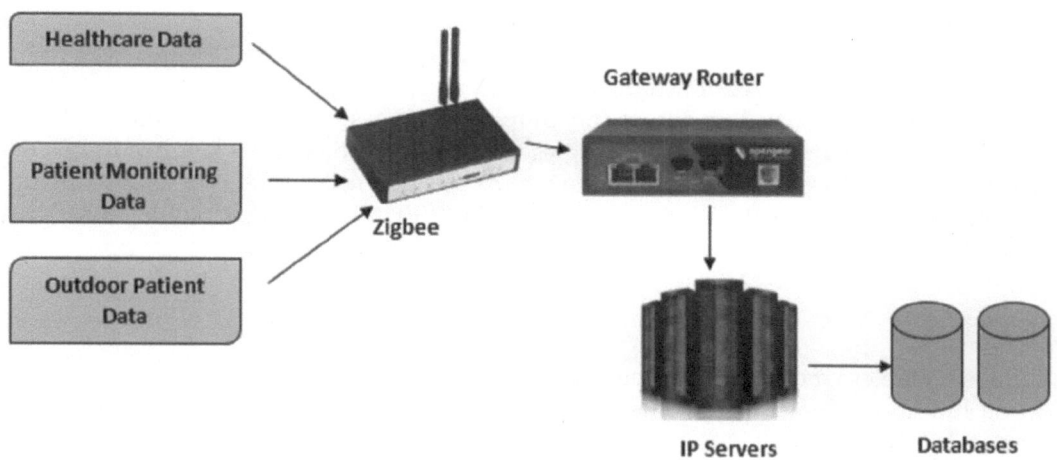

FIGURE 16.5
Storage and Monitoring of Data Through IoT.

paves the way for the user to learn and determine where the faults occurred within the specified system.

To execute any process or to monitor a system, physical devices should interact with each other as given in Figure 16.5, and internet connectivity is an essential requirement in order to communicate with each other. The devices are represented by an IP address, but unfortunately very limited addresses are available in the naming registry. This IP naming will not be sufficient due to the rising number of devices getting involved and an alternative system is to be identified to represent the physical device. Usage of electronic devices is huge in almost every field and this massive usage is going to generate a very vast amount of big data. Since IoT relies heavily on sensors to collect data especially in real time, the data storage and retrieval is to be framed in a meaningful setup. The technology developed based on IoT may deal with services that may be offered with more a advanced level and it entirely transformed the lifestyle of people.

16.4 Communication Options

16.4.1 ZIGBEE

ZigBee plays the major role in wireless personal area networking (WPAN), where it brings the radio connection in digital between the computers and the related systems. ZigBee, or low-ate WPAN, is responsible for devices with certain specifications like low power consumption, low data rate, and a long lifetime of battery as well. Completely controlled and networked homes are made possible with the help of ZigBee technology wherein all the devices communicate with a single unit. The ZigBee was defined by a standard body called ZigBee Alliance, that publishes applications and also allows OEM vendors in multiples to develop interoperable products. The devices provisioned with ZigBee technology are equipped with standard parameters like 2.4 GHz which is the worldwide accepted ISM band and throughput rate of 250 Kbps which is against Bluetooth which has 1 Mbps.

ZigBee has been explored in many controls ranging from household devices with less power like smoke alarms to centralized control units [5].

The focus of ZigBee standards towards the network applications include features such as a dense network, less power consumption, and ease of implementation with less cost. Compared to the wireless technologies existing, the protocol stack size of the ZigBee is 32 Kb which is three times less than the others. The potential benefits of ZigBee include data integrity and security. ZigBee influences the IEEE 802.15.4 MAC sublayer security model [6].

16.4.2 LoRa

LoRa is a wireless modulation scheme used in the physical layer to develop a communication link in the long range. Other wireless technologies applied in the physical layer include WCDMA and OFDAMA, which may be used for LTE and UMTS (3G) networks. LoRaWAN is a short form of a low-power wide-area network (LPWAN), established to connect the sensors and IoT devices for the heavy volume of deployment. LoRa is implemented to provide certain key benefits such as extended battery in terms of lifetime with several years and also deployment cost effective with extended range. LoRa uses a modulation scheme such as chirp spread spectrum modulation to significantly increase the communication range. The main reason to prefer LoRa compared to other wireless technologies is its long-range communication capability. A base station or a single gateway can extend its service to hundreds of kilometers or a complete city [7,8].

16.4.3 Wi-Fi

Wi-Fi is the abbreviated of wireless fidelity. It falls under the IEEE 802.11 family of standards and Wi-Fi is predominantly a LAN technology intended to provide broadband coverage within a building structure. Current Wi-Fi systems use 54 Mbps as physical layer data rate and provide approximately 100 feet of indoor coverage. Remarkably, Wi-Fi offers a peak data rate compared to a 3G systems and operates on 20 MHz larger bandwidth, but it was not designed to provide high mobility speed.

According to the Wi-Fi guide

> Wi-Fi has become the de facto standard for last mile broadband connectivity in homes, offices, and public hotspot locations. Systems can typically provide a coverage range of only about 1,000 feet from the access point. The Wi-Fi standards define a fixed channel bandwidth of 25 MHz for 802.11b and 20 MHz for either 802.11a or g networks [9].

One significant benefit of Wi-Fi against WiMAX and 3G is its wide accessibility of terminal devices. A majority of laptops distributed today have a built-in Wi-Fi interface. Wi-Fi interfaces are being built into a variety of devices, including cell phones, personal data assistants (PDAs), cameras, cordless phones, and media players.

16.4.4 Low-Power, Short Range IoT Networks

This paradigm talks about the networks and the communication that is to be built either in short range or long range wireless connectivity. The present wireless radio community provides longer range connectivity with IoT and the internet. The short range could connect wearable devices to a smart phone that may be used to control the home environment.

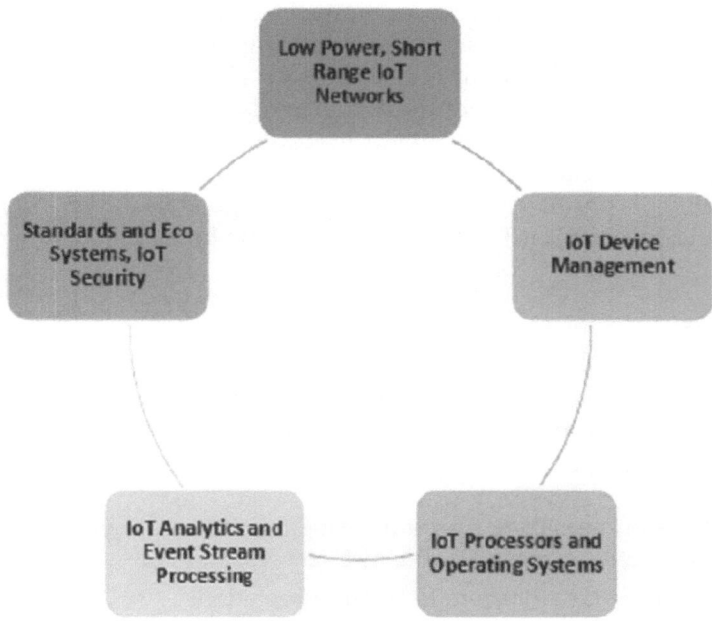

FIGURE 16.6
Building Blocks of IoT Technologies.

The essential, and one of the predominant requirements of, IoT devices are batteries that range from low-power to advanced batteries.

Short range communication established with devices like Bluetooth, Zigbee, sensors and actuators, and gateways are monitored with transmit and receive power, and resource availability decides the massive concurrent accesses of data. Many IoT applications are in need of wide area coverage with good battery lifetime, meager hardware and operating cost, and low bandwidth, but these technical features may not deliver properly by the traditional cellular networks [10]. Emerging standards of IoT may provide high connection density and are likely to dominate the cellular networks.

The following are the emerging IoT technologies identified by Nick Jones, vice president and distinguished analyst at Gartner: the roadmap of his analysis on IoT technologies is given in Figure 16.6.

16.5 IoT Device Management

The tools involved in IoT applications should be capable enough to manage and monitor thousands of devices. Since the deployment of IoT devices significantly increases day to day, it is quintessential to monitor and track connected devices [11]. After deployment, the devices should be ensured to work properly like to detect any issues and troubleshoot problems and manage software updates.

Since the number of chronic diseases is growing exponentially day by day due to the aging population, healthcare services are costlier than in the older days. It may not be possible to reduce the population or to eradicate the chronic diseases with technology, but it can offer accessible healthcare easier. The right diagnosis for any disease is an essential need and it ingests a large amount of clinical bills. The routines of technology shall be moved from the hospital-centric to the home-centric with the help of technology. This makes the medical centres work competently to acquire the enhanced treatment. Healthcare systems based on the technology have tremendous benefits that might improve the efficiency and quality of consulting and treatments to get good health for the patients [12,13].

Wearables and sensors in IoT play a major role. The following are the devices used in healthcare:

16.5.1 Indoor Air Quality Meter

An indoor air quality meter permits the user to carry out the measurements that are essential to verify and tune the air conditioning and ventilation systems in addition to the quality assessment of indoor air. The potential uses are as follows:

- A wide range of measuring functions using possible accessories (e.g., moisture sensors, anemometers, degree of turbulence sensors, and lux)
- Superior memory size for up to 10,000 interpretations
- Exhibits dew point difference like, maximum, minimum, and mean values
- Has system software for investigating and recording measuring data

The advantages are the CO_2 level, indoor pollution level, a temperature sensor and humidity sensor, and it practices Wi-Fi for outdoor air quality data. Data might also be available on a smartphone or tablet. Information is beneficial to turn on air purifiers and to open or close windows [14].

16.5.2 Wearables

A few wearable sensing devices are listed in the following Table 16.1:

TABLE 16.1

Wearable Sensing Devices

S. No	IoT Type	Description	Uses
1	Bio strap	It is a wristband and display clip to show blood oxygen level, monitor heart rate, movement of body while sleep. Charging wirelessly and battery can resist for five days	Used for health charts, exercise tracking, activity logging and timers
2	Medical alert system	Pre-programmed mobile calls triggered by two medical alert pendants. A mobile-based station also possesses an alert button	Can call up to three mobile numbers. Useful for disabled, elderly, or if living alone
3	Smart watch with heart rate and GPS	Enabled spark cardio system with uploaded music, supports as a GPS fitness watch, heart rate monitor, and provided three GB music storage (small, black)	Takes up to three weeks between uploading and recharging the health data. Smartphone is not essential. Audio provided with Bluetooth [15]

16.5.3 Sensors

Sensors are used to gather data without depending on the kinds of tasks allocated by various organizations. Depending on the type of data to be fetched, a vast number of sensors of different kinds are involved in this aspect. Sensors are taken to a different level through the emergence of IoT. The gathered data are piled up together, and are then shared with a whole network of related devices. The accumulated data functions the devices and acquires the decision automatically, thus it is becoming 'smarter and efficient' day by day [16]. Devices share the data among each other by clubbing the communication network and a set of sensors that will improve the efficiency and functionality of data. Let us have a look at some of the sensors especially used in the IoT world shown in Figure 16.7.

- *Temperature Sensors*
  It is a device used to detect a notable physical transformation in temperature from a specific cause. It measures the amount of heat energy and converts this data for a user or any device. Temperature sensors are used in most of the IoT projects across the planet for the organization of the data gathering process and to give healthier and firm care for patients.
  Usage of medical gadgets is becoming complex in nature since the number of chronic diseases are increasing rapidly. The need for cheaper and lower persistent equipment that are sterilized in either steam or chemicals are making it more complicated to use electronics, plastics, and the devices themselves [17].

- *Proximity Sensors*
  To identify the presence of any nearby object, proximity sensors are used. Its main purpose is to find the presence of the object and the properties of that object and thus convert this pack of information into a signal. Any user of the electronic equipment

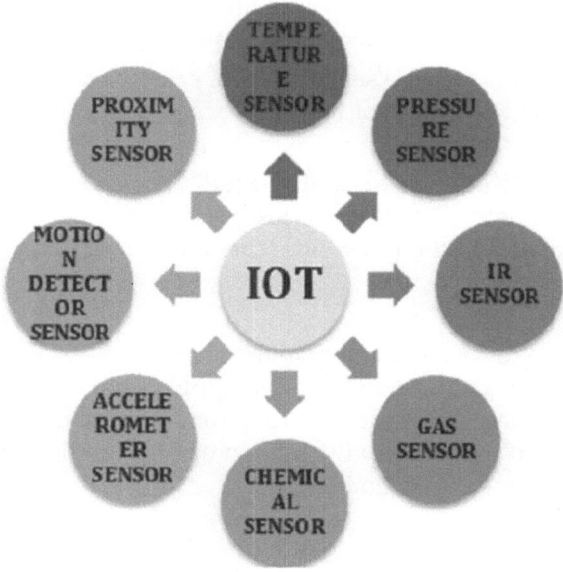

FIGURE 16.7
Kinds of Sensors Used in IoT Healthcare System.

can read this signal without any trouble and also contacting with the object can be avoided.

Proximity sensors are sub-categorized as follows:

- Inductive proximity sensors can work at higher speeds compared to the mechanical switches and also these sensors are utilized to discover the occurrence of metal objects by means of an electromagnetic radiation rays or an electromagnetic field.
- In a very large portion of the target, capacitive sensors are used to sense the presence of tiny objects. Hence, capacitive sensors are normally used in very difficult, and most complicated, applications.
- Photoelectric sensors utilizes a beam of light to identify the absence or presence of any object since it is made up of light sensitive parts. It is the suitable replacement for inductive sensors and its main centre of attention is on long distance sensing.
- *Motion Detector Sensors*

The device is used to sense the physical movement in a particular location and the detected information obtained undergoes conversion to produce an electric signal. This takes place in the method of motion called a motion detector. The movement of human beings or the motion of any object can be easily detected with the help of this device.

The raising requirement for the motion detector sensor is in security industry. Its primary usage was identified in the healthcare sector in the laboratory and the identification of the motion of new personnel at the unusual time inside the hospitals. The primary role is ensured with the intrusion detection system developed in the hospital data sectors, smart cameras that may recognize/capture the motion of an object or through video recording, automatic door control, and so forth.

Some of the widely used motion sensors are the following:

- Ultrasonic Sensors: The ultrasonic waves are received as pulses, track the speed of sound waves, and measure reflection from a motion object.
- Passive Infrared Sensors: These sensors play a primary role in a data collection centre where it detects body heat of an object since it responds only to the infrared energy.
- Microwave Sensors: These sensors emit the radio wave pulses and detects the reflection of an object while moving. These sensors cover a wider area than the ultrasonic and passive infrared. On the other side, the device cost is high and susceptible to electrical interferences.

- *Pressure Sensors*
  The sensors used to sense the pressure in the phenomena and convert it to an electric signal. Depending on the level of pressure applied, the amount of electrical signal will be generated. These sensors are useful for monitoring the water heating systems in laboratories and if there is any fluctuation or drop in the indicated level, the IoT system should be immediately notified.

- *IR Sensors*
  Infrared (IR) sensors are the foremost applied sensor in the IoT field and it is mainly utilized to sense certain specific individualities of its atmospheres by sensing or

discharging infrared radioactivity. These sensors are mainly used for determining the temperature produced by objects.

Though IR sensors are used in wide variety of smart devices like smartphones, smartwatches, and so on, its usage was highly recognized and essentially utilized in monitoring the blood pressure and blood flow of the body. Its other applications include wearable health devices, blood flow of the body, analysis of breathing, temperature-based measurements, and optical communiqué.

- *Accelerometer Sensors*

These sensors work based on the principle of 'rate of conversion of speed over time'. This is a transducer mainly focused to detect quantifiable acceleration due to any external forces practiced by an object and this acceleration is to be converted into an electrical output.

   Accelerometer sensors are widely used in the detection of vibrations. Medical equipment that is built with the inclusion of these sensors is smart enough to monitor and raise an alarm/alert signal if a patient is disturbed by any external forces.

- *Chemical and Gas Sensors*

These sensors are used to identify the presence of various gases and the nature of sensing data is similar to chemical sensors. It plays an essential role in healthcare and pharmaceutical sectors. The composition of the types of gases used for a purpose will be shown through an IoT device and it is made available to the people who are involved. Some of the common gas sensors are hydrogen sensors, breath analyzer, carbon monoxide and carbon dioxide detectors, and oxygen sensors [18].

16.5.4 Storage Options for IoT

In this era, to make an efficient use of IoT in domain applications, data plays the major role and the challenging part is how to collect, organize, and store the data. Cloud computing is a healthier choice for data storage and dynamic integration of data and resources with IoT. The cloud environment is highly constructive for IoT to utilize the enormous amount of resources that are available.

Cloud computing is a technology that drastically changes the whole way of how technologies and data can be accessed, processed, and modified. It is broadly agreed that this can be used as a utility service for future domain applications. The actual fact of using this technology is that it has been involved in various technologies such as virtualization, networking the legacy software systems, and grid computing. And it shares services and computing resources across the globe through internet. IoT generally involves a number of objects with little storage and processing capacity, efficient operations are highly dependent on the services provided by the cloud. The controversial part is that the cloud is highly interoperable whereas IoT concentrates more on variety rather than interoperability [19,20].

Based on resources utilized from the cloud, cloud deployment models are categorized as:

- Public cloud
- Private cloud

- Hybrid cloud
- Communal cloud

Based on the services:

- Software as a service (SaaS)
- Platform as a service (PaaS)
- Infrastructure as a service (IaaS)

Generally, the deployment models provided by cloud computing come under the category of public, where all the resources are made accessible to customers through the internet. Mostly this will be owned by profitable organizations like Amazon. In contrast, if any single organization provides infrastructure to serve the purpose of a particular customers, this is termed as a private cloud. A private cloud assures more security and a higher level of control over services. To overcome the flaws in a public and private cloud, a hybrid cloud came into effect. Infrastructure shared to different concerns for the same purpose are called communal cloud.

16.5.4.1 Storage Options

Since most devices use IoT for smart communication, it may generate an enormous amount of information that is generally called data. Habitually these data maybe unstructured or semi-structured data. Pertaining to the characteristics of data such as type of data, frequency of generation of data and size, big data plays a role. To deal with this massive amount of data in a cost-effective and timely without a storage location and in a secure manner, cloud computing is the appropriate solution. Probabilities of aggregating, integrating, and sharing the organized data with customers are done in an elegant way by using this technology [21–23].

Figure 16.8 shows the cloud-oriented IoT architecture, which is comprised of three layers. To be precise, the layers are application, view, and the cloud layer. The application layer affords the meeting point for different set of services, the view layer detects objects to gather data from the adjacent nature, and the cloud layer is responsible to transfer the gathered data and to make it available to everyone in the cloud. Its contributions to the healthcare area are highly substantial like intelligent drug management, medicine control and monitoring, and equipment monitoring systems.

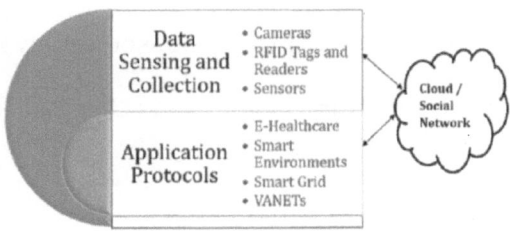

FIGURE 16.8
Cloud-Based IoT Architecture.

16.6 IoT Analytics

The information or data collected by 'things' are the major backbone to make up any decision in the system. To bring a solution to any issue, IoT business models plays an extensive role which will demand new algorithms and analytic tools very often as the volumes of data have grown exponentially in the last few years.

In a multi-faceted life, many decisions that we make may not be concrete in nature and decision making is an essential requirement to solve any complex issues. In day-to-day life, the data gathered from people are to be streamed properly using fuzzy sets and algorithms since the IoT contributes to the collecting the data and synergizing the equipment, processes, and public. The accuracy of health monitoring devices such as, CYCORE, glucose monitor, connected inhalers, and insulin pens are entirely dependent on the data which are analyzed and streamed properly to the cloud.

According to Daniel Harris [24], even though new sensors and wireless and mobile technologies are the driving force for the evolution of IoT, data analytics creates the significant value of IoT rather than the software and hardware used.

16.6.1 Uniqueness of IoT Analytics

The specialty of analytics lies in IoT datasets that may be collected from various sensors and it may be represented clearly with use cases at a lower cost. Rather than collecting data from sensors all the time, other kinds of data also involved in IoT projects include the potential use of sensors that were showcased with their capability to bring together data from the physical phenomena, which will be evaluated or pooled with different types of data to perceive patterns. A senior vice president of research and head of the advanced analytics lab, Dr. Joachim Schaper, briefs about sensor data that can be used to comprehend public as well as things. This gives the importance of data received from healthcare domain.

Rather than collecting data from sensors all the time, other kinds of data also involved in IoT projects include the following:

- Data from video
- Geo-location data from mobile
- Any product usage data, gathered from various sources other than sensors
- Data from social media
- Log records (operations executed and stored in computer-generated records and events in networks and software applications)

Cameras are the rich data sources that combine sensors and one can analyze the same data with different perspectives due to the frequent set of captured images at different angles. So IoT includes cameras and expands its services widely.

The healthcare industry faces pressures and challenges from nearly every aspect of operation. From managing equipment, inventory and time to tracking patients, the level of accountability is high. A good example is medical inventories. Many hospitals overstock certain inventories to prevent running out during an emergency [24].

Putting a wealth of complex devices together is problematic on several levels. One in particular involves standards. IoT will rely on even greater standardization of communications protocols in the future. Work is ongoing to create guidelines for wireless communications between monitoring devices that share data with care [20,25].

Yet, open issues remain that can slow down adoption. They include:

(i) limited interoperability between vertical silos,

(ii) difficulty to guarantee deterministic service provisioning using interoperable architectures,

(iii) limited flexibility and adaptability of hard real-time service models and related IoT deployments,

(iv) lack of definitive information-centric networking approaches to inherent data-centric IoT applications,

(v) fragmentation of design guidelines and definitions, which are tightly tied to each single application domain,

(vi) efficient and scalable service and resource discovery,

(vii) and mobility support and self-configuration for smart objects and mobile devices [26,27].

The challenges and opportunities arising from a proper integration of IoT with mobile computing and industrial systems create a fascinating research field that deserves in-depth investigations. On the one hand, mobile devices are expected to become the joining link between connected smart objects, the web, and end users. This enables IoT technologies to play a significant role for the quick convergence between operational management systems (OMS) and distributed ICT sensing and actuation platforms [28].

16.7 Conclusion

From the affirmative study of tools and technologies, it is clearly evident that IoT offers significant contributions to the healthcare industry. To migrate from the traditional approach to a modernized approach to automate the tasks and to identify the health issues in an early stage, IoT tools play the major role. The IoT wearables simplify the task of doctors and sensors are quintessential to make these wearables to sense data. The heart of the IoT devices are microcontrollers. Raspberry Pi, ARM processors, and the Arduino board are the basic building blocks of IoT and are capable enough to manage any communication protocols. These devices are structured with RTOS and communicate with one another with the help of LORA, Wi-Fi, and ZigBee. Low-power short-range wireless networks are mainly required to make devices send data from one location to another.

In the operational environment, data generated by the IoT devices are an enormous amount and it needs to be processed, organized, and stored with limited storage capability. Cloud computing is the suitable choice for the aforementioned requirements. IoT and cloud computing are the two giant technologies that come together to bring out the meaningful data with limited utilization of devices and maximum use of cloud services and infrastructures. To process the huge amount of data, data analytics comes into play. To analyze the cause for a particular disease, data analytics are required. To inculcate the complete acceptance of the IoT in healthcare domain, it is life-threatening to recognize and examine distinctive features of IoT in terms of security and privacy, including vulnerabilities, security requirements, and countermeasures from the healthcare viewpoint.

Even though these different technologies and analytical tools are used to bring significant improvements in the healthcare sector, it is highly dependent on the cost of obtaining these services from hospitals and the cooperation of the patient. Too much of awareness through the internet makes people confused on how to make use of the services. Apart from availability, one should understand the purpose of these data. This chapter provides the roadmap of the tools and technologies used in the healthcare industry and may create research avenues with the technologies applied.

References

[1] Chakrabarty, Ankush, Stamatina Zavitsanou, Tara Sowrirajan, Francis J. Doyle, and Eyal Dassau. 2019. *Getting IoT-Ready*. Elsevier.

[2] https://www.codemag.com/Article/1607071/Introduction-to-IoT-Using-the-Raspberry-Pi/

[3] https://www.hackster.io/PatelDarshil/things-you-should-know-before-using-esp8266-wifi-module-784001

[4] Dewangan, Kiran, and Mina Mishra. 2018. "Internet of Things for Healthcare: A Review." *International Journal of Advanced in Management, Technology and Engineering Sciences* 8 (3): 526–534.

[5] https://zigbeealliance.org/

[6] Waher, Peter. 2015. *Learning Internet of Things*. Packt Publishing Ltd.

[7] http://www.3glteinfo.com/LoRa/

[8] Holler, Jan, Vlasios Tsiatsis, Catherine Mulligan, Stamatis Karnouskos, et al. 2014. *Machine-to-Machine to the Internet of Things*. Elsevier.

[9] https://www.tutorialspoint.com/wi-fi/wifi_quick_guide.htm

[10] Uddin Ahmed, Mobyen, Shahina Begum, and Wasim Raad. 2016. *Internet of Things Technologies for HealthCare*. Springer.

[11] Illegems, Janni. 2017. "The Internet of Things In Health Care." Master's Dissertation.

[12] Miller, Lawrence. 2017. *CISSP. Internet of Things for Dummies*. Qorvo Special Edition. Wiley Publishers.

[13] https://www.networkworld.com/article/3258812/the-future-of-iot-device-management.html

[14] https://www.testo.com/en-IN/testo-435-2/p/0563-4352

[15] https://www.techradar.com/in/wearables

[16] https://www.finoit.com/blog/top-15-sensor-types-used-iot/

[17] https://enterpriseiotinsights.com/20170706/healthcare/the-role-of-temperature-sensors-in-medical-devices-tag27

[18] https://www.digikey.com/en/articles/techzone/2014/jul/the-role-of-sensors-in-iot-medical-and-healthcare-applications

[19] https://www.sciencedirect.com/topics/computer-science/cloud-deployment-model

[20] Rajesh Kumar, D., and A. Shanmugam. 2017. "A Hyper Heuristic Localization Based Cloned Node Detection Technique Using GSA Based Simulated Annealing in Sensor Networks." In *Cognitive Computing for Big Data Systems Over IoT*. 307–335.

[21] Atlam, Hany F., Ahmed Alenezi, Abdulrahman Alharthi, et al. 2017. "Integration of Cloud Computing with Internet of Things: Challenges and Open Issues." In *IEEE International Conference on Internet of Things (iThings) and IEEE Green Computing and Communications. IEEE* 105: 670–675.

[22] Kumar, D. R., T. A. Krishna, and A. Wahi. 2018. "Health Monitoring Framework for in Time Recognition of Pulmonary Embolism Using Internet of Things." *Journal of Computational and Theoretical Nanoscience* 15 (5): 1598–1602.

[23] Ahmadi, Hossein, Goli Arji, Leila Shahmoradi, Reza Safdari, Mehrbakhsh Nilashi, and Mojtaba Alizadeh. 2018. "The Application of Internet of Things in Healthcare: A Systematic Literature Review and Classification." *Universal Access in the Information Society.* 1–33.

[24] https://www.sam-solutions.com/blog/internet-of-things-iot-protocols-and-connectivity-options-an-overview/

[25] Sathish, R., and D. R. Kumar. 2013. *"Dynamic Detection of Clone Attack in Wireless Sensor Networks."* In *2013 International Conference on Communication Systems and Network Technologies. Presented at the 2013 International Conference on Communication Systems and Network Technologies (CSNT 2013).*

[26] Al-Sarawi, Shadi, Mohammed Anbar, Kamal Alieyan, and Mahmood Alzubaidi. 2017. *"Internet of Things (IoT) Communication Protocols: Review."* In *8th International Conference on Information Technology (ICIT).*

[27] Nikoukar, Ali, Saleem Raza, Angelina Poole, Mesut Güneş, and Behnam Dezfouli. 2018. "Low-Power Wireless for the Internet of Things: Standards and Applications." *IEEE Access* 6: 67893–67926.

[28] Zhang, Zhi-Kai, Michael Cheng Yi Cho, Chia-Wei Wang, Chia-Wei Hsu, and Chong-Kuan Chen. 2014. *"IoT Security: Ongoing Challenges and Research Opportunities."* In *IEEE 7th International Conference on Service-Oriented Computing and Applications.*

Index

Page numbers in *italic* indicate figures. Page numbers in **bold** indicate tables.